Solutions Manual to Accompa

D0948024

Inorganic Chemistry

Seventh Edition

Alen Hadzovic
University of Toronto

OXFORD
UNIVERSITY PRESS

OXFORD
UNIVERSITY PRESS

Great Clarendon Street, Oxford, OX2 6DP,
United Kingdom

Oxford University Press is a department of the University of Oxford.
It furthers the University's objective of excellence in research, scholarship,
and education by publishing worldwide. Oxford is a registered trade mark of
Oxford University Press in the UK and in certain other countries

Fourth edition 2006
Fifth edition 2010
Sixth edition 2014

Impression: 4

Published in the United States of America by Oxford University Press
198 Madison Avenue, New York, NY 10016, United States of America

British Library Cataloguing in Publication Data
Data available

ISBN 978–0–19–881468–9

Printed in Great Britain by
CPI Group (UK) Ltd, Croydon CR0 4YY

TABLE OF CONTENTS

PREFACE

This *Solutions Manual* accompanies *Inorganic Chemistry*, Seventh Edition by Mark Weller, Tina Overton, Jonathan Rourke, and Fraser Armstrong. Within its covers, you will find the detailed solutions for all self-tests and end of chapter exercises. New to this edition of the Solutions Manual is the inclusion of guidelines for the selected tutorial problems—those problems for which the literature reference is not provided—for the majority of chapters. Many solutions include figures specifically prepared for the solution, and not found in the main text. As you master each chapter in *Inorganic Chemistry*, this manual will help you not only to confirm your answers and understanding but also to expand the material covered in the textbook.

The Solutions Manual is a learning aid—its primary goal is to provide you with means to ensure that your own understanding and your own answers are correct. If you see that your solution differs from the one offered in the Solutions Manual, do not simply read over the provided answer. Go back to the main text, reexamine and reread the important concepts required to solve that problem, and then, with this fresh insight, try solving the same problem again. The self-tests are closely related to the examples that precede them. Thus, if you had a problem with a self-test, read the preceding text and analyze the worked example. The solutions to the end of chapter exercises direct you to the relevant sections of the textbook, which you should reexamine if the exercise proves challenging to you.

Inorganic chemistry is a beautiful, rich, and exciting discipline, but it also has its challenges. The self-tests, exercises, and tutorial problems have been designed to help you test your knowledge and meet the challenges of inorganic chemistry. The Solutions Manual is here to help you on your way, provide guidance through the world of chemical elements and their compounds and, together with the text it accompanies, take you to the very frontiers of this world.

With a hope you will find this manual useful,
Alen Hadzovic

ACKNOWLEDGMENTS

I would like to thank the authors Mark Weller, Tina Overton, Jonathan Rourke, and Fraser Armstrong for their insightful comments, discussions, and valuable assistance during the preparation of the seventh edition of the Solutions Manual. I would also like to express my gratitude to Roseanne Levermore, Editor for Oxford University Press, for all of her efforts and dedication to the project.

Chapter 1 Atomic Structure

Self-Tests

S1.1 **a)** For the Paschen series $n_1 = 3$ and $n_2 = 4, 5, 6, \ldots$ The second line in the Paschen series is observed when $n_2 = 5$. Hence, starting from equation 1.1, we have

$$\frac{1}{\lambda} = R\left(\frac{1}{n_1^2} - \frac{1}{n_2^2}\right) = 1.097 \times 10^7\,\text{m}^{-1}\left(\frac{1}{3^2} - \frac{1}{5^2}\right) = 1.097 \times 10^7\,\text{m}^{-1} \times 0.071 = 779967\,\text{m}^{-1}.$$

The wavelength is the reciprocal value of the above-calculated wavenumber: $\dfrac{1}{779967\,\text{m}^{-1}} = 1.28 \times 10^{-6}\,\text{m}$ or 1280 nm.

b) For the Lyman series $n_1 = 1$ is set while n_2 has to be calculated. We can re-write equation 1.1 as

$$\frac{1}{\lambda} = R\left(\frac{1}{n_1^2} - \frac{1}{n_2^2}\right) = \frac{R}{n_1^2} - \frac{R}{n_2^2}.$$

From here:

$$\frac{R}{n_2^2} = \frac{R}{n_1^2} - \frac{1}{\lambda} = \frac{1.097 \times 10^7\,\text{m}^{-1}}{1^2} - \frac{1}{103\,\text{nm} \times 10^{-9}\,\dfrac{\text{m}}{\text{nm}}} = 1.097 \times 10^7\,\text{m}^{-1} - 9.71 \times 10^6\,\text{m}^{-1} = 1.26 \times 10^6\,\text{m}^{-1}.$$

And

$$n_2^2 = \frac{R}{1.26 \times 10^6\,\text{m}^{-1}} = \frac{1.097 \times 10^7\,\text{m}^{-1}}{1.26 \times 10^6\,\text{m}^{-1}} = 8.71.$$

Finally

$$n_2 = \sqrt{8.71} = 2.95.$$

Since n values are integer, our n_2 equals 3.

S1.2 **a)** The third shell is given by $n = 3$, and the subshell for $l = 2$ consists of the d orbitals. Therefore, the quantum numbers $n = 3$, $l = 2$ define a 3d set of orbitals. For $l = 2$, there are $2l + 1 = 5$ m_l values: +2, +1, 0, –1, –2. Thus, there are five orbitals in this set.

b) The value of quantum number $n = 5$ is given. The value of quantum number l can be determined from the letter designation f as $l = 3$. The number of orbitals in this set is given by the number of possible m_l values: $2l + 1 = 2 \times 3 + 1 = 7$. The allowed m_l values in this case are +3, +2, +1, 0, –1, –2 and –3.

S1.3 **a)** The number of radial nodes is given by the expression: $n - l - 1$. For the 5s orbital, $n = 5$ and $l = 0$. Therefore: $5 - 0 - 1 = 4$. Thus, there are four radial nodes in a 5s orbital. Remember, the first occurrence of a radial node for an s orbital is the 2s orbital, which has one radial node, the 3s has two, the 4s has three, and finally the 5s has four. If you forget the expression for determining radial nodes, just count by a unit of one from the first occurrence of a radial node for that particular "shape" of orbital. Figure 1.9 shows the radial wavefuntions of 1s, 2s, and 3s hydrogenic orbitals. The radial nodes are found where the radial wavefunction has a value of zero (i.e., it intersects the x-axis).

b) In this case we have to determine the value of quantum number n for which a p orbital ($l = 1$) would have two radial nodes. Thus:

$$n - l - 1 = 2$$
$$n - 1 - 1 = 2$$
$$n = 4.$$

All three 4p orbitals have two radial nodes.

S1.4 There is no figure showing the radial distribution functions for 3p and 3d orbitals, so you must reason by analogy. In the example, you saw that an electron in a p orbital has a smaller probability of close approach to the nucleus than in an s orbital, because an electron in a p orbital has a greater angular momentum than in an s orbital. Visually, Figure 1.12 shows this. The area under the graph represents where the electron has the highest probability of being found. The origin of the graph is the nucleus, so one can see that the 2s orbital, on average, spends more time closer to the nucleus than a 2p orbital. Similarly, an electron in a d orbital has a greater angular momentum than in a p orbital: l(d) > l(p) > l(s). Therefore, an electron in a p orbital has a greater probability than in a d orbital of close approach to the nucleus.

S1.5 **a)** To start with, we write down the ground state electronic configuration of Si atom:

$$1s^2 2s^2 2p^6 3s^2 3p^2.$$

From this electronic configuration we see that silicon's outermost electron resides in one of three 3p orbitals. Now we group the orbitals as described in the text and in Example 1.5:

$$(1s^2)(2s^2 2p^6)(3s^2 3p^2).$$

Next, we analyse the groups as follows:

- There are three additional electrons in the same group as our outermost electron (two in 3s and one more in 3p orbital). Each of them contributes 0.35 to the screening constant.
- There are eight electrons in $n - 1$ group (two in 2s and six in 2p orbitals). Each contributes 0.85 to the screening constant.
- Finally, there are just two electrons (1s electrons) in one lower shell/group contributing 1.00 to the screening constant.

Adding all this together we have:

$$\sigma = (3 \times 0.35) + (8 \times 0.85) + (2 \times 1.00) = \mathbf{8.95}.$$

b) Recall that effective nuclear charge is given by:

$$Z_{eff} = Z - \sigma.$$

The full nuclear charge, Z, is also the atomic number, and for chlorine it is 17. Now we have to determine s following previous examples.

For Cl atom:

$$(1s^2)(2s^2 2p^6)(3s^2 3p^5)$$

$$\sigma = (6 \times 0.35) + (8 \times 0.85) + (2 \times 1.00) = \mathbf{10.9}.$$

We can now calculate Z_{eff}:

$$Z_{eff} = Z - \sigma = 17 - 10.9 = \mathbf{6.1}.$$

S1.6 **a)** The configuration of the valence electrons, called the valence configuration, is as follows for the four atoms in question:

Li: $2s^1$	B: $2s^2 2p^1$
Be: $2s^2$	C: $2s^2 2p^2$.

We can use the effective nuclear charge values from Table 1.2 in the textbook to analyse the trends. Thus, when an electron is added to the 2s orbital on going from Li to Be, Z_{eff} increases by 0.63. When an electron is added to an empty p orbital on going from B to C, Z_{eff} increases by 0.72. The s electron already present in Li repels the incoming electron more strongly than the p electron already present in B repels the incoming p electron, because the incoming p electron goes into a new orbital. Therefore, Z_{eff} increases by a smaller amount on going from Li to Be than from B to C. However, extreme caution must be exercised with arguments like this because the effects of electron–electron repulsions are very subtle. This is illustrated in period 3, where the effect is opposite to that just described for period 2.

b) Al is below B in the same group of the periodic table. In this case Z significantly increases (from $Z = 5$ for B to $Z = 13$ for Al) because of the added protons as we move through the elements. However, the increase in Z_{eff} is smaller (from $Z_{eff} = 2.42$ for B to $Z_{eff} = 4.07$ for Al) because the Al atoms contain one extra filled electronic shell of eight electrons ($2s^2 2p^6$) that shield the outermost electrons from the full nuclear charge.

S1.7 **a)** Following the example, for an atom of Ni with $Z = 28$ the electron configuration is:

$$\text{Ni: } 1s^2 2s^2 2p^6 3s^2 3p^6 3d^8 4s^2 \text{ or } [Ar]3d^8 4s^2.$$

Once again, the 4s electrons are listed last because the energy of the 4s orbital is higher than the energy of the 3d orbitals. Despite this ordering of the individual 3d and 4s energy levels for elements past Ca (see Figure 1.19), interelectronic repulsions prevent the configuration of an Ni atom from being [Ar]$3d^{10}$. For an Ni^{2+} ion, with two fewer electrons than an Ni atom but with the same Z as an Ni atom, interelectronic repulsions are less important. Because of the higher energy 4s electrons as well as smaller Z_{eff} than the 3d electrons, the 4s electrons are removed from Ni to form Ni^{2+}, and the electron configuration of the ion is:

$$Ni^{2+}: 1s^22s^22p^63s^23p^63d^8 \text{ or } [Ar]3d^8.$$

b) The ground state electronic configuration of Cu atom is:

$$1s^22s^22p^63s^23p^63d^{10}4s^1 \text{ or } [Ar]3d^{10}4s^1.$$

Recall from the textbook Section 1.5(b) that copper atom is one exception to the expected electronic configuration: it is not $3d^94s^2$ (as would be anticipated following nickel's $3d^84s^2$) but rather $3d^{10}4s^1$. Note that this leaves 4s orbital half-filled and 3d orbitals filled.
Taking this electronic configuration of the neutral Cu, we can easily obtain the electronic configurations for ions by removing the outermost electrons:

$$Cu^{+}: 1s^22s^22p^63s^23p^63d^{10} \text{ or } [Ar]3d^{10}$$

and

$$Cu^{2+}: 1s^22s^22p^63s^23p^63d^9 \text{ or } [Ar]3d^9.$$

S1.8 **a)** The valence electrons are in the $n = 4$ shell. Therefore, the element is in period 4 of the periodic table. It has two valence electrons that are in a 4s orbital, indicating that it is in Group 2. Therefore, the element is calcium, Ca.

b) In this case the highest value for the quantum number n is 5 placing the element in period 5. Note that there is an incomplete d subshell in shell 4: $4d^4$. This means that our element belongs in d block and has six valence electrons in total (four in 4d and two in 5s). The element is in group 6. This place in the periodic table belongs to molybdenum (Mo).

S1.9 **a)** When considering questions like these, it is always best to begin by writing down the electron configurations of the atoms or ions in question. If you do this routinely, a confusing comparison may become more understandable. In this case the relevant configurations are:

$$F: 1s^22s^22p^5 \text{ or } [He]2s^22p^5$$

$$Cl: 1s^22s^22p^63s^23p^5 \text{ or } [Ne]3s^23p^5.$$

The electron removed during the ionization process is a 2p electron for F and a 3p electron for Cl. The principal quantum number, n, is lower for the electron removed from F ($n = 2$ for a 2p electron), so this electron is bound more strongly by the F nucleus than a 3p electron in Cl is bound by its nucleus.

A general trend: within a group, the first ionization energy decreases down the group because in the same direction the atomic radii and principal quantum number n increase. There are only a few exceptions to this trend, and they are found in Groups 13 and 14.

b) It is important to recall that s orbitals are somewhat lower in energy than the p orbitals with the same quantum number n. Looking at the relevant electronic configurations

$$Mg: 1s^22s^22p^63s^2$$
$$Al: 1s^22s^22p^63s^23p^1$$

we see that ionization removes a 3s electron from Mg atom and the 3p electron from Al atom. Greater stability of s electrons requires greater ionization energy, and magnesium's I_1 is greater than aluminium's.

S1.10 When considering questions like these, look for the highest jump in energies. This occurs for the fifth ionization energy of this element: $I_4 = 6229$ kJ mol^{-1}, while $I_5 = 37838$ kJ mol^{-1}, indicating breaking into a complete subshell after the removal of the fourth electron. Therefore, the element is in Group 14 (C, Si, Ge, etc.).

S1.11 **a)** The electron configurations of these two atoms are:

$$C: [He]2s^22p^2 \text{ and } N: [He]2s^22p^3.$$

An additional electron can be added to the empty 2p orbital of C, and this is a favourable process ($E_a = 122$ kJ mol^{-1}). However, all 2p orbitals of N are already half occupied, so an additional electron added to N would experience sufficiently strong interelectronic repulsions. Therefore, the electron-gain process for N is unfavourable (E_a = –8 kJ mol^{-1}). This is despite the fact that the 2p Z_{eff} for N is larger than the 2p Z_{eff} for C (see Table 1.2). This tells you that attraction to the nucleus is not the only force that determines electron affinities (or, for that matter, ionization energies). Interelectronic repulsions are also important.

b) Recall from section 1.7(c) "Electron affinity" that negative values of E_a indicate lower stability of A^-(g) than A(g). The fact that all Group 18 elements have negative E_a values indicates that their anions are less stable than neutral atoms in the gas phase. A new electron would occupy completely new electronic shell for the Group 18 elements (for example He(g) would accept an electron in 2s orbital to become He^-(g)). These are well shielded from the nucleus by completely filled shells of neutral atoms and are consequently very weakly bound making an electron gain for all these elements endothermic.

S1.12 **a)** According to Fajan's rules, small, highly charged cations have polarizing ability. Cs^+ has a larger ionic radius than Na^+. Both cations have the same charge, but because Na^+ is smaller than Cs^+, Na^+ is more polarizing.

b) The important difference between NaF and NaI is the size of the anion: I^- is significantly larger than F^-. Recall from Fajan's rules that larger anions are easier to polarize, and (following Example 1.12 as well), I^- is easier to polarize. Consequently, Na^+ can distort the electron distribution of I^- more than that of F^- pulling electron cloud closer and making NaI bond more covalent in character.

Exercises

E1.1 The energy of a hydrogenic ion, like He^+ or Be^{3+}, is defined by equation 1.3:

$$E_n = -\frac{Z^2 Rhc}{n^2}.$$

Both He^+ and Be^{3+} have ground state electronic configuration $1s^1$; thus, for both, the principal quantum number n in the above equation equals 1. For the ratio $E(He^+)/E(Be^{3+})$, after cancelling all constants, we obtain:

$$E(He^+)/E(Be^{3+}) = Z^2(He^+)/Z^2(Be^{3+}) = 2^2/4^2 = 0.25.$$

E1.2 **a)** The ground state of hydrogen atom is $1s^1$. The wavefunction describing 1s orbital in H atom is given with $\Psi = \frac{1}{(\pi a_0^3)^{1/2}} e^{-r/a_0}$. The values of this function for various values of r/a_0 are plotted on Figure 1.8. From the plot, we can see that the most probable location of an electron in this orbital is at nucleus because that is where this function has its maximum.

b) The most probable distance from the nucleus for a 1s electron in H atom can be determined from the radial distribution function, $P(r)$, for 1s orbital. The radial distribution function for 1s orbital is given by $P(r) = 4\pi r^2 R(r)^2 = 4\pi r^2 \times \frac{1}{\pi a_0^3} e^{-2r/a_0} = \frac{4r^2}{a_0^3} e^{-2r/a_0}$. Remember that the wavefunction for 1s orbitals depends only on radius, not on angles, thus $R_{1s}(r) = \Psi_{1s}$. To find the most probable distance we must find the value of r where the value for the radial distribution function is maximum. This can be done by finding the first derivative of $P(r)$ by r, making this derivative equal to zero and solving the equation for r:

$$\frac{P(r)}{dr} = \frac{8r}{a_0^3} e^{-2r/a_0} - \frac{8r^2}{a_0^4} e^{-2r/a_0} = \frac{8}{a_0^3}\left(2r - \frac{2r^2}{a_0}\right) e^{-2r/a_0} = 0.$$

The exponential part of this derivative is never zero, so we can take the part in the brackets, make it equal to zero, and solve it for r:

$$\left(2r - \frac{2r^2}{a_0}\right) = 0$$

$$2r = \frac{2r^2}{a_0} \Rightarrow r = a_0.$$

Thus, the most probable distance from the nucleus is exactly at Bohr radius, a_0. This value and the value determined in part a) of this question are different because the radial distribution function gives the probability of finding an electron anywhere in a shell of thickness dr at radius r regardless of the direction, whereas the wavefunction simply describes the behaviour of electron.

c) Similar procedure is followed as in b) but using the $R(r)$ function for 2s electron. The result is $3+\sqrt{5}a_0$.

E1.3 The expression for E given in Equations 1.3 and 1.4 (see Exercise 1.1 as well) can be used for a hydrogen atom as well as for hydrogenic ions. For the ratio $E(\text{H}, n = 1)/E(\text{H}, n = 6)$, after cancelling all constants, we obtain:

$$E(\text{H}, n = 1)/E(\text{H}, n = 6) = (1/1^2)/(1/6^2) = 36.$$

The value for $E(\text{H}, n = 1)$ has been given in the problem (13.6 eV). From this value and above ratio we can find $E(\text{H}, n = 6)$ as: $E(\text{H}, n = 6) = (E(\text{H}, n = 1))/36 = 13.6\text{ eV}/36 = 0.378\text{ eV}$, and the difference is:

$$E(\text{H}, n = 1) - E(\text{H}, n = 6) = 13.6\text{ eV} - 0.378\text{ eV} = 13.2\text{ eV}.$$

E1.4 Both rubidium and silver are in period 5; hence their valence electrons are in their 5s atomic orbitals. If we place hydrogen's valence electron in 5s orbital, the ionization energy would be:

$$I = -E_{5s} = \frac{hcRZ^2}{n^2} = \frac{13.6\,\text{eV}}{5^2} = 0.544\,\text{eV}.$$

This is significantly lower value than for either Rb or Ag. First the difference for H atom only: the energy of an electron in a hydrogenic atom is inversely proportional to the square of the principal quantum number n. Hence, we can expect a sharp decrease in I with an increase in n (i.e., from 13.6 eV for $n = 1$ to 0.544 eV for $n = 5$). The difference between H(5s^1) on one side and Rb and Ag on the other lies in Z (atomic number or nuclear charge; or better Z_{eff}–effective nuclear charge); as we increase Z (or Z_{eff}) the ionization energy is exponentially increasing (Z(Rb) = 37 and Z(Ag) = 47). And finally, the difference in ionization energies for Rb and Ag is less than expected based on the difference in their nuclear charges; we would expect significantly higher I value for Ag than is actually observed. The discrepancy is due to the shielding effect and Z_{eff}; in comparison to Rb, silver has an additional ten 4d electrons placed between the nucleus and its 5s valence electron. These 4d electrons shield the 5s electron from the nucleus (although, being d electrons, not very efficiently).

E1.5 When a photon emitted from the helium lamp collides with an electron in an atom, one part of its energy is used to ionize the atom while the rest is converted to the kinetic energy of the electron ejected in the ionization process. Thus, the total energy of a photon ($h\nu$) is equal to the sum of the first ionization energy (I_1) and the kinetic energy of an electron ($m_e v_e^2/2$):

$$h\nu = I_1 + \frac{m_e v_e^2}{2}.$$

From here the ionization energy is:

$$I_1 = h\nu - \frac{m_e v_e^2}{2}.$$

Since both krypton and rubidium atoms are ionized with the same radiation, we can calculate the energy of one photon emitted from the helium discharge lamp:

$$h\nu = h \times \frac{c}{\lambda} = 6.626\times10^{-34}\,\text{J s} \times \frac{2.998\times10^{8}\,\text{m s}^{-1}}{58.4\times10^{-9}\,\text{m}} = 3.40\times10^{-18}\,\text{J}.$$

Now we can calculate the first ionization energies using given velocities of respective electrons:

$$I_1^{Kr} = h\nu - \frac{m_e v_{e,Kr}^2}{2} = 3.40\times10^{-18}\,\text{J} - \frac{9.109\times10^{-31}\,\text{kg}\times\left(1.59\times10^{6}\,\text{m s}^{-1}\right)^2}{2} = 2.25\times10^{-18}\,\text{J}$$

$$I_1^{Rb} = h\nu - \frac{m_e v_{e,Rb}^2}{2} = 3.40\times10^{-18}\,\text{J} - \frac{9.109\times10^{-31}\,\text{kg}\times\left(2.45\times10^{6}\,\text{m s}^{-1}\right)^2}{2} = 6.68\times10^{-18}\,\text{J}.$$

Note that calculated energies are the first ionization energies per one atom because we used only one photon. To calculate the first ionization energies in eV as asked in the exercise, the above energies must be multiplied by the Avogadro's constant (to obtain the energies in J mol^{-1}) and then divided by the conversion factor 96485 J mol^{-1} eV^{-1} to obtain the values in eV:

$$I^{Kr}_{1,eV} = 2.25\times10^{-18}\,\text{J}\times\frac{6.022\times10^{23}\text{mol}^{-1}}{96485\,\text{J mol}^{-1}\,\text{eV}^{-1}} = 14.0\,\text{eV}$$

$$I^{Rb}_{1,eV} = 6.68\times10^{-18}\,\text{J}\times\frac{6.022\times10^{23}\text{mol}^{-1}}{96485\,\text{J mol}^{-1}\,\text{eV}^{-1}} = 4.16\text{ eV}.$$

E1.6 To solve this exercise we are going to recall equation 1.1 and substitute the given values for n_1 and n_2:

$$\frac{1}{\lambda} = R\left(\frac{1}{n_1^2}-\frac{1}{n_2^2}\right) = 1.097\times10^7\,\text{m}^{-1}\left(\frac{1}{1^2}-\frac{1}{3^2}\right) = 1.097\times10^7\,\text{m}^{-1}\times0.899 = 9752330\,\text{m}^{-1}.$$

The corresponding wavelength is $(9752330\text{ m}^{-1})^{-1}$ = 102.5 nm.

The energy is: $h\nu = h\times\frac{c}{\lambda} = 6.626\times10^{-34}\,\text{J s}\times2.998\times10^8\,\text{m s}^{-1}\times9752330\,\text{m}^{-1} = 1.937\times10^{-18}\,\text{J}$.

E1.7 The visible region starts when n_1 = 2. The next transition is for n_2 = 3. This can be determined using the Rydberg equation (Equation 1.1).

$$\frac{1}{\lambda} = R\left(\frac{1}{2^2}-\frac{1}{3^2}\right) = 1.524\times10^{-3}\,\text{nm}^{-1}.$$

And from the above result, λ = 656.3 nm.

E1.8 The version of the Rydberg equation which generates the Lyman series is:

$$\frac{1}{\lambda} = R\left(\frac{1}{1^2}-\frac{1}{n^2}\right) \quad R = 1.097\times10^7\text{ m}^{-1}.$$

Where n is a natural number greater than or equal to n = 2 (i.e., n = 2, 3, 4 ,..., ∞). There are infinitely many spectral lines, but they become very dense as n approaches infinity, so only some of the first lines and the last one appear. If we let $n = \infty$, we get an approximation for the first line:

$$\frac{1}{\lambda} = R\left(\frac{1}{1^2}-\frac{1}{\infty^2}\right) = 1.0974\times10^7\,\text{m}^{-1}$$

and $\lambda = 91.124\,\text{nm}$.

The next line is for n = 4:

$$\frac{1}{\lambda} = R\left(\frac{1}{1^2}-\frac{1}{4^2}\right) = 1.0288\times10^7\,\text{m}^{-1}$$

and $\lambda = 97.199\,\text{nm}$.

The next line is for n = 3:

$$\frac{1}{\lambda} = R\left(\frac{1}{1^2}-\frac{1}{3^2}\right) = 9.7547\times10^6\,\text{m}^{-1}$$

$$\lambda = 102.52\,\text{nm}.$$

The final line is for n = 2:

$$\frac{1}{\lambda} = R\left(\frac{1}{1^2}-\frac{1}{2^2}\right) = 8.2305\times10^6\,\text{m}^{-1}$$

$$\lambda = 121.499\,\text{nm}.$$

These numbers predict the Lyman series within a few significant figures.

E1.9 The principal quantum number n labels one of the shells of an atom. For a hydrogen atom or a hydrogenic ion, n alone determines the energy of all orbitals contained in a given shell (since there are n^2 orbitals in a shell, these would be n^2-fold degenerate). For a given value of n, the angular momentum quantum number l can assume all integer values from 0 to $n - 1$.

E1.10 For the first shell ($n = 1$), there is only one orbital, the 1s orbital. For the second shell ($n = 2$), there are four orbitals, the 2s orbital and the three 2p orbitals. For $n = 3$, there are 9 orbitals, the 3s orbital, three 3p orbitals, and five 3d orbitals. The progression of the number of orbitals so far is 1, 4, 9, which is the same as n^2 (e.g., $n^2 = 1$ for $n = 1$, $n^2 = 4$ for $n = 2$, etc.). As a further verification, consider the fourth shell ($n = 4$), which, according to the analysis so far, should contain $4^2 = 16$ orbitals. Does it? Yes; the fourth shell contains one 4s orbital, three 4p orbitals, five 4d orbitals, and seven 4f orbitals, and $1 + 3 + 5 + 7 = 16$.

E1.11 Completed table:

n	l	m_l	Orbital designation	Number of orbitals
2	**1**	**+1, 0, –1**	2p	**3**
3	2	**+2, +1, ..., –2**	**3d**	**5**
4	**0**	**0**	4s	**1**
4	**3**	+3, +2, ..., –3	**4f**	**7**

(Note: the table entries in bold are the sought solutions.)

E1.12 When $n = 5$, $l = 3$ (for the f orbitals) and $m_l = +3, +2, +1, 0, -1, -2, -3$, which represent the seven orbitals that complete the 5f subshell. The 5f orbitals represent the start of the actinoids, starting with Th and ending with Lr.

E1.13 The plots of R (the radial part of the wavefunction ψ) vs. r shown in Figures 1.8 and 1.9 are plots of the radial parts of the total wavefunctions for the indicated orbitals. Notice that the function R(2s) can take both positive and negative values (Figure 1.8), requiring that for some value of r the wavefunction R(2s) = 0 (i.e., the wavefunction has a node at this value of r; for a hydrogen atom or a hydrogenic ion, R(2s) = 0 when $r = 2a_0/Z$). Notice also that the plot of R(2p) vs. r is positive for all values of r (Figure 1.9). Although a 2p orbital does have a node, it is not due to the radial part of the total wavefunction but rather due to the angular part, Y.

The radial distribution function is $P(r) = r^2R^2$ (for the s orbitals this expression is the same as $4\pi r^2\psi^2$). The plot of r^2R^2 vs. r for a 1s orbital in Figure 1.10 is a radial distribution function. Figure 1.12 provides plots of the radial distribution functions for the hydrogenic 2s and 2p orbitals.

Comparing the plots for 1s (Figure 1.10) and 2s (Figure 1.12) orbitals we should note that the radial distribution function for a 1s orbital has a single maximum, and that for a 2s orbital has two maxima and a minimum (at $r = 2a_0/Z$ for hydrogenic 2s orbitals). The presence of the node at $r = 2a_0/Z$ for R(2s) requires the presence of the two maxima and the minimum in the 2s radial distribution function. Using the same reasoning, the absence of a radial node for R(2p) requires that the 2p radial distribution function has only a single maximum, as shown in Figure 1.12.

E1.14 Your sketch should be similar to the one shown below:

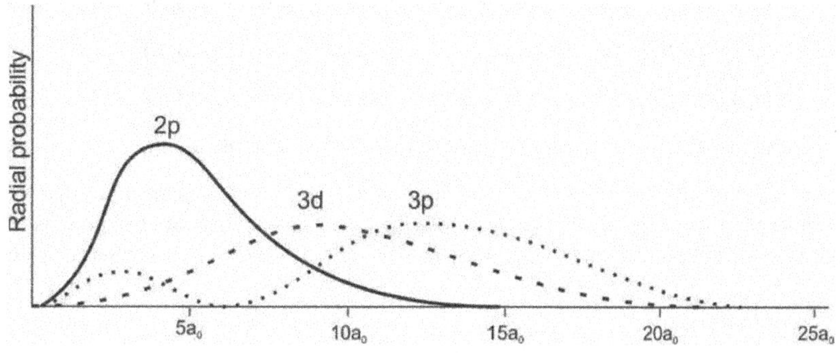

An important feature to note is that a 3p orbital has a local maximum closer to the nucleus (i.e., to the origin of the plot). That indicates that 3p electrons can approach the atomic nucleus closer than 3d electrons. Therefore, 3p electrons are better stabilized by a positive nuclear charge and thus have lower energy than 3d electrons.

E1.15 An orbital defined with the quantum number l has l nodal planes (or angular nodes). For a 4p set of atomic orbitals $l = 1$, and each 4p orbital has one nodal plane. The number of radial nodes is given by $n - l - 1$; and 4p orbitals have $4 - 1 - 1 = 2$ radial nodes. The total number of nodes is thus 3. (Note that the total number of nodes, angular and radial, is given by $n - 1$.)

E1.16 The two orbitals are d_{xy} and $d_{x^2-y^2}$. The sketches, mathematical functions, and labelled pairs of Cartesian coordinates are provided below.

Orbital	Sketch	Function	Label
d_{xy}	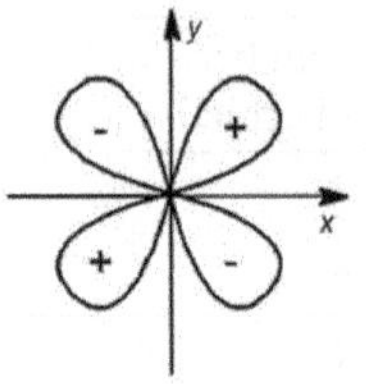	$xyR(r)$	xy
$d_{x^2-y^2}$	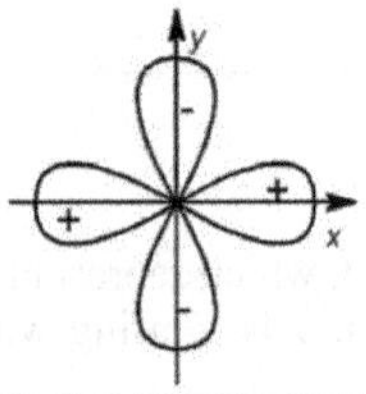	$\frac{1}{2}(x^2 - y^2)R(r)$	$x^2 - y^2$

Note: $R(r)$ for 3d orbitals is $\frac{1}{81\sqrt{30}}\left(\frac{Z}{a_0}\right)^{3/2}\frac{4Z^2r^2}{a_0^2}e^{-Zr/3a_0}$.

E1.17 In an atom with more than one electron, as beryllium, the outer electrons (the 2s electrons in this case) are simultaneously attracted to the positive nucleus (the protons in the nucleus) and repelled by the negatively charged electrons occupying the same orbital. The two electrons in the 1s orbital on average are statically closer to the nucleus than the 2s electrons, thus the 1s electrons "feel" more positive charge than the 2s electrons. The 1s electrons also shield that positive charge from the 2s electrons, which are further out from the nucleus than the 1s electrons. Consequently, the 2s electrons "feel" less positive charge than the 1s electrons for beryllium.

E1.18 Follow Slater's rules outlined in the text to calculate the values for shielding constants, σ.

Li: $1s^22s^1$; $(1s^2)(2s^1)$, $\sigma = 2 \times 0.85 = \mathbf{1.70}$

Be: $1s^22s^2$; $(1s^2)(2s^2)$, $\sigma = (2 \times 0.85) + (1 \times 0.35) = \mathbf{2.05}$

B: $1s^22s^22p^1$; $(1s^2)(2s^22p^1)$, $\sigma = (2 \times 0.85) + (2 \times 0.35) = \mathbf{2.40}$

C: $1s^22s^22p^2$; $(1s^2)(2s^22p^2)$, $\sigma = (2 \times 0.85) + (3 \times 0.35) = \mathbf{2.75}$

N: $1s^22s^22p^3$; $(1s^2)(2s^22p^3)$, $\sigma = (2 \times 0.85) + (4 \times 0.35) = \mathbf{3.10}$

O: $1s^22s^22p^4$; $(1s^2)(2s^22p^4)$, $\sigma = (2 \times 0.85) + (5 \times 0.35) = \mathbf{3.45}$

F: $1s^22s^22p^5$; $(1s^2)(2s^22p^5)$, $\sigma = (2 \times 0.85) + (6 \times 0.35) = \mathbf{3.80}$

The trend is as expected, the shielding constant increases going across a period because of increased number of electrons entering the same shell. For example, the one 2s electron in Li atom is shielded by two 1s electrons; but one valence electron in Be atom is shielded by both 1s electrons as well as the second 2s electron.

E1.19 Chromium second ionization energy is higher than expected because the second electron is removed from a 3d orbital. In the case of manganese, the second electron is removed from a 4s orbital. Since 3d is lower in energy (more stable) than 4s, I_2 for Cr is higher than that for Mn.

E1.20 The first ionization energies of calcium and zinc are 6.11 and 9.39 eV, respectively (these values can be found in the Resource Section 2 of your textbook). Both atoms have an electron configuration that ends with $4s^2$: Ca is $[Ar]4s^2$ and Zn is $[Ar]3d^{10}4s^2$. An atom of zinc has 30 protons in its nucleus and an atom of calcium has 20, so clearly zinc has a higher nuclear charge than calcium. Remember, though, that it is *effective* nuclear charge (Z_{eff}) that directly affects the ionization energy of an atom. Since I(Zn) > I(Ca), it would seem that Z_{eff}(Zn) > Z_{eff}(Ca). How can you demonstrate that this is as it should be? The actual nuclear charge can always be readily determined by looking at the periodic table and noting the atomic number of an atom. The effective nuclear charge cannot be directly determined, that is, it requires some interpretation on your part. Read Section 1.4 *Penetration and shielding* again and pay attention to the orbital shielding trends. Also, study the trend in Z_{eff} for the period 2 p-block elements in Table 1.2. The pattern that emerges is that not only Z but also Z_{eff} rises from boron to neon. Each successive element has one additional proton in its nucleus and one additional electron to balance the charge. However, the additional electron never completely shields the other electrons in the atom. Therefore, Z_{eff} rises from B to Ne. Similarly, Z_{eff} rises through the d block from Sc to Zn, and that is why Z_{eff}(Zn) > Z_{eff}(Ca). As a consequence, I_1(Zn) > I_1(Ca).

E1.21 The first ionization energies of strontium, barium, and radium are 5.69, 5.21, and 5.28 eV. Normally, atomic radius increases and ionization energy decreases down a group in the periodic table. However, in this case I(Ba) < I(Ra). Study the periodic table, especially the elements of Group 2 (the alkaline earth elements). Notice that Ba is 18 elements past Sr, but Ra is 32 elements past Ba. The difference between the two corresponds to the fourteen 4f elements (the lanthanoids) between Ba and Lu. Therefore, radium has a higher first ionization energy because it has such a large Z_{eff} due to the insertion of the lanthanides.

E1.22 The second ionization energies of the elements calcium through manganese increase from left to right in the periodic table with the exception that I_2(Cr) > I_2(Mn). The electron configurations of the elements are:

Ca	Sc	Ti	V	Cr	Mn
$[Ar]4s^2$	$[Ar]3d^14s^2$	$[Ar]3d^24s^2$	$[Ar]3d^34s^2$	$[Ar]3d^54s^1$	$[Ar]3d^54s^2$

Both the first and the second ionization processes remove electrons from the 4s orbital of these atoms, with the exception of Cr. In general, the 4s electrons are poorly shielded by the 3d electrons, so Z_{eff}(4s) increases from left to right and I_2 also increases from left to right. While the I_1 process removes the sole 4s electron for Cr, the I_2 process must remove a 3d electron. The higher value of I_2 for Cr relative to Mn is a consequence of the special stability of half-filled subshell configurations.

E1.23 See Example and Self-Test 1.7 in your textbook (if necessary review Section 1.5(a) of your textbook).

(a) C? Carbon is the sixth element in the periodic table and we have to sort six electrons in atomic orbitals obeying the build-up principle and the Hund's rule. Starting from the bottom we place two electrons in 1s, two in 2s (this fills both 1s and 2s) and finally two remaining electrons are placed in two different 2p orbitals with parallel spins. This gives us the configuration $1s^22s^22p^2$. Recognizing that $1s^2$ is in fact the electronic configuration of preceding noble gas helium, the short-hand configuration is $[He]2s^22p^2$. Follow the same principle for questions (b)–(f).

(b) F? Atomic number (Z) is 9, and nine electrons have to be sorted: $1s^22s^22p^5$ or $[He]2s^22p^5$.

(c) Ca? Z = 20, and 20 electrons: $1s^22s^22p^63s^23p^64s^2$ or $[Ar]4s^2$. Note that we have placed 19th and 20th electron in 4s and not 3d orbitals. Review section 1.4 of your textbook and pay attention to Figure 1.20 if you need further explanation for this apparent reversal in expected filling order.

(d) Ga^{3+}? Z = 31, Ga has 31 electrons, but we have to sort 28 because we are looking at a 3+ cation: $1s^22s^22p^63s^23p^63d^{10}$ or $[Ar]3d^{10}$. Note that Ga is just past the first row of the d-block so the 3d orbitals are filled.

(e) Bi? Z = 83 and 83 electrons. We have to be a bit careful here: before Bi in the periodic table, we find d-block elements (La to Hg) of the sixth row, as well as elements Ce to Lu from the f-block. So Bi would have filled 5d and 4f atomic orbitals. Starting from the bottom we would have: $1s^22s^22p^63s^23p^63d^{10}4s^24p^64d^{10}4f^{14}5s^25p^65d^{10}6s^26p^3$ and $[Xe]4f^{14}5d^{10}6s^26p^3$. Note that Xe ends period 5 of the periodic table and has 5d and 4f subshells empty.

(f) Pb^{2+}? $Z = 82$, but 80 electrons (because it is a 2+ cation): $1s^22s^22p^63s^23p^63d^{10}4s^24p^64d^{10}4f^{14}5s^25p^65d^{10}6s^2$ or $[Xe]4f^{14}5d^{10}6s^2$.

E1.24 Following instructions for Question 1.23:

(a) Sc? $Z = 21$: $1s^22s^22p^63s^23p^63d^14s^2$ or $[Ar]3d^14s^2$. Note the order of the subshells, with Sc 3d subshell is lower in energy than 4s. The same applies to (b)–(g) parts of this exercise.

(b) V^{3+}? $Z = 23$ but 3+ charge leaves us with 20 electrons to sort: $1s^22s^22p^63s^23p^63d^2$ or $[Ar]3d^2$. Note that it might seem tempting to place the last two electrons in 4s orbital (just like for Ca that has the same number of electrons as V^{3+}). However, V is a member of the d-block and, inside this block, 3d orbitals are lower in energy than 4s orbitals and thus are filled first. Another way to check this (and some might find it easier) is to start with the ground state configuration for neutral V atom ($1s^22s^22p^63s^23p^63d^34s^2$) and then remove the necessary electrons—in this case three. We first start removing electrons from the subshell highest in energy, in this case 4s, and then if necessary move to lower subshells. Removing three electrons from V atom would leave 4s empty and take away one electron from the 3d subshell. This note and strategy applies to other cases in this and next exercise.

(c) Mn^{2+}? $Z = 25$ but 2+ charge leaves us with 23 electrons to sort; $1s^22s^22p^63s^23p^63d^5$ or $[Ar]3d^5$.

(d) Cr^{2+}? $Z = 24$ but 2+ charge leaves us with 22 electrons to sort: $1s^22s^22p^63s^23p^63d^4$ or $[Ar]3d^4$.

(e) Co^{3+}? $Z = 27$ but 3+ charge leaves us with 24 electrons to sort; $1s^22s^22p^63s^23p^63d^6$ or $[Ar]3d^6$.

(f) Cr^{6+}? $Z = 24$, but with a +6 charge it has the *same* number of electrons (18) and electron configuration as Ar: $1s^22s^22p^63s^23p^6$ or [Ar]. Sometimes inorganic chemists will write the electron configuration as $[Ar]3d^0$ to emphasize that there are no d electrons for this d-block metal ion in its highest oxidation state.

(g) Cu? $Z = 29$, and 29 electrons: $1s^22s^22p^63s^23p^63d^{10}4s^1$ or $[Ar]3d^{10}4s^1$. Recall Section 1.5(b) and Self-test 1.7(b) about copper's ground state electronic configuration being an exception.

(h) Gd^{3+}? $Z = 64$ but 3+ charge leaves us with 61 electrons to sort. Gd belongs to Period 6, f-block of the periodic table where 4f subshell is being filled. Note that 3d and 4d levels have already been filled in the d-block. $1s^22s^22p^63s^23p^63d^{10}4s^24p^64d^{10}4f^75s^25p^6$ or $[Xe]4f^7$. Note where 4f orbital is in the configuration.

E1.25 Following instructions for Question 1.23:

(a) W? $Z = 74$. If you assumed that the configuration would resemble that of chromium, you would write $[Xe]4f^{14}5d^56s^1$. It turns out that the actual configuration is $[Xe]4f^{14}5d^46s^2$. The configurations of the heavier d- and f-block elements show some exceptions to the trends for the lighter d-block elements.

(b) Rh^{3+}? $Z = 45$ but 3+ charge leaves us with 42 electrons to sort: $1s^22s^22p^63s^23p^63d^{10}4s^24p^64d^6$ or $[Kr]4d^6$.

(c) Eu^{3+}? $Z = 63$ but 3+ charge leaves us with 61 electrons to sort (similar to Gd^{3+} in the previous Exercise) $1s^22s^22p^63s^23p^63d^{10}4s^24p^64d^{10}4f^75s^25p^6$ or $[Xe]4f^6$.

(d) Eu^{2+}? This will have one more electron than Eu^{3+}. Therefore, the ground-state electron configuration of Eu^{2+} is $[Xe]4f^7$.

(e) V^{5+}? $Z = 23$ but with its 5+ charge it has the *same* number of electrons (18) and electron configuration as Ar: $1s^22s^22p^63s^23p^6$ or $[Ar]3d^0$. See the note for Exercise 1.24(f) above.

(f) Mo^{4+}? $Z = 42$ but 4+ charge leaves us with 38 electrons to sort: $1s^22s^22p^63s^23p^63d^{10}4s^24p^64d^2$ or $[Kr]4d^2$.

E1.26 See Example and Self-Test 1.8 in your textbook.
(a) S
(b) Sr
(c) V
(d) Tc
(e) In
(f) Sm

E1.27 See Figure 1.22 and the inside front cover of this book. You should start learning the names and positions of elements that you do not know. Start with the alkali metals and the alkaline earths. Then learn the elements in the p-block. A blank periodic table can be found on the inside back cover of this book. You should make several photocopies of it and should test yourself from time to time, especially after studying each chapter. This will equip you with a very important skill for mastering inorganic chemistry: being able to navigate through the periodic table is essential for many topics covered in this book.

E1.28 The following values were taken from Tables 1.5, 1.6, and 1.7:

Element	Configuration	I_1(eV)	E_a (eV)	χ
Na	[Ne]$3s^1$	5.14	0.548	0.93
Mg	[Ne]$3s^2$	7.64	–0.4	1.31
Al	[Ne]$3s^23p^1$	5.98	0.441	1.61
Si	[Ne]$3s^23p^2$	8.15	1.385	1.90
P	[Ne]$3s^23p^3$	11.0	0.747	2.19
S	[Ne]$3s^23p^4$	10.36	2.077	2.58
Cl	[Ne]$3s^23p^5$	13.10	3.617	3.16
Ar	[Ne]$3s^23p^6$	15.76	–1.0	

In general, I_1, E_a, and χ all increase from left to right across period 3 (or from top to bottom in the table above). All three quantities reflect how tightly an atom holds on to its electrons, or how tightly it holds on to additional electrons. The cause of the general increase across the period is the gradual increase in Z_{eff}, which itself is caused by the incomplete shielding of electrons of a given value of *n* by electrons with the same *n*. The exceptions are explained as follows: I_1(Mg) > I_1(Al) and E_a(Na) > E_a(Al)—both of these are due to the greater stability of 3s electrons relative to 3p electrons; E_a(Mg) and E_a(Ar) < 0—filled subshells impart a special stability to an atom or ion (in these two cases the additional electron must be added to a higher energy subshell (for Mg) or shell (for Ar)); I_1(P) > I_1(S) and E_a(Si) > E_a(P)—the loss of an electron from S and the gain of an additional electron by Si both result in an ion with a half-filled p subshell, which, like filled subshells, imparts a special stability to an atom or ion.

E1.29 To follow the answer for this question you should have a copy of the periodic table of elements in front of you. If you look at Table 1.3 in your textbook, you will see that this is a general trend. Normally, the period 6 elements would be expected to have larger metallic radii than their period 5 vertical neighbours; only Cs and Ba follow this trend; Cs is larger than Rb and Ba is larger than Sr. Lutetium, Lu, is significantly smaller than yttrium, Y, and Hf is almost the same size as Zr. The same similarity in radii between the period 5 and period 6 vertical neighbours can be observed for the rest of the d-block. This phenomenon can be explained as follows. There are no intervening elements between Sr and Y, but there are 14 intervening elements, the lanthanides, between Ba and Lu. A contraction of the radii of the elements starting with Lu is due to incomplete shielding by the 4f electrons. By the time we pass the lanthanides and reach Hf and Ta, the atomic radii have been contracted so much that the d-block period 6 elements have almost identical radii to their vertical neighbours in the period 5.

E1.30 Recall from Section 1.7(c) *Electron affinity*, that the frontier orbitals are the highest occupied and the lowest unoccupied orbitals of a chemical species (atom, molecule, or ion). Since the ground-state electron configuration of a beryllium atom is $1s^22s^2$, the frontier orbitals are the 2s orbital (highest occupied) and three 2p orbitals (lowest unoccupied). Note that there can be more than two frontier orbitals if either the highest occupied and/or lowest unoccupied energy levels are degenerate. In the case of beryllium, we have four frontier orbitals (one 2s and three 2p).

E1.31 Mulliken defined electronegativity as an average value of the ionization energy (*I*) and electron affinity (E_a) of the element, that is, $\chi_M = ½ (I + E_a)$, thus making electronegativity an atomic property (just like atomic radius or ionization energy). Since both *I* and E_a should have units eV for Mulliken electronegativity scale, the data in Tables 1.6 and 1.7 have to be converted from J/mol to eV (these can also be found in the Resource Section 2 of your textbook). The graph below plots variation of $I_1 + E_a$ (vertical axis) across the second row of the periodic table. The numbers next to the data points are the values of Mulliken electronegativity for the element shown on the x-axis. As we can see, the trend is almost linear, supporting Mulliken's proposition that electronegativity values are proportional to $I + E_a$. There are, however, two points on the plot, namely for B and O, that significantly deviate from linearity. Boron (B) should have significatly higher electronegativity than lithium (Li) because: (1) its I_1 is higher than that for Li and (2) B's E_a is more negative than Li's. This discrepancy is due to the fact that Mulliken's electronegativity scale does not use *I* and E_a values for the atomic ground-states (which are the values of I_1 and E_a used to construct the above graph). It rather uses the values for valence states; for boron this state is $2s^12p^12p^1$ (an elctronic state that allows B atom to have its normal valency 3). It is easier to remove the electron from B's valence configuration and the I_1 drops resulting in lower-than-expected electronegativity for B. Similarly, the graph uses E_{a1} for oxygen (which is positive).

Oxygen's normal valency, however, is 2, and its E_{a2} (which should also be taken into consideration) has negative value.

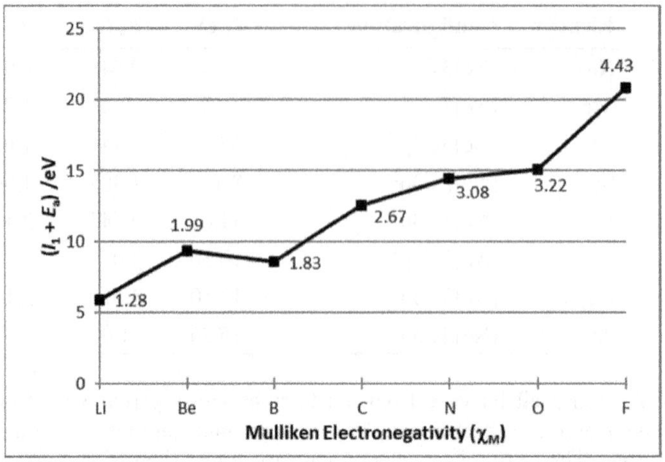

Guidelines for Selected Tutorial Problems

T1.2 Your coverage of early proposals for the periodic table should at least include Döbereiner's triads, Newlands' Law of Octaves, and Meyer's and Mendeleev's tables. From the modern designs (post-Mendeleev) you should consider Hinrichs' spiral periodic table, Benfey's oval table, Janet's left-step periodic table, and Dufour's Periodic Tree.

A good starting reference for this exploration is Eric R. Scerri. (1998). The Evolution of the Periodic System. *Scientific American,* 279, 78–83 and references provided at the end of this article. The website "The Internet Database of Periodic Tables" (http://www.meta-synthesis.com/webbook/35_pt/pt_database.php) is also useful.

T1.6 The variation of atomic radii is one of the periodic trends, and as such could be used to group the elements. Look up the atomic radii of the six elements and see which grouping, (a) or (b), better follows the expected trends in atomic radii. After that you should extend your discussion to the chemical properties of the elements in question and see if your choice makes chemical sense as well.

Chapter 2 Molecular Structure and Bonding

Self-Tests

S2.1 **a)** One phosphorus and three chlorine atoms supply 5 + (3 × 7) = 26 valence electrons. Since P is less electronegative than Cl, it is likely to be the central atom, so the 13 pairs of electrons are distributed as shown below. In this case, each atom obeys the octet rule. Whenever it is possible to follow the octet rule without violating other electron counting rules, you should do so.

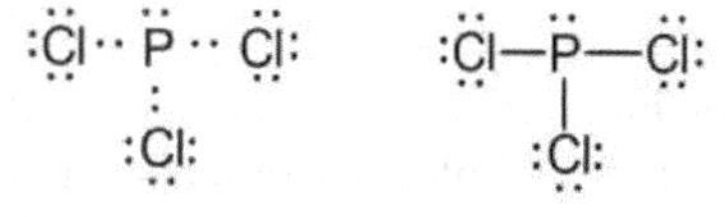

b) BF_3 has in total 3 + 3 × 7 = 24 valence electrons or 12 electron pairs. B atom is less electronegative and it is a likely central atom. The 12 electron pairs can be distributed as follows:

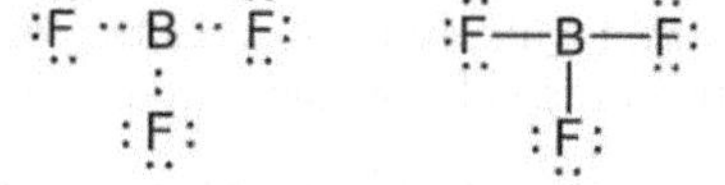

S2.2 The Lewis structures and the shapes of H_2S, XeO_4, and SOF_4 are shown below. According to the VSEPR model, electrons in bonds and in lone pairs can be thought of as charge clouds that repel one another and stay as far apart as possible. First, write a Lewis structure for the molecule, and then arrange the lone pairs and atoms around the central atom, such that the lone pairs are as far away from each other as possible.

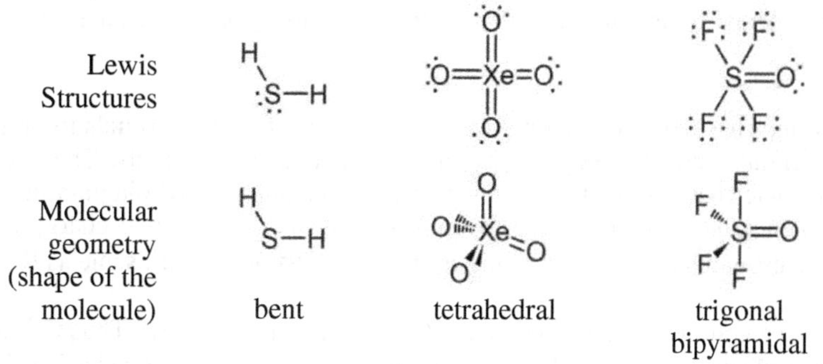

All atoms in both structures have formal charges of zero. The actual structure of XeO_4 is tetrahedral; it is a highly unstable colourless gas. Note that the oxygen atom in SOF_4 molecules lies in one plane with two fluorine atoms. This is because the double bonds contain higher density of negative charge and cause higher repulsion in comparison to single bonds. Placing O atom and four electrons forming S=O bond in equatorial plane reduces repulsion.

S2.3 The Lewis structures and molecular shapes for XeF_2 and ICl_2^+ are shown. The XeF_2 Lewis structure has an octet for the 4 F atoms and an expanded valence shell of 10 electrons for the Xe atom, with the 8 + (2 × 7) = 22 valence electrons provided by the three atoms. The five electron pairs around the central Xe atom will arrange themselves at the corners of a trigonal bipyramid (as in PF_5). The three lone pairs will be in the equatorial plane, to minimize lone pair–lone pair repulsions. The resulting shape of the molecule, shown at the right, is linear (i.e., the F–Xe–F bond angle is 180°).

Two chlorine atoms and one iodine atom in total have 21 electrons. However, we have a cationic species at hand, so we have to remove one electron and start with a total of 20 valence electrons for ICl_2^+. That gives a Lewis structure in which iodine is a central atom (being the more electropositive of the two) and is bonded to two chlorines with a single bond to each. All atoms have a precise octet. Looking at iodine, there are two bonding electron pairs and two lone electron pairs making overall tetrahedral electron pair geometry. The lone pair–

bonding pair repulsion is going to distort the ideal tetrahedral geometry lowering the Cl–I–Cl angle to less than 109.5°, resulting in bent molecular geometry.

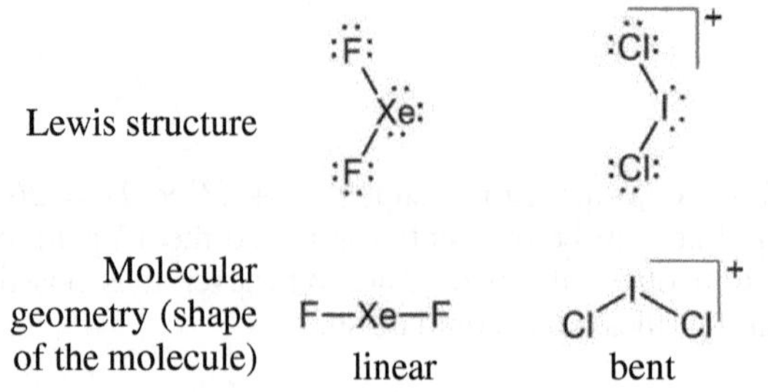

S2.4 **(a)** Figure 2.17 gives the distribution of electrons in molecular orbitals of O_2 molecule. We see that $1\pi_g$ level is a set of two degenerate (of same energy) molecular orbitals. Thus, following the Hund's rule, we obtain the $1\pi_g^2$ electronic configuration of O_2 molecule by placing one electron in each $1\pi_g$ orbital with spins parallel. This gives two unpaired electrons in O_2 molecule.

If we add one electron to O_2 we obtain the next species, the anion O_2^-. This extra electron continues to fill $1\pi_g$ level but it has to have an antiparallel spin with respect to already present electron. Thus, after O_2 molecule receives an electron, one electron pair is formed and only one unpaired electron is left.

Addition of second electron fills $1\pi_g$ set, and O_2^{2-} anion has no unpaired electrons.

(b) The first of these two anions, S_2^{2-}, has the same Lewis structure as peroxide, O_2^{2-}. It also has a similar electron configuration to that of peroxide, except for the use of sulfur atom valence 3s and 3p atomic orbitals instead of oxygen atom 2s and 2p orbitals. There is no need to use sulfur atom 3d atomic orbitals, which are higher in energy than the 3s and 3p orbitals, because the 2(6) + 2 = 14 valence electrons of S_2^{2-} will not completely fill the stack of molecular orbitals constructed from sulfur atom 3s and 3p atomic orbitals. Thus, the electron configuration of S_2^{2-} is $1\sigma_g^2 2\sigma_u^2 3\sigma_g^2 1\pi_u^4 2\pi_g^4$. The Cl_2^- anion contains one more electron than S_2^{2-}, so its electron configuration is $1\sigma_g^2 2\sigma_u^2 3\sigma_g^2 1\pi_u^4 2\pi_g^4 4\sigma_u^1$.

S2.5 ClO^- anion is isoelectronic (has the same number of electrons) with ICl. The orbitals to be used are chlorine's 3s and 3p valence shell orbitals and the oxygen's 2s and 2p valence shell orbitals. The bonding orbitals will be predominantly O in character being that O is more electronegative but the MO diagram of ClO^- will be similar to ICl. We have seven valence electrons from Cl, six from O and one for negative charge giving us a total of 14 electrons. Therefore, the ground-state electron configuration is $1\sigma^2 2\sigma^2 3\sigma^2 1\pi^4 2\pi^4$, same as ICl.

S2.6 **a)** The number of valence electrons for C_2^{2-} is equal to 10 (4 + 4 + 2 (for charge)). Thus C_2^{2-} is isoelectronic with N_2 (which has 10 valence electrons as well). The configuration of C_2^{2-} would be $1\sigma_g^2 1\sigma_u^2 1\pi_u^4 2\sigma_u^2$. The bond order would be ½[2 – 2 + 4 + 2] = 3. C_2^{2-} has a triple bond.

b) The number of valence electrons for Ne_2 is 16. The electronic configuration is $1\sigma_g^2 1\sigma_u^2 2\sigma_g^2 1\pi_u^4 1\pi_g^4 2\sigma_u^2$. The bond order is ½[2 – 2 + 2 + 4 – 4 – 2] = 0. In this case the bond order is 0 because the bonding and antibonding MOs are equally populated. Consequently, the molecule Ne_2 does not exist.

S2.7 In general, the more bonds you have between two atoms, the shorter the bond length and the stronger the bond. Therefore, the ordering for bond length going from shortest to longest is C≡N, C=N, and C–N. For bond strength, going from strongest to weakest, the order is C≡N > C=N > C–N.

S2.8 **a)** The synthesis reaction is:

$$H_2 + F_2 \longrightarrow 2HF.$$

Looking at the reaction we see that 1 mol of H–H and 1 mol of F–F bonds must be broken. Therefore, we must use 436 kJ mol^{-1} for H_2 and 155 kJ mol^{-1} for F_2 to break the bonds. On the right side of the reaction we form 2 mols of HF, indicating that 2 × 565 kJ mol^{-1} has been released. Accounting the use and release of energies with plus and minus signs respectively, we have:

$$\Delta_f H = +436 \text{ kJ mol}^{-1} + 155 \text{ kJ mol}^{-1} - (2 \times 565 \text{ kJ mol}^{-1}) = -539 \text{ kJ mol}^{-1}.$$

b) You can prepare H_2S from H_2 and S_8 in the following reaction:

$$^1/_8 S_8 + H_2 \rightarrow H_2 S.$$

On the left side, you must break one H–H bond and produce one sulfur atom from cyclic S_8. Since there are eight S–S bonds holding eight S atoms together, you must supply the mean S–S bond enthalpy *per S atom*. On the right side, you form two H–S bonds. From the values given in Table 2.8, you can estimate:

$$\Delta_f H = +436 \text{ kJ mol}^{-1} + 264 \text{ kJ mol}^{-1} - (2 \times 338 \text{ kJ mol}^{-1}) = 24 \text{ kJ mol}^{-1}.$$

This estimate indicates a slightly endothermic enthalpy of formation, but the experimental value, -21 kJ mol^{-1}, is slightly exothermic.

S2.9   **(a)** The Pauling electronegativity values from Table 1.7 for Be and F are 1.57 and 3.98 respectively. The difference in electronegativities is:

$$\Delta\chi = 3.98 - 1.57 = 2.41.$$

The average electronegativity is:

$$\chi_{\text{mean}} = (3.98 + 1.57)/2 = 2.77.$$

These values place BeF_2 within the ionic bonding region.

(b) Following the same procedure and taking the electronegativity values $\chi(N) = 3.04$ and $\chi(O) = 3.44$ we get:

$$\Delta\chi = 3.44 - 3.04 = 0.40 \text{ and } \chi_{\text{mean}} = (3.44 + 3.04)/2 = 3.24.$$

These two values place NO within the covalent bonding region.

Exercises

E2.1   **a)** NO^+ has in total 10 valence electrons (5 from N + 6 from O and -1 for a positive charge). The most feasible Lewis structure is the one shown below with a triple bond between the atoms and an octet of electrons on each atom.

$$[:N{\equiv}O:]^+$$

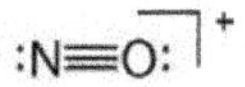

b) ClO^- has in total 16 valence electrons (7 from Cl + 6 from O + 1 for a negative charge). The most feasible Lewis structure is the one shown below with a single Cl–O bond and an octet of electrons on both atoms.

$$[:\ddot{\underset{..}{Cl}}-\ddot{\underset{..}{O}}:]^-$$

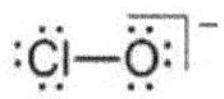

c) H_2O_2 has 14 valence electrons and the most feasible Lewis structure is shown below. Note that the structure requires O–O bond.

$$H-\ddot{\underset{..}{O}}-\ddot{\underset{..}{O}}-H$$

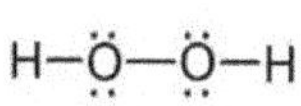

d) The most feasible Lewis structure for CCl_4 is shown below. Each element has a precise octet.

$$\begin{array}{c} :\ddot{\underset{..}{Cl}}: \\ | \\ :\ddot{\underset{..}{Cl}}-C-\ddot{\underset{..}{Cl}}: \\ | \\ :\ddot{\underset{..}{Cl}}: \end{array}$$

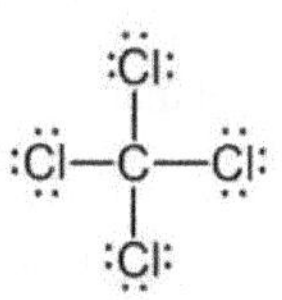

e) The most feasible Lewis structure for HSO_3^- is shown below. Being the most electropositive of the three elements, sulfur is the central atom.

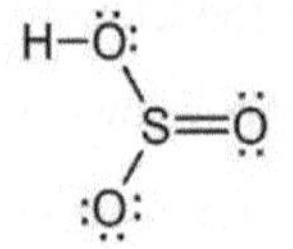

E2.2 The resonance structures for CO_3^{2-} are shown below. Keep in mind that the resonance structures differ only in allocation of electrons while the connectivity remains the same.

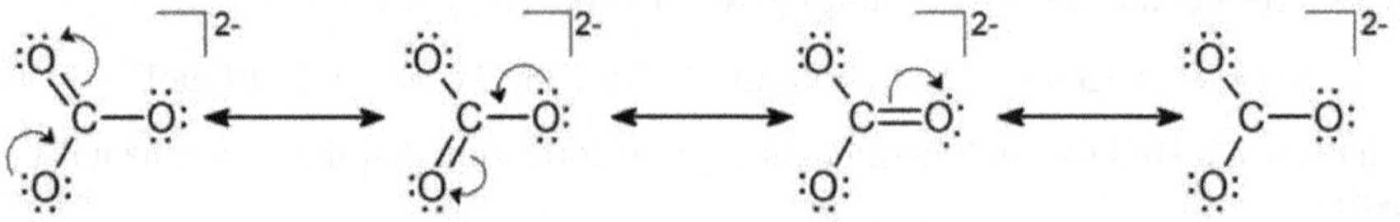

E2.3 **a)** The Lewis structure for hydrogen selenide, H_2Se, is shown below. The shape would be expected to be bent with the H–Se–H angle less than 109°. However, the angle is actually close to 90°, indicative of considerable p character in the bonding between S and H.
b) The BF_4^- structure is shown below (for the Lewis structure see Example 2.1 of your textbook). The shape is tetrahedral with all angles 109.5°.
c) The NH_4^+ structure of the ammonium ion is shown below. Again, the shape is tetrahedral with all angles 109.5°.

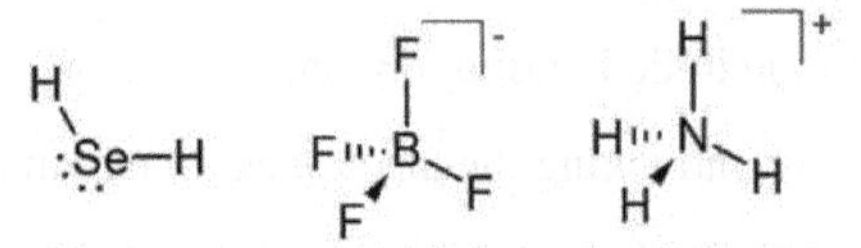

bent tetrahedral tetrahedral

E2.4 **a)** The Lewis structure of sulfur trioxide, SO_3, is shown below. With three σ bonds and no lone pairs, you should expect a trigonal-planar geometry (like BF_3). The shape of SO_3 is also shown below.

b) The Lewis structure of sulfite ion, SO_3^{2-} is shown below. With three σ bonds and one lone pair, you should expect a trigonal pyramidal geometry such as NH_3. The shape of SO_3^{2-} is also shown below.

c) The Lewis structure of iodine pentafluoride, IF_5, is shown below. With five σ bonds and one lone pair, you should expect a square pyramidal geometry. The shape of IF_5 is also shown below.

The Lewis structures and molecular shapes for each species a)–c) are summarized in the table below:

	SO_3	**SO_3^{2-}**	**IF_5**
Lewis structure	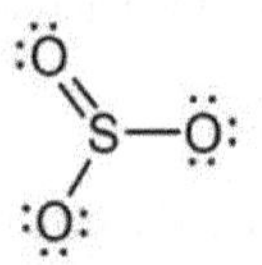	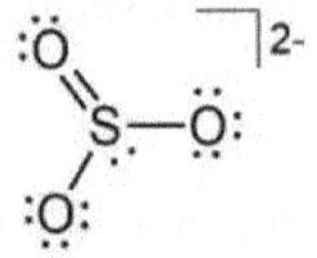	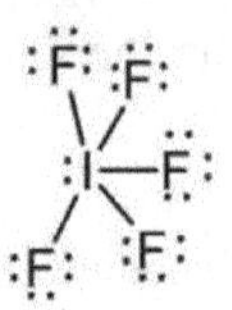
Molecular geometry (shape of the molecule)	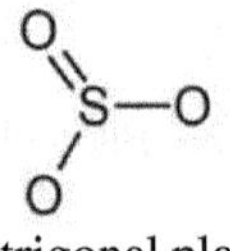 trigonal planar	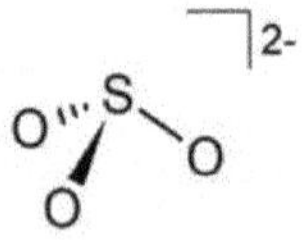 trigonal pyramidal	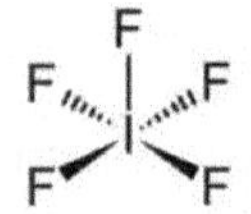 square pyramidal

Please note that for SO_3 and SO_3^{2-} only one possible resonance structure is shown. If you compare the electronic structures of these two species and CO_3^{2-} from Exercise 2.2, you can easily derive the other possible resonance structures. Note that the molecular geometry always remains the same regardless of which resonance structure is shown.

E2.5 **a)** The central iodine atom in IF_6^+ is bonded to the six fluorine atoms through sigma bonds and has no lone electron pairs. Thus, the six bonding electron pairs repel each other equally and the cation has octahedral geometry.

b) In this case the central iodine atom forms three sigma bonds with three fluorine atoms and has two lone electron pairs. These lone electron pairs and one fluorine atom would form a trigonal plane with two more fluorine atoms residing above and below that plane. Overall electron group geometry would then be trigonal bipyramidal but molecular geometry would be T-shaped.

c) The central xenon atom forms single bonds with four fluorine atoms and a double bond with one oxygen atom. It also has a lone electron pair. Thus, the electron group geometry would be octahedral, but the molecular geometry is square pyramidal.

The Lewis structures and molecular shapes for each species a)–c) are summarized in the table below:

	IF_6^+	IF_3	$XeOF_4$

Lewis structure

Molecular geometry (shape of the molecule)

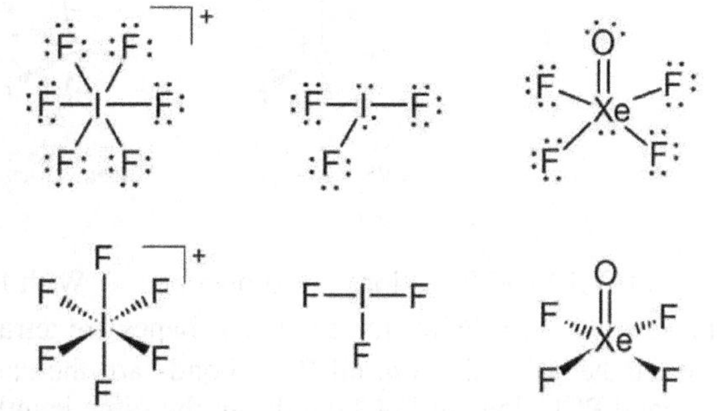

Octahedral | T-shaped | Square pyramidal

E2.6 **a)** The chlorine atom in ClF_3 is bonded to the three fluorine atoms through sigma bonds and has two nonbonding electron lone pairs. Both lone pairs occupy equatorial positions (the largest angles in a trigonal bipyramid), resulting in a T-shape for the molecule.

b) The iodine atom in ICl_4^- is bonded to the four chlorine atoms through sigma bonds and has two sets of lone pairs. The lone pairs are opposite each other, occupying the axial sites of an octahedron. The overall shape of the molecule is square planar.

c) The iodine atom is bonded to two other iodine atoms through sigma bonds and has three sets of lone pairs. The lone pairs occupy the equatorial sites of a trigonal bipyramid. The overall shape of the molecule is linear.

The Lewis structures and molecular shapes for each species a)–c) are summarized in the table below:

	ClF_3	ICl_4^-	I_3^-

Lewis structure

Molecular geometry (shape of the molecule)

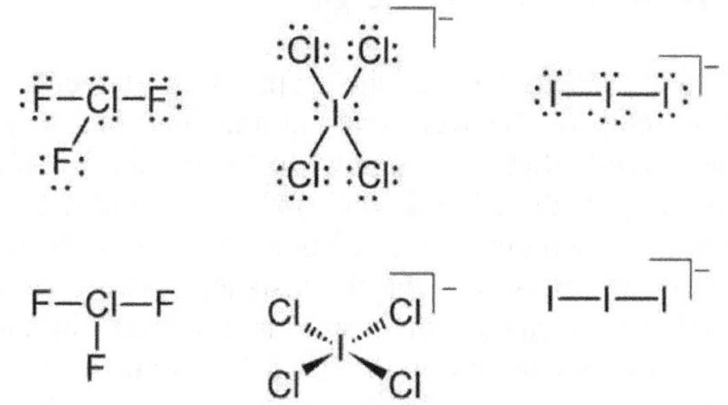

T-shaped | Square planar | Linear

E2.7 The Lewis structure of ICl_6^- and its geometry based on VSEPR theory is shown. The central iodine atom is surrounded by six bonding and one lone pair. However, this lone pair in the case of ICl_6^- is a *stereochemically* *inert* lone electron pair. This is because of the size of iodine atom—large size of this atom spread around all seven electron pairs resulting in a minimum repulsion between electron pairs. Therefore, all bond angles are expected to be equal to 90° with overall octahedral geometry.

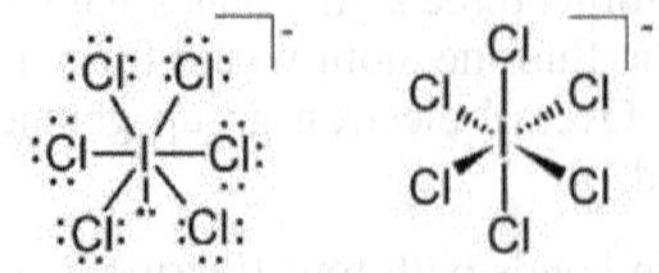

The VSEPR model and actual structure of SF_4 are shown below. Remember, lone pairs repel bonding regions, read Section 2.3(b). The VSEPR theory predicts a see-saw structure with a bond angle of 120° between the S and equatorial F's and a bond angle of 180° between the S and the axial F's. The bond angle is actually 102° for the S and equatorial Fs and 173° for the S and the axial Fs. This is due to the equatorial lone pair repelling the bonding S-F atoms.

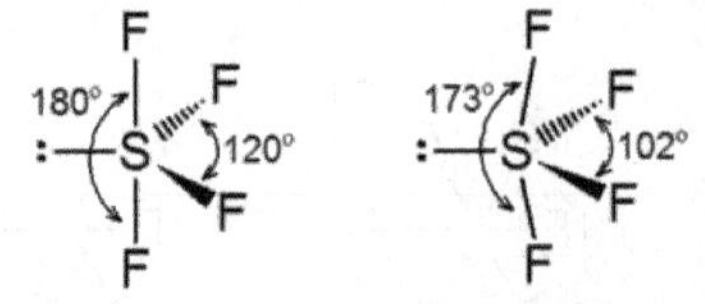

VSEPR model Actual structure

E2.8 The Lewis structures of PCl_4^+ and PCl_6^- ions are shown below. With four σ bonds and no lone pairs for PCl_4^+, and six σ bonds and no lone pairs for PCl_6^-, the expected shapes are tetrahedral (like CCl_4) and octahedral (like SF_6), respectively. In the tetrahedral PCl_4^+ ion, all P–Cl bonds are the same length and all Cl–P–Cl bond angles are 109.5°. In the octahedral PCl_6^- ion, all P–Cl bonds are the same length and all Cl–P–Cl bond angles are either 90° or 180°. The P–Cl bond distances in the two ions would not necessarily be the same length.

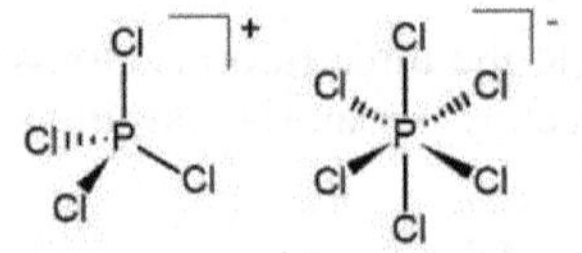

E2.9 **a)** From the covalent radii values given in Table 2.7, 77 pm for single-bonded C and 99 pm for Cl, the C–Cl bond length in CCl_4 is predicted to be 77 pm + 99 pm = 176 pm. The agreement with the experimentally observed value of 177 pm is excellent.

b) The covalent radius for Si is 118 pm. Therefore, the Si–Cl bond length in $SiCl_4$ is predicted to be 118 pm + 99 pm = 217 pm. This is 8% longer than the observed bond length, so the agreement is not as good as in the case of C–Cl.

c) The covalent radius for Ge is 122 pm. Therefore, the Ge–Cl bond length in $GeCl_4$ is predicted to be 221 pm. This is 5% longer than the observed bond length.

E2.10 Table 2.7 lists selected covalent radii of the main group elements. As we can see from the table, the trends in covalent radii follow closely the trends in atomic and ionic radii covered in Chapter 1 (Section 1.7(a)). Considering first the horizontal periodic trends we can see that the single bond covalent radii decrease if we move from left to right in the periodic table. Recall that the atomic radii decrease while the Z_{eff} increases in the same direction. Because of these two trends, the valence electrons are held tighter and closer to the atomic nucleus. This means that, as we move from left to right, the distance between two non-metallic atoms has to decrease in order for the covalent bond to be formed. This results in a shorter internuclear separation (i.e., shorter bond) and as a consequence decrease in covalent radii in the same direction.

If we consider the vertical trends, that is, the trends within one group of the elements, we observe an increase in covalent radii. Recall again that the atomic radii increase in the same direction. As we go down a group the valence electrons are found in the atomic orbitals of higher principal quantum number. There are more core (or non-valence) electrons between the valence electrons and the nucleus that shield the valence electrons from the influence of the nucleus. Therefore, the valence electrons are further away from the nucleus. This means that when a covalent bond is formed, two atomic nuclei are further apart as we go down the group and the covalent radii increase.

E2.11 You need to consider the enthalpy difference between one mole of Si=O double bonds and two moles of Si–O single bonds. The difference is:

$$2(\text{Si–O}) - (\text{Si=O}) = 2 \times (466 \text{ kJ mol}^{-1}) - (640 \text{ kJ mol}^{-1}) = 292 \text{ kJ mol}^{-1}.$$

Therefore, the two single bonds will always be better enthalpically than one double bond. If silicon atoms only have single bonds to oxygen atoms in silicon-oxygen compounds, the structure around each silicon atom will be tetrahedral: each silicon will have four single bonds to four different oxygen atoms.

E2.12 Diatomic nitrogen has a triple bond holding the atoms together, whereas six P–P single bonds hold together a molecule of P_4. If nitrogen were to form N_4 molecules with the P_4 structure, then two N≡N triple bonds would be traded for six N–N single bonds, which are weak. The net enthalpy change can be estimated from the data in Table 2.8 to be $2 \times (945 \text{ kJ mol}^{-1}) - 6 \times (163 \text{ kJ mol}^{-1}) = 912 \text{ kJ mol}^{-1}$, which indicates that the tetramerization of nitrogen is *very* unfavourable. On the other hand, multiple bonds between period 3 and larger atoms are not as strong as two times the analogous single bond, so P_2 molecules, each with a P≡P triple bond, would not be as stable as P_4 molecules, containing only P–P single bonds. In this case, the net enthalpy change for $2P_2 \rightarrow P_4$ can be estimated to be $2 \times (481 \text{ kJ mol}^{-1}) - 6 \times (201 \text{ kJ mol}^{-1}) = -244 \text{ kJ mol}^{-1}$.

E2.13 Consider the reaction:

$$2H_2(g) + O_2(g) \rightarrow 2H_2O(g).$$

Since you must break two moles of H–H bonds and one mole of O=O bonds on the left-hand side of the equation and form four moles of O–H bonds on the right-hand side, the enthalpy change for the reaction can be estimated as:

$$\Delta H = 2 \times (436 \text{ kJ mol}^{-1}) + 497 \text{ kJ mol}^{-1} - 4 \times (463 \text{ kJ mol}^{-1}) = -483 \text{ kJ mol}^{-1}.$$

The experimental value is –484 kJ, which is in closer agreement with the estimated value than ordinarily expected. Since Table 2.8 contains average bond enthalpies, there is frequently a small error when comparing estimates to a specific reaction.

E2.14 Recall from the Section 2.10(b) *Bond correlations* that bond enthalpy increases as bond order increases, and that bond length decreases as bond order increases. Thus, we have also to think about bond order to answer this question. As you can see from the table, C_2 molecule has the highest dissociation energy. The reason for this lies in the fact that C_2 molecule has a bond order of 3 and relatively short C–C distance. The next highest bond dissociation energy is found in O_2 molecule. There is significant difference in bond dissociation energies between C_2 and O_2. This can be attributed to the lower bond order in O_2 (bond order 2) vs. C_2 (bond order 3), particularly considering that the O–O bond length is very similar to the C–C bond length. Following O_2 we have BeO as third in bond dissociation energy. The bond order in BeO is the same as in O_2, that is, 2; thus, the difference here lies in the bond length: significantly longer Be–O bond (133.1 pm vs. 120.7 pm in O_2) makes this bond weaker than the one in O_2. After BeO comes BN molecule. Considering that this bond is shorter than the one in BeO, we can safely suggest that the difference in the bond dissociation enthalpy is due to the lower bond order in B–N, that is, bond order of 1. The last molecule in the list is NF with a very similar bond length to BeO molecules but significantly lower bond dissociation energy. The reason for this difference again lies in the bond order: 1.

The atoms that obey the octet rule are O in BeO, F in NF, and O in O_2. Keep in mind that this is only based on the Lewis structures, not on the results of more accurate MO theory.

E2.15 Consider the first reaction:

$$S_2^{2-}(g) + {}^1/_4S_8(g) \rightarrow S_4^{2-}(g).$$

Hypothetically, two S–S single bonds (of S_8) are broken to produce two S atoms, which combine with S_2^{2-} to form two new S–S single bonds in the product S_4^{2-}. Since two S–S single bonds are broken and two are made, the net enthalpy change is zero.

Now consider the second reaction:

$$O_2^{2-}(g) + O_2(g) \rightarrow O_4^{2-}(g).$$

Here there is a difference. A mole of O=O double bond of O_2 is broken, and two moles of O–O single bonds are made. The overall enthalpy change, based on the mean bond enthalpies in Table 2.7, is:

$$O{=}O - 2 \times (O{-}O) = 497 \text{ kJ mol}^{-1} - 2 \times (146 \text{ kJ mol}^{-1}) = 205 \text{ kJ mol}^{-1}.$$

The large positive value indicates that this is not a favourable process.

E2.16 **a)** Since we have an anion with overall –2 charge, the sum of oxidation numbers must equal –2. The oxidation number most commonly assigned to oxygen atom is –2 (SO_3^{2-} does not belong to the classes in which oxygen has oxidation number different from –2). Thus, if we denote the oxidation state of S atom as X, we have X + 3 × (–2) = –2. From here X = +4, that is, the oxidation number of sulfur in SO_3^{2-} is +4.

b) Following the same procedure as outlined above with X being oxidation number of nitrogen atom and overall charge being +1, we have X + (–2) = +1 and oxidation number of nitrogen is +3.

c) In this case there are two chromium atoms each having oxidation state X. Thus we have 2X + 7 × (–2) = –2. The oxidation state of Cr atoms is +6.

d) Again, there are two V atoms, but the species is neutral; thus the sum of oxidation numbers of all elements must equal zero. We have 2X + 5×(–2) = 0. The oxidation state of vanadium atoms from here is +5.

e) PCl_5 is neutral and we can assign –1 oxidation state to Cl because it is bound to an atom of lower electronegativity (phosphorus). We have X + 5 × (–1) = 0, X = 5. The oxidation state of phosphorus atom is +5.

E2.17 The differences in electronegativities are AB 0.5, AD 2.5, BD 2.0, and AC 1.0. The increasing covalent character is AD < BD < AC < AB.

E2.18 **a)** Using the Pauling electronegativity values in Table 1.7 and the Ketelaar triangle in Figure 2.28, the $\Delta\chi$ for BCl_3 = 3.16 – 2.04 = 1.12 and χ_{mean} = 2.60. This value places BCl_3 in the covalent region of the triangle.

b) $\Delta\chi$ for KCl = 3.16 – 0.82 = 2.34 and χ_{mean} = 1.99. This value places KCl in the ionic region of the triangle.

c) $\Delta\chi$ for BeO = 3.44 – 1.57 = 1.87 and χ_{mean} = 2.51. This value places BeO in the ionic region of the triangle.

E2.19 **a)** BCl_3 has a trigonal planar geometry, according to Table 2.4, the most likely hybridization would be sp^2.

b) NH_4^+ has a tetrahedral geometry, so the most likely hybridization would be sp^3.

c) SF_4 has distorted see-saw geometry, with the lone pair occupying one of the equatorial sites, see Exercise 2.4 above. Therefore, it would be sp^3d or spd^3.

d) XeF_4 has a square planar molecular geometry with two lone pairs on the central Xe atom, so its hybridization is sp^3d^2.

E2.20 **a) O_2^-?** You must write the electron configurations for each species, recalling Figure 2.17, and then apply the Pauli exclusion principle to determine the situation for incompletely filled degenerate orbitals. In this case the electron configuration is $1\sigma_g^2 1\sigma_u^2 2\sigma_g^2 1\pi_u^4 1\pi_g^3$. With three electrons in the pair of $1\pi_g$ molecular orbitals, one electron must be unpaired. Thus, the superoxide anion has a single unpaired electron.

b) O_2^+? The configuration is $1\sigma_g^2 1\sigma_u^2 2\sigma_g^2 1\pi_u^4 1\pi_g^1$, so the oxygenyl cation also has a single unpaired electron.

c) BN? You can assume that the energy of the 3σ molecular orbital is *higher* than the energy of the 1π orbitals, because that is the case for CO (see Figures 2.22 and 2.23). Therefore, the configuration is $1\sigma^2 1\sigma^2 1\pi^4$, and, as observed, this diatomic molecule has no unpaired electrons.

d) NO^-? The molecular orbitals of NO are based on the molecular orbitals of CO. Nitrogen atom, however, has one more valence electron than carbon and the total number of valence electrons in NO is 11. Thus, the electronic configuration of NO is $1\sigma^2 2\sigma^2 1\pi^4 3\sigma^2 2\pi^1$. An additional electron would go as unpaired electron in the second π

orbital of the doubly degenerate 2π set. Consequently, this anion has two unpaired electrons. See Figures 2.22 and 2.23.

E2.21 **a) Be_2^+?** Two Be atoms have in total four valence electrons. We have to remove one to account for a positive charge of this diatomic cation. This gives the electron configuration $1\sigma_g^2 1\sigma_u^1$. The HOMO (highest occupied molecular orbital) for Be_2 is a σ ungerade antibonding orbital formed from two 2s atomic orbitals, shown below. (The black spheres are atomic nuclei; different shading represents two opposite phases of the orbital wave functions. The same applies for the solutions b)–d).)

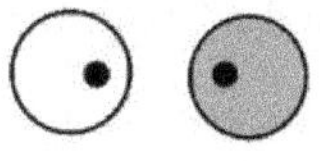

b) B_2^-? Two boron atoms have six valence electrons in total. We have to add one more to account for the negative charge. The electron configuration is $1\sigma_g^2 1\sigma_u^2 1\pi_u^3$. The HOMO for B_2 is a π ungerade bonding MO, shown below.

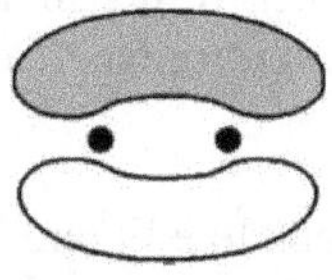

c) C_2^-? The total number of valence electrons is eight, but we again have to add one electron to account for the negative charge. The electron configuration is $1\sigma_g^2 1\sigma_u^2 1\pi_u^4 2\sigma_g^1$. The HOMO for C_2^- is a σ bonding MO formed from mixing two 2p atomic orbitals, shown below.

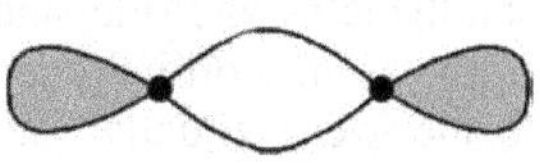

(d) F_2^+? The total number of valence electrons is 14 (seven from each F atom) but we must subtract one to account for the plus one charge. The electron configuration is $1\sigma_g^2 1\sigma_u^2 2\sigma_g^2 1\pi_u^4 1\pi_g^3$. The HOMO for F_2^+ is a π gerade antibonding MO, shown below.

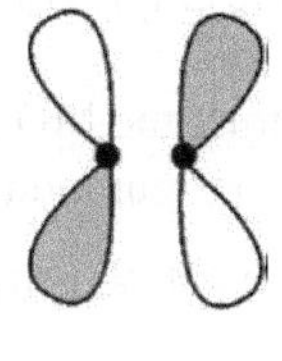

E2.22 Recall from Section 2.4 of your textbook that σ bond has cylindrical symmetry around the internuclear axis. Also, Section 2.5 defines the internuclear axis (the line that connects two bonded nuclei) as the *z*-axis. If you have a closer look at Figure 1.15 you will see that d_{z2} atomic orbital already has a cylindrical symmetry around *z*-axis. Interaction of two d_{z^2} orbitals along *z*-axis would thus produce a molecular orbital of cylindrical symmetry, i.e., a σ molecular orbital:

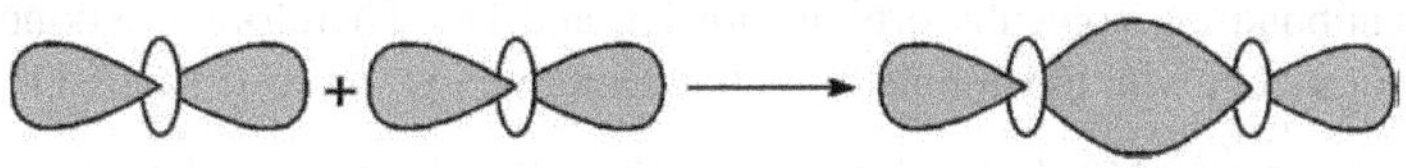

Section 2.5 also describes the formation of π orbital: π orbital is formed when atomic orbitals approach side by side, if the π orbital is rotated around the nuclear axis, the signs of orbital lobes are interchanged. Again, referring to Figure 1.15 we see that four d orbitals have this type of symmetry (all except d_{z^2} used above). To narrow our choices, we take *z*-axis as the internuclear axis. Then two possibilities emerge: overlap of two d_{xz} or two d_{yz} orbitals. Note in the sketch below that the lobe signs are indicated by shading.

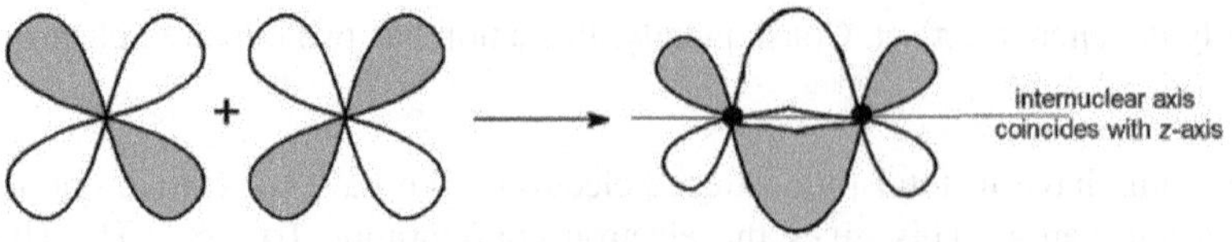

E2.23 The configuration of C_2^{2-} would be $1\sigma_g^2 1\sigma_u^2 1\pi_u^4 2\sigma_g^2$. The bond order would be ½[2 – 2 + 4 + 2] = 3. So C_2^{2-} has a triple bond, as discussed in the Self Test S2.6, above. The HOMO is in a sigma bonding orbital.

The configuration for the neutral C_2 would be $1\sigma_g^2 1\sigma_u^2 1\pi_u^4$ (see Figure 2.17). The bond order would be ½[2 – 2 + 4] = 2.

E2.24 Your molecular orbital diagram should look like Figure 2.18 which shows the MO diagram for Period 2 homonuclear diatomic molecules from Li_2 to N_2. Each carbon atom has four valence electrons, thus a total of eight electrons has to be placed in the molecular orbitals on the diagram. Keep in mind that you still follow the Hund's rule and the Pauli exclusion principle when filling molecular orbitals with electrons. See the solution for E2.23 above for electronic configuration and bond order of this species. Figures 2.15 and 2.16 provide you with the appropriate sketches of each MO.

E2.25 Your molecular orbital diagram should look like Figure 2.22 showing the MO diagram for CO molecule. There are several important differences between this diagram and that for C_2 from Exercise 2.24. First, the energies of B and N atomic orbitals are different. The atomic orbitals on N atom are lower in energy than corresponding atomic orbitals on B atom. Recall from Chapter 1 that N should have lower atomic radius and higher Z_{eff} than B. These differences result in higher stability (lower energy) of N atomic orbitals. Also, nitrogen is more electronegative than boron. Thus, B atom should be on the left-hand side, and N should be on the right-hand side of your diagram. Second, because of this difference in the atomic orbital energies, the MO are localized differently. For example, while all MO orbitals in C_2 molecule were equally localized over both nuclei (because the molecule is homonuclear), in the case of BN, 1σ is more localized on N while 2σ is more localized on B. BN molecule has a total of eight valence electrons (three from B atom and five from N atom). Thus, after we fill the MOs with electrons we should get the following electronic configuration $1\sigma^2 2\sigma^2 1\pi^4$. This means that a degenerate set of two 1π molecular orbitals is HOMO of BN while empty 3σ are LUMO of this molecule.

E2.26 **a)** The orbitals that would be used to construct the MO diagram are the 5p and 5s of I, and the 4p and 4s of Br. IBr is isoelectronic to ICl, so the ground state configuration is the same, $1\sigma^2 2\sigma^2 3\sigma^2 1\pi^4 2\pi^4$.

b) The bond order would be ½[2 + 2 – 2 + 4 – 4] = 1; there would be single bond between I and Br.

c) IBr^- would have the ground state configuration of $1\sigma^2 2\sigma^2 3\sigma^2 1\pi^4 2\pi^4 4\sigma^1$, with a bond order of ½[2 + 2 – 2 + 4 – 4 – 1] = ½, it would not be very stable.. While IBr^- would have the ground state configuration of $1\sigma^2 2\sigma^2 3\sigma^2 1\pi^4 2\pi^4 4\sigma^2$, with a bond order of ½[2 + 2 – 2 + 4 – 4 – 2] = 0, it would not exist, there is no bond between the two atoms.

E2.27 **a)** The electron configuration of S_2 molecule is $1\sigma_g^2 1\sigma_u^2 2\sigma_g^2 1\pi_u^4 1\pi_g^2$. The bonding molecular orbitals are $1\sigma_g$, $1\pi_u$, and $2\sigma_g$, while the antibonding molecular orbitals are $1\sigma_u$ and $1\pi_g$. Therefore, the bond order is ½((2 + 4 + 2) – (2 + 2)) = 2, which is consistent with the double bond between the S atoms as predicted by Lewis theory.

b) The electron configuration of Cl_2 is $1\sigma_g^2 2\sigma_u^2 3\sigma_g^2 1\pi_u^4 2\pi_g^4$. The bonding and antibonding orbitals are the same as for S_2, above. Therefore, the bond order is (1/2)((2 + 4 + 2) – (2 + 4)) = 1 again as predicted by Lewis theory.

c) The electron configuration of NO^+ is $1\sigma^2 1\sigma^2 2\sigma^2 1\pi^4$. The bond order for NO^+ is ½((2 + 2 + 4) – 2) = 3, as predicted by the Lewis theory.

E2.28 **a)** $\mathbf{O_2 \rightarrow O_2^+ + e^-}$**?** The molecular orbital electron configuration of O_2 is $1\sigma_g^2 1\sigma_u^2 2\sigma_g^2 1\pi_u^4 1\pi_g^2$. The two $1\pi_g$ orbitals are π antibonding molecular orbitals, so when one of the $1\pi_g$ electrons is removed, the oxygen-oxygen bond order increases from 2 to 2.5. Since the bond in O_2^+ becomes stronger, it should become shorter as well.

b) $N_2 + e^- \rightarrow N_2^-$? The molecular orbital electron configuration of N_2 is $1\sigma_g^2 1\sigma_u^2 1\pi_u^4 2\sigma_g^2$. The next electron must go into the $1\pi_g$ orbital, which is π antibonding (refer to Figures 2.17 and 2.18). This will decrease the nitrogen-nitrogen bond order from 3 to 2.5. Therefore, N_2^- has a weaker and longer bond than N_2.

c) $NO \rightarrow NO^+ + e^-$? NO has similar molecular orbitals to CO. Hence, the electronic configuration of NO molecule is $1\sigma^2 2\sigma^2 1\pi^4 3\sigma^1 2\pi^1$. Notice that NO has one electron more than CO because nitrogen atom has five valence electrons. Removal of one electron from 2π antibonding orbital will increase the bond order from 2.5 to 3, making the bond in NO^+ stronger.

E2.29 Recall from Section 2.8(b) that molecular frontier orbitals are lowest unoccupied molecular orbital (LUMO) and highest occupied molecular orbital (HOMO). It is also good to have Figures 2.12, 2.17 and 2.18 in front while analysing this exercise.

For C_2^{2-}: In Exercise 2.23 we have seen that the electronic configuration of C_2^{2-} is $1\sigma_g^2 1\sigma_u^2 1\pi_u^4 2\sigma_g^2$. From this configuration, we see that the HOMO of this anion is $2\sigma_g$. This MO is bonding and formed from overlap of two p_z atomic orbitals. (If you do not know this already, refer to Figure 2.18 and keep in mind that σ orbitals must have cylindrical symmetry. Note that this orbital also has some contribution form 2s atomic orbitals; but for this type of analysis, this contribution can be neglected because $2\sigma_g$ is close in energy to atomic 2p orbitals.). The HOMO can be sketched as

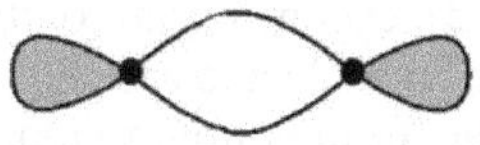

Now refer to Figure 2.17. The next orbital, one step higher in energy, is $1\pi_g$. It is a set of two degenerate, unoccupied MOs forming a LUMO frontier orbital. These are formed in π-overlap from p_x and p_y atomic orbitals. One of the $1\pi_g$ is sketched below.

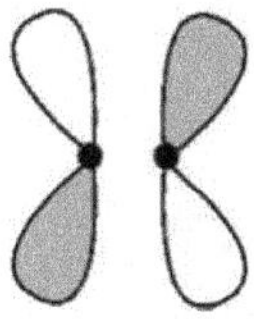

LUMO is the last bonding MO for C_2^{2-}; higher in energy are empty antibonding orbitals. Adding electrons to HOMO would decrease the bond order and weaken the C–C bond in the anion.

For N_2: Note that N_2 and C_2^{2-} are isolelectronic (they have the same number of electrons). Thus, the same discussion applies for N_2 and the frontier orbitals would look the same. The major difference between the MOs of these two chemical species is in relative energies of molecular orbitals. The atomic orbitals on N atom are lower in energy than corresponding orbitals on C atom. Consequently, the MOs of N_2 molecule are lower in energy than C_2^{2-}. The same sketches as above as well as chemical consequences.

For CO: Refer to the Figure 2.23 of your textbook. From the diagram, we can see that the frontier orbitals are 3σ (HOMO) and doubly degenerate set of 2π MOs. The figure also shows simple sketches of these orbitals. Note that the smaller orbital lobes are located on O atom because oxygen atomic orbitals are lower in energy than carbon's. Because of this difference and electron configuration, CO molecule is polar with negative side of the dipole oriented towards C atom. Adding an electron to CO molecule would decrease CO bond order, elongate the bond and weaken it.

For O_2: Referring to figure 2.17 again we see that HOMO of O_2 is doubly degenerate set of $1\pi_g$ while LUMO is $2\sigma_u$. LUMO orbitals are formed from p_x and p_y atomic orbitals and their sketch looks like the one for C_2^{2-} above. The HOMO is antibonding σ_u MO formed from p_z atomic orbitals and can be sketched as:

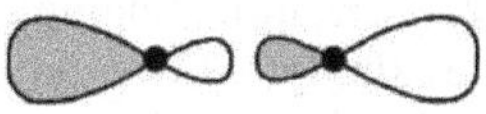

Both HOMO and LUMO are antibonding MOs and adding electrons would significantly weaken O–O bond. At the same time the paramagnetic character of this molecule would decrease as well (see Example 2.4).

E2.30 Ultraviolet (UV) radiation is energetic enough to be able to ionize molecules in the gas phase (you can see this from the figure—the *x*-axis shows the ionization energies, *I*, of the molecule). Thus, the UV photoelectron spectra of a kind shown on Figure 2.31 can tell us a lot about the orbital energies. As you can see from the figure, the major regions of the spectra have been already assigned. The line at about 15.2 eV corresponds to excitation of the electrons from 3σ molecular orbital (the HOMO of CO). The group of four peaks between 17 and about 17.8 eV correspond to the excitation of 1π electrons, and finally the last peak (highest in energy) corresponds to the excitation of 2σ electrons. Note that, as expected, the least energy is required to remove an electron from HOMO molecular orbital. Also, the fine structure seen for the 1π is due to excitation of different vibrational modes in the cation formed by photoejection of the electrons.

To predict the appearance of the UV photoelectron spectrum of SO we have to determine its electronic configuration. The MO diagram for this molecule would look somewhat similar to that of CO, but the electronic configuration would be different because the total number of valence electrons is different: 10 for CO but 12 for SO. The electronic configuration of SO would then be $1\sigma^2 1\sigma^2 2\sigma^2 1\pi^4 3\sigma^2 2\pi^2$. Therefore, the UV photoelectron spectrum should have one more line at even lower energies corresponding to the ionization from 2π degenerate set of molecular orbitals.

E2.31 Four atomic orbitals can yield four independent linear combinations, each giving one molecular orbital. The four relevant ones for a hypothetical linear H_4 molecule are shown on the left, and for a hypothetical square H_4 on the right. The orbital energies are increasing from bottom to top. The most stable orbital has the fewest nodes, the next orbital in energy has only one node, and so on to the fourth and highest energy orbital. Note that for the square geometry two linear combinations have one node only and, consequently, have the same energy—they form a doubly degenerate set.

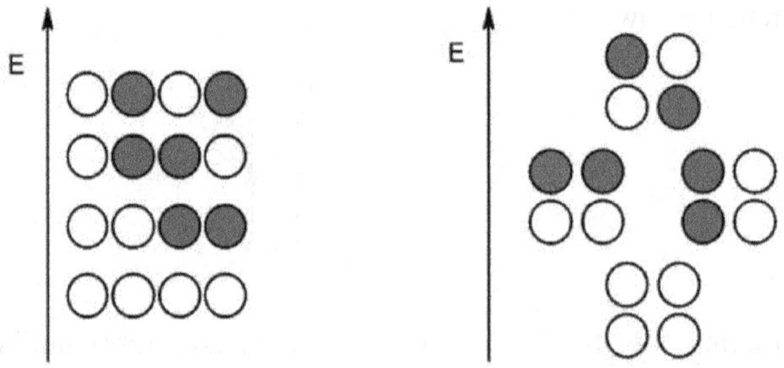

Both linear and square geometry produce the same number of MOs and must accommodate the same number of electrons—four (one from each H atom). Looking at the linear molecule (left-hand diagram) we can place two electrons in the lowest MO and two electrons in the next MO. Thus, all electrons are paired and the linear H_4 molecule is diamagnetic. The square geometry also accepts two electrons in the lowest MO, but the degenerate set of two orbitals that come next accept one electron each (recall Hund's rule). So, square H_4 has two unpaired electrons and is paramagnetic.

E2.32 **Molecular orbitals of linear $[HHeH]^{2+}$?** The three atoms of $[HHeH]^{2+}$ will form a set of three molecular orbitals; one bonding, one nonbonding, and one antibonding. They are shown below. You should note that helium's 1s atomic orbital is lower in energy than 1s in hydrogen atom. Therefore, He 1s orbital contributes more to the bonding MO than hydrogen's 1s. Since $[HHeH]^{2+}$ has two electrons, only the bonding orbital is filled.

Helium, being a noble gas and unreactive, holds tight to its electrons and is not keen on losing them or sharing them; thus it is unlikely that $[HHeH]^{2+}$ would last long and would decompose to $2H^+$ and He. In isolation the cation should be only moderately stable. There are two reasons for this low stability: (1) Although only the bonding MO is filled, it spans over three nuclei, making the average H–He bond order about 0.5 (note that bond order 1 requires two electrons shared between two nuclei; here we have two electrons shared between three nuclei—situation referred to as 3 centre-2 electron bonding); (2) H and (particularly) He are small atoms. Bonding them together brings three nuclei in proximity: +/2+/+ repulsion between H–He–H nuclei is going to be strong particularly because it is "diluted" by only two electrons.

In solution it would be unstable particularly with respect to proton transfer to another chemical species that can act as a base, such as the solvent or counterion. Removal of one proton would produce HeH^+, a species with lower positive charge and H–He bond order of 1. Any substance is more basic than helium.

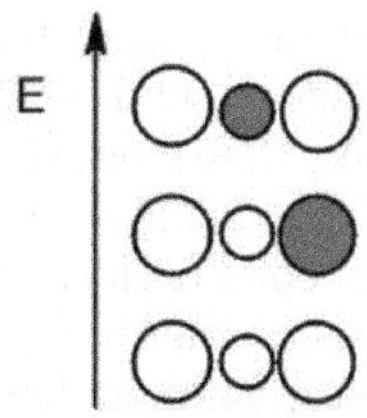

E2.33 The molecular orbital diagram for HeHe* is shown below.

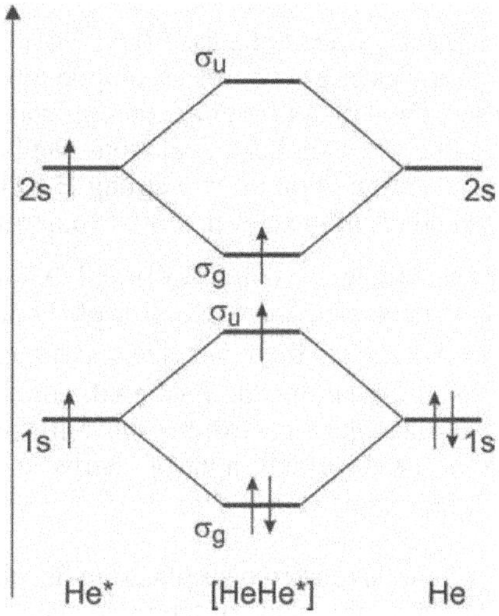

Note that this molecule has a bond order 1: ½(2 – 1 + 1) = 1.

E2.34 Your molecular orbital diagram for the N_2 molecule should look like the one shown on Figure 2.18, whereas the diagrams for O_2 and NO should be similar to those shown on Figures 2.12 and 2.22 respectively. The variation in bond lengths can be attributed to the differences in bond orders. The electronic configuration of N_2 molecule is $1\sigma_g^2 1\sigma_u^2 1\pi_u^4 2\sigma_g^2$, and the bond order of N_2 is b = ½(2 – 2 + 4 + 2) = 3. The electronic configuration of O_2 molecule is $1\sigma_g^2 1\sigma_u^2 2\sigma_g^2 1\pi_u^4 2\pi_g^2$, and the bond order is b = ½(2 – 2 + 2 + 4 – 2) = 2. Thus, the lower bond order in O_2 results in longer O–O bond in comparison to N–N bond. The electronic configuration of NO is $1\sigma^2 2\sigma^2 1\pi^4 3\sigma^2 2\pi^1$ and the b = ½(2 + 2 + 4 – 2 – 1) = 2.5. This intermediate bond order in NO is reflected in an intermediate bond distance.

E2.35 **(a) Square H_4^{2+}?** The drawing below shows a square array of four hydrogen atoms. Clearly, each line connecting any two of the atoms is not a typical bond in which two electrons are shared between two nuclei because this molecular ion has only two electrons and is electron deficient. It is a hypothetical example of (4c,2e) bonding. We cannot write a Lewis structure for this species. It is not likely to exist; it should be unstable with respect to two separate H_2^+ diatomic species.

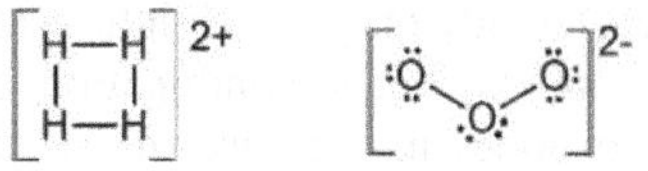

(b) Angular O_3^{2-}? A proper Lewis structure for this 20-electron ion is shown above. Therefore, it is electron precise. It could very well exist.

Guidelines to Selected Tutorial Problems

T2.2 The substances common in the Earth's crust are silicate minerals. These contain Si–O bonds with no Si–Si bonds. On the other hand, biosphere contains living systems that are based on biologically important organic molecules that contain C–C and C–H bonds. To understand this "segregation" of Si and C (both elements of Group 14) we have to analyse the mean bond enthalpies, *B*, for these bonds listed in Table 2.8. If we concentrate on silicon first, we see that the bond enthalpy for Si–Si bond is about ½ of the Si–O bond (226 kJ mol^{-1} vs. 466 kJ mol^{-1}). Also, Si–H bond enthalpy is about 2/3 of Si–O bond (318 kJ mol^{-1} vs. 466 kJ mol^{-1}). Consequently, any Si–Si and Si–H bonds will be converted to Si–O bonds (and water in the case of Si–H). Although it appears that the bond enthalpy of O_2 molecule is high (497 kJ mol^{-1}), keep in mind that two Si–O bonds are formed if we break one molecule of O_2. Thus, we are considering the following general reaction and associated changes in enthalpy:

$$O_2 + \text{Si–Si} \rightarrow 2\text{Si–O}$$

$$-2B(\text{Si–O}) + B(O_2) + B(\text{Si–Si}) = -(2 \times 466 \text{ kJ mol}^{-1}) + 497 \text{ kJ mol}^{-1} + 226 \text{ kJ mol}^{-1} = -209 \text{ kJ mol}^{-1}$$

Positive signs above indicate that the energy has to be used (i.e., we have to use up the energy to break O=O and Si–Si bonds) while the negative sign indicates energy released upon a bond formation (i.e., energy liberated when Si–O bond is formed). As we can see, the process is very exothermic and as long as there is O_2, Si will be bonded to O rather than to another silicon atom. Very similar calculations can be performed to show the preference of Si–O bond over Si–H bond, but in this case formation of very strong H–O bonds (about 463 kJ mol^{-1} per H–O bond) has to be taken into account. This would result in an even more exothermic process.

Carbon on the other hand, forms much stronger bonds with H and O in comparison to silicon. It also forms much stronger C–C double and triple bonds in comparison to Si–Si triple bonds. (Recall that as we descend a group the strength of π bonds decrease due to the increased atomic size and larger separation of p atomic orbitals in space. This larger separation prevents efficient orbital overlap.) You will note that the formation of C–O and O–H bonds from organic material (C–C bonds) is still somewhat exothermic, but a high bond enthalpy in O_2 molecule presents a high activation barrier. Thus, we do need a small input of energy to get the reaction going—a spark to light a piece of paper or burning flame to light a camp fire.

T2.6 The planar form of NH_3 molecule would have a trigonal planar molecular geometry with the lone pair residing in nitrogen's p orbital that is perpendicular to the plane of the molecule. This structure belongs to a D_{3h} point group (for discussion of molecular symmetry and point groups refer to Chapter 6). Now we can consult Resource Section 5. Table RS5.1 provides us with *the symmetry classes* of atomic orbitals on a central atom of AB_n molecules. In our case, the central atom is N with its 2s and 2p valence orbitals. From the table, we look at the symmetry classes of s and p orbitals (table rows) for the D_{3h} point group (in columns). We can see that the s orbital is in A_1' class (a non-degenerate orbital), p_x and p_y in E′ class (a doubly degenerate orbital), and finally p_z is in A_2'' class (again a single degenerate orbital).

Recall from Section 2.9(a) *Polyatomic molecular orbitals* that molecular orbitals are formed as linear combinations of atomic orbitals of the same symmetry and similar in energy. We find first a linear combination of atomic orbitals on peripheral atoms (in this case H) and the coefficient c, and then combine the combinations of appropriate symmetry with atomic orbitals on the central atom. The linear combinations of atomic orbitals of peripheral atoms (also called symmetry-adapted linear combinations—SALC) are given after Table RS5.1 in the Resource Section 5. Thus, we have to find linear combinations in D_{3h} point group that belong to symmetry classes A_1', A_2'', and E′. We also have to keep in mind that H atom has only one s orbital.

We can find one A_1' and two E′ linear combinations that are based only on s orbitals (all other given combinations in the D_{3h} group with the same symmetry class are based on p orbitals—this can be concluded based on the orbital shape). This means that p_z atomic orbital on nitrogen atom (from A_2'' class) does not have a symmetry match in the linear combinations of hydrogen's s orbitals and will remain essentially a nonbonding orbital in the planar NH_3 molecule. The plot is given below. Keep in mind the relative energies of atomic orbitals given in the text of the problem. These are important because only orbitals of right symmetry and similar energies will combine to make molecular orbitals. Further note that A_1' linear combination of H atomic orbitals is lower in energy than E′ combinations because there are no nodes in the case of A_1'' combinations. (In the plot below AOs stands for "atomic orbitals" and SALC stands for "symmetry-adapted linear combinations." Both H atomic orbitals and SALCs formed from them are sketched.)

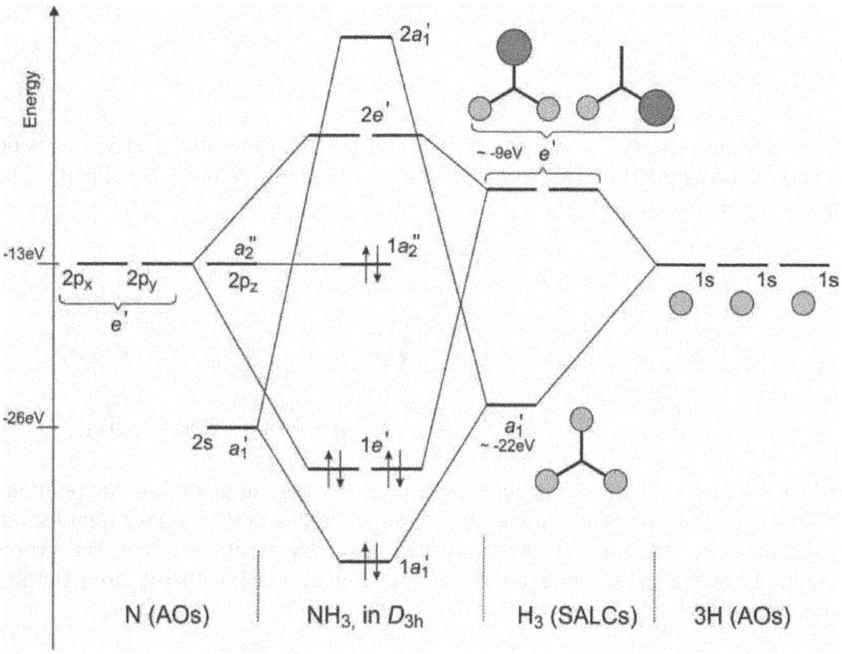
Energy
-13eV
-26eV
2px
2py
e'
a2''
2pz
2s
a1'
2a1'
2e'
1a2''
1e'
1a1'
~ -9eV
e'
a1'
~ -22eV
1s
1s
1s
N (AOs)
NH3, in D3h
H3 (SALCs)
3H (AOs)

Chapter 3 Molecular Symmetry

Self-Tests

S3.1 See Figure 3.7 for the sketch of S_4 in a structurally similar CH_4 molecule. The S_4 axis is best shown by separating it into its two components: rotation by 90° (or ¼ of a turn around the axis and reflection (see the figure below). This ion has three S_4 axes.

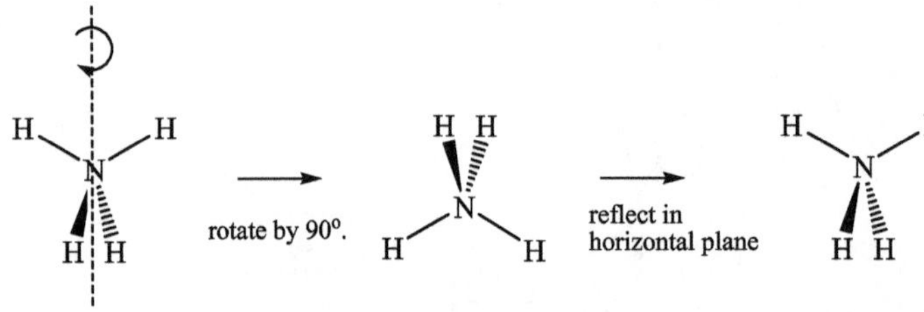

S3.2 Using the decision tree in Figure 3.9 is generally the easiest way to determine the point group of a molecule or ion. You might find it helpful to make a model of the molecule or ion if you have a molecular modelling kit. This can help you visualize the structure. If the molecular geometry is not given in the problem (i.e., you have been provided with a molecular formula only), you can use VSEPR theory to determine the geometry of the molecule/ion in question.

a) BF_3 point group: According to the decision tree on Figure 3.9, the first step is answering the question "Is the molecule/ion linear?" Since BF_3 is a planar molecule, the answer is NO. Following the "N" line from "Linear?" we find the next question we have to answer: "Are there two or more C_n with n > 2?" The answer is again NO: BF_3 possesses only one three-fold rotation axis (C_3) which passes through B atom and is perpendicular to the plane of the molecule. Keep in mind that the highest order axis, in this case C_3, is the principal axis. The molecule should be repositioned in such a way that the principal axis (i.e., C_3 here) is vertical (thus, it coincides with the z coordinate axis). We have to follow "N" line again to the next step: "Are there any rotational axes (C_n)?" The answer is YES: BF_3 has one C_3 (already mentioned above) and three C_2 symmetry axes. We follow the "Y" path and look if the three C_2 axes are perpendicular to the principal axis C_3. Since C_2 axes coincide with three B–F bonds, they are perpendicular to the principal axis—so the answer is YES. Again, we follow the "Y" path and then ask "Are there any horizontal mirror planes (σ_h)?" The answer is YES: there is a mirror plane perpendicular to C_3, the principal axis. This mirror plane coincides with the plane of BF_3 molecule. Finally, the "Y" path leads us to the D_{nh} point group where n in the subscript stands for the order of the principal axis. Since the order of the principal axis in our case is 3, the point group to which BF_3 belongs is D_{3h}.

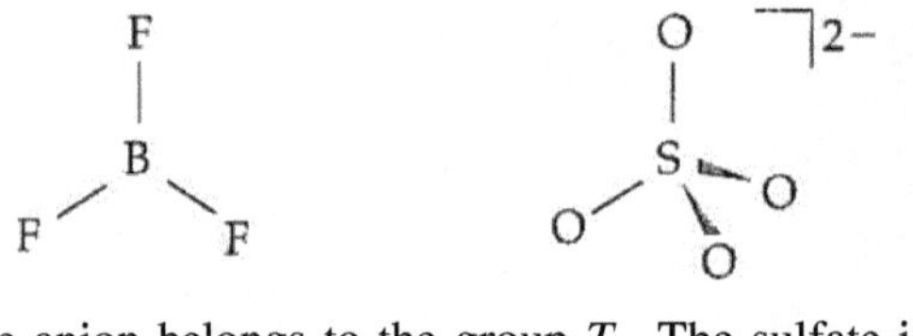

b) SO_4^{2-} point group: The sulfate anion belongs to the group T_d. The sulfate ion (i) is nonlinear; (ii) possesses four three-fold axes (C_3), like NH_4^+ (see the answer to S6.1), and (iii) does not have a centre of symmetry. The sequence of "no, yes, no" on the decision tree leads to the conclusion that SO_4^{2-} belongs to the T_d point group.

S3.3 H_2S molecule is in the same point group as H_2O, C_{2v}. For the location of symmetry elements of H_2S we can refer to Figure 3.10, and for the C_{2v} character table to Table 3.4 in your textbook. The convention we always follow is to place the z-axis along the highest order axis, in this case C_2 symmetry axis, and the mutually perpendicular mirror planes σ_v and σ_v' in planes of xz and yz axes respectively. Now check how all d orbitals behave under symmetry operations of C_{2v} group. (To successfully work through this problem, you should know the shapes, orientation, and signs of d orbital lobes with respect to the coordinate axes (x, y, and z)—refer to Figure 1.13 to refresh your memory if necessary.) Keep in mind that all d orbitals will remain unchanged by E (as a matter of fact, everything remains unchanged by E, hence 1 under its column) so we can concentrate on remaining symmetry elements. The d_{z^2} atomic orbital remains unchanged by C_2, hence it has character 1 for this symmetry element. It also remains unchanged by σ_v and σ_v', thus for both the character would be 1. Thus, the characters of this orbital are (1, 1, 1, 1) for E, C_{2v}, σ_v and σ_v'. The only row in Table 3.4 that has these four characters is A_1—therefore the d_{z^2} orbital has symmetry species A_1 (i.e., this orbital is totally symmetric under C_{2v}). The $d_{x^2-y^2}$ orbital remains unchanged by all symmetry operations as well; it also has characters (1, 1, 1, 1) and symmetry species A_1. The d_{xy} orbital, with its lobes between x and y axes remains unchanged by C_2 (hence character 1) but it would change the sign by both σ_v and σ_v'. For σ_v and σ_v', d_{xy} has characters –1 and –1. Overall, this orbital has (1, 1, –1, –1) as its characters, which corresponds to the symmetry species A_2. The d_{yz} orbital, once rotated around C_2 (i.e., around z-axis for 180°) would change its sign, thus the C_2 character for this orbital is –1. It would also change its sign after reflection through σ_v (which we have placed in the xz plane)—the character is again –1. And finally, the reflection through σ_v' would leave this orbital unchanged for character 1. Therefore, d_{yz} orbital has characters (1, –1, –1, 1) and has symmetry species B_2. Following similar analysis, we can find that d_{zx} orbital has symmetry species B_1.

S3.4 The SF_6 molecule has octahedral geometry and belongs to the O_h space group. If you refer to the character table for this group, which is given in Resource Section 4, you find that there are characters of 1, 2, *and* 3 in the column headed by the identity element, E. Therefore, the *maximum* possible degree of degeneracy of the orbitals in SF_6 is 3 (although nondegenerate and twofold degenerate orbitals are also allowed as revealed by 1 and 2 characters).

S3.5 Like the pentagonal prismatic (or eclipsed) configuration, the pentagonal antiprismatic (or staggered) conformation of ruthenocene has a C_5 axis passing through the centroids of the C_5H_5 rings and the central Ru atom. It also has five C_2 axes that pass through the Ru atom but are perpendicular to the principal C_5 axis, so it belongs to one of the D point groups. The staggered conformation, however, lacks the σ_h plane of symmetry that the pentagonal prismatic structure has. Thus, it cannot be in the D_{5h} space group. It does have five dihedral mirror planes, located between C_2 symmetry axes. Thus, the pentagonal antiprismatic conformation of ruthenocene belongs to the D_{5d} point group. Since this point group has a C_5 axis *and* perpendicular C_2 axes, it is not polar (see Section 3.3). You may find it difficult to find the n C_2 axes for a D_{nd} structure. However, if you draw the mirror planes, the C_2 axes lie between them. In this case, one C_2 axis interchanges the front vertex of the top ring with one of the two front vertices of the bottom ring, while a second C_2 axis, rotated exactly 36° from the first one, interchanges the same vertex on top with the other front bottom one.

S3.6 Except for the identity, E, the only element of symmetry that this conformation of hydrogen peroxide possesses is a C_2 axis that passes through the midpoint of the O–O bond and bisects the two O–O–H planes (these are *not* mirror planes of symmetry). Hence this form of H_2O_2 belongs to the C_2 point group, and it is chiral because this group does not contain any S_n axes. In general, any structure that belongs to a C_n or D_n point group is chiral, as are molecules that are asymmetric (C_1 symmetry). However, considering the free rotation around O–O bond we would not be able to observe optically active H_2O_2 under ordinary conditions. Optically active H_2O_2 might be observable, even transiently, if bound in a chiral host such as the active site of an enzyme. Or if the rotation is prevented in any other way (e.g., very low temperature in solid state).

S3.7 The Lewis structure of linear nitrous oxide molecule is shown below. However, unlike similarly linear CO_2 with which it is isoelectronic, N_2O does not have a centre of symmetry. Therefore, the exclusion rule does not apply, and a band that is IR active *can* be Raman active as well.

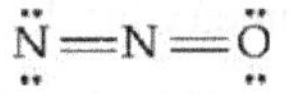

S3.8 All of the operations of the D_{2h} point group leave the displacement vectors unchanged during the symmetric stretching of Pd–Cl bonds in the *trans* isomer. Therefore, all the operations have a character of 1. This corresponds to the first row in the D_{2h} character table, which is the A_g symmetry type.

S6.9 The reducible representation is obtained as follows:

D_{4h}	E	$2C_4$	C_2	$2C_2'$	$2C_2''$	i	$2S_4$	σ_h	$2\sigma_v$	$2\sigma_d$
Γ_{3N}	4	0	0	2	0	0	0	4	2	0

This reduces to $A_{1g} + B_{1g} + E_u$.

A_{1g} and B_{1g} transform as $x^2 + y^2$, z^2 and $x^2 - y^2$ respectively and are thus Raman active but not IR active. E_u is IR active but not Raman active as it transforms as (x, y).

S3.10 The molecule CH_4 has T_d symmetry, and the given symmetry-adapted linear combination (SALC) of H atom's 1s orbitals must also have T_d symmetry. This is true because each time the set of four H atom orbitals is subjected to an operation in the T_d point group, the set changes into itself. The sketch below shows the 1s orbitals around carbon atom and location of one of the three C_3 axes. You can make a similar sketch and confirm that all symmetry elements leave the set unchanged and thus have a character 1. This places SALC in A_1 symmetry species.

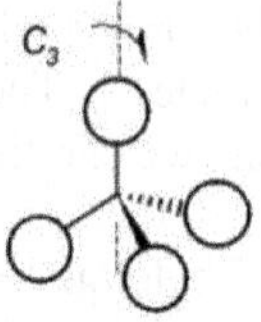

S3.11 You must adopt some conventions to answer this question. First, you assume that the combination of H atom 1s orbitals given looks like the figure below showing four H atoms arranged with D_{4h} symmetry. Note that shaded circles correspond to the H atoms whose 1s orbital did not change the sign (A1s and C1s) while open circles represent the H atoms whose 1s orbital did change the sign (B1s and D1s)—the change of sign is indicated by a minus sign in front of the wavefunction Ψ. Inspection of the character table for this group, which is given in Resource Section 4, reveals that there are three different types of C_2 rotation axes, that is, there are three columns labelled C_2, C_2', and C_2''. The first of these is the C_2 axis that is coincident with the C_4 axis; the second type, C_2', represents two axes in the H_4 plane that pass through pairs of opposite H atoms; the third type, C_2'', represents two axes in the H_4 plane that do not pass through H atoms but rather between them (see Figure 3.3 in the textbook). Now, instead of applying operations from all ten columns to this array, to see if it changes into itself (i.e., the +/– signs of the lobes stay the same) or if it changes sign, you can make use of a shortcut. Notice that the array changes into itself under the inversion operation through the centre of symmetry. Thus, the character for this operation, i, is 1. This means that the symmetry label for this array is one of the first four in the character table, A_{1g}, A_{2g}, B_{1g}, or B_{2g}. Notice also that for these four, the symmetry type is uniquely determined by the characters for the first five columns of operations, which are:

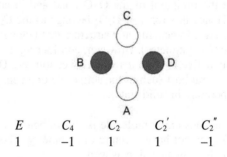

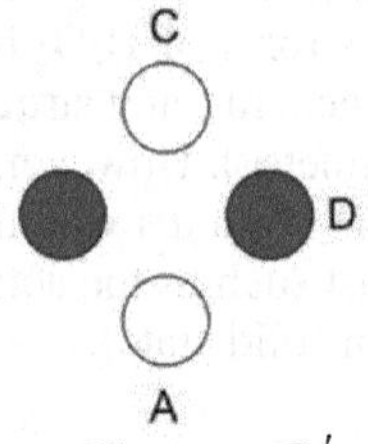

E	C_4	C_2	C_2'	C_2''
1	−1	1	1	−1

These match the characters of the B_{1g} symmetry label.

Note that this SALC also looks like $d_{x^2-y^2}$, thus the above SALC and $d_{x^2-y^2}$ have to have the same symmetry—look at the last column of D_{4h} character table for the B_{1g}.

S3.12 By consulting Resource Section 5 and the D_{4h} character table, we note that Pt's 5s and $4d_{z^2}$ have A_{1g} symmetry so they would combine with A_{1g} SALCs; the $d_{x^2-y^2}$ has B_{1g} symmetry and can combine with B_{1g} SALCs. Finally, $5p_x$ and $5p_y$ with E_u symmetry can combine with E_u SALCs. Therefore, these Pt orbitals with matching symmetries can be used to generate SALCs.

S3.13 Recall that symmetry types with the same symmetry as the function $x^2 + y^2 + z^2$ are Raman active, not IR active. On the other hand, symmetry types with the same symmetry as the functions x, y, or z are IR active, not Raman active.

SF_6 has O_h symmetry. Analysis of the stretching vibrations leads to (see Example 3.13):

$$A_{1g} \text{ (Raman, polarized)} + E_g \text{ (Raman)} + T_{1u} \text{ (IR)}.$$

Trans-SF_4Cl_2 belongs to the D_{4h} point group. (Note: *trans* means that the two Cl atoms are located on the opposite sides of the molecule, see the structure below). Analysis of the stretching vibrations leads to:

$$A_{2u} \text{ (IR)} + 2E_u \text{ (IR)} + A_{1g} \text{ (Raman, polarized)} + B_{1g} \text{ (Raman)} + B_{2g} \text{ (Raman)}.$$

Thus, for *trans*-SF_4Cl_2 we will observe two IR and three Raman stretching absorptions, whereas for SF_6 only one IR and two Raman absorptions.

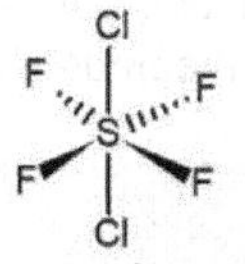

The structure of *trans*-SF_4Cl_2

S3.14 For the molecule or ion with D_{4h} symmetry,

D_{4h}	E	$2C_4$	C_2	$2C_2'$	$2C_2''$	i	$2S_4$	σ_h	$2\sigma_v$	$2\sigma_d$
Γ_{3N}	15	1	−1	−3	−1	−3	−1	5	3	1

This reduces to $A_{1g} + A_{2g} + B_{1g} + B_{2g} + E_g + 2A_{2u} + B_{2u} + 3E_u$.

The translations span A_{2u} and E_u and the rotations span A_{2g} and E_g. Subtracting these terms gives:

$$A_{1g} + B_{1g} + B_{2g} + A_{2u} + B_{2u} + 2E_u$$

as the symmetries for the vibration modes.

S3.15 $A_{1g} + E_g + T_{1u}$, see Resource Section 5.

Exercises

E3.1 The analysis here is going to focus on the topic of this chapter, i.e., molecular symmetry. The VSEPR model has been covered in quite a lot of detail in the previous chapter and only the basic overview is going to be given here. If you have trouble determining the geometries of molecules and ions in this exercise, you should revisit the relevant parts of Chapter 2.

To predict the structure of **BrF_4^-** using VSEPR model, we have to take care of 36 valence electrons. The correct Lewis structure places four bonding and two lone electron pairs on the central Br atom resulting in an octahedral electron geometry but square planar molecular geometry. The principal rotation axis in this case is C_4 perpendicular to the square plane. The ion also has one horizontal mirror plane and two vertical mirror planes (only one shown below) as well as the centre of inversion located at Br atom (plus two diagonal mirror planes). For illustration showing all symmetry elements of a square planar molecule/ion, please refer to Figure 3.3 in your textbook.

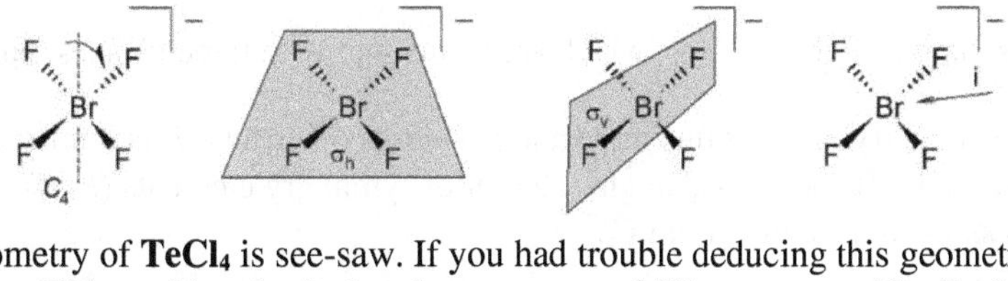

The molecular geometry of **$TeCl_4$** is see-saw. If you had trouble deducing this geometry, review Example 2.3 from the previous chapter. This problem looked at the geometry of SF_4, more specifically the effect of lone electron pair on central S atom on the overall geometry. Note that Te and Cl are in the same group of the periodic table as S and F respectively. That means the total number of valence electrons in $TeCl_4$ matches the one for SF_4. Since both Te and Cl have larger atomic radii than S and F respectively, the effect of lone electron pair on geometry is going to be diminished and overall $TeCl_4$ is going to be very close to ideal see-saw geometry. $TeCl_4$ molecule has a C_2 axis as a principal axis and two vertical mirror planes. It does not have *i*. Note that ideally the molecule should be rotated so that C_2 axis, as a principal axis in this case, is vertical.

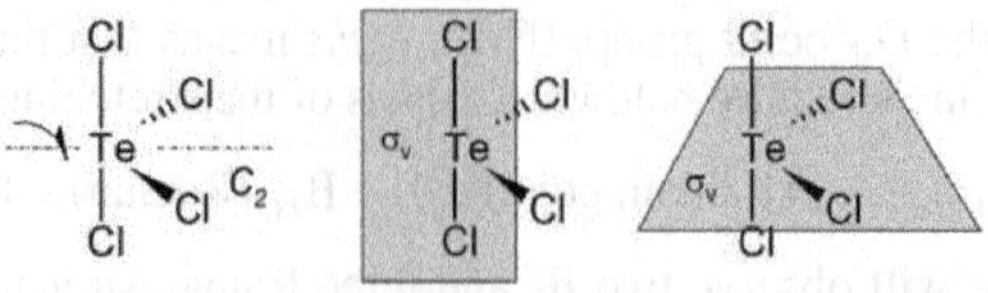

The $\mathbf{I_2Cl_6}$ molecule is planar. It is a dimer consisting of two ICl_3 monomers bridged by two Cl atoms. There is one principal C_2 axis (there are two more C_2 axes), one horizontal mirror plane and two vertical mirror planes (only one shown). The centre of inversion lies in the middle of I–Cl–I–Cl rhombohedron.

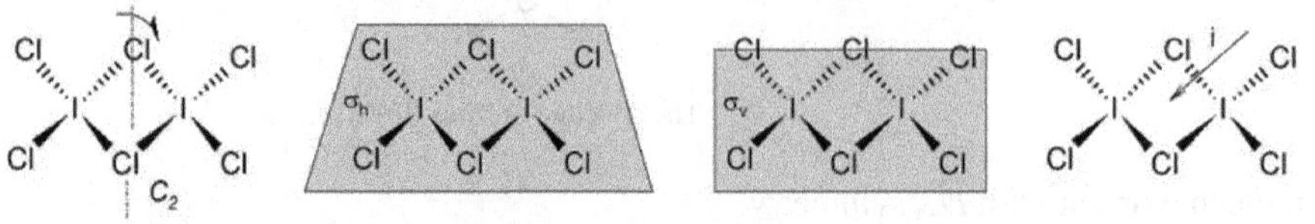

The anion $\mathbf{ClO_3^-}$ has 26 valence electrons with trigonal pyramidal molecular geometry. The principal axis is C_3, there are no horizontal mirror planes but there are three vertical ones. The anion does not possess a centre of inversion.

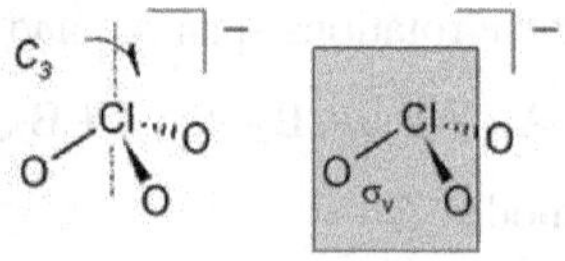

E3.2 a) A C_3 axis and a σ_v plane in the NH_3 molecule:

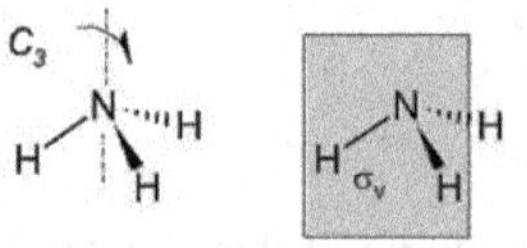

b) A C_4 axis and a σ_h plane in the square-planar $[PtCl_4]^{2-}$ ion:

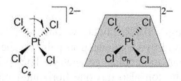

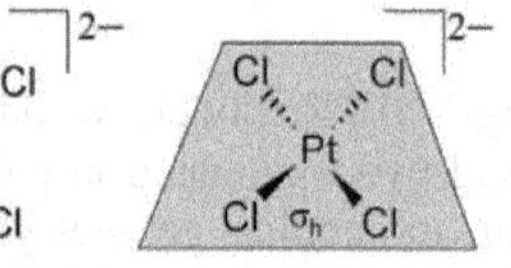

E3.3 a) $\mathbf{CO_2}$: This linear molecule has both a centre of inversion, *i*, and S_4.

b) $\mathbf{C_2H_2}$: This linear molecule also has both *i* and S_4.

c) $\mathbf{BF_3}$: This molecule possesses neither *i* nor S_4.

d) $\mathbf{SO_4^{2-}}$:This ion has three different S_4 axes, which are coincident with three C_2 axes, but there is no *i*.

E3.4 a) $\mathbf{NH_2Cl}$: The only symmetry element this molecule possesses other than *E* is a mirror plane that contains the N and Cl atoms and bisects the H–N–H bond angle. The set of symmetry elements (*E*, σ), following the decision tree on Figure 3.9, corresponds to the point group C_s.

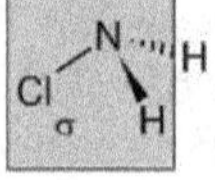

b) CO_3^{2-}: The carbonate anion is planar, so it possesses at least one plane of symmetry. Since this plane is perpendicular to the principal rotation axis, C_3, it is called σ_h. In addition to the C_3 axis, there are three C_2 axes coinciding with the three C–O bond vectors. There are also other mirror planes and improper rotation axes, but the elements listed so far (E, C_3, σ_h, $3C_2$) uniquely correspond to the D_{3h} point group according to the decision tree on Figure 3.9. CO_3^{2-} has the same symmetry as BF_3; if you have problems with this exercise, review Self-test 3.2(a). A complete list of symmetry elements is E, C_3, $3C_2$, S_3, σ_h, and $3\sigma_v$.

c) SiF_4: This tetrahedral molecule has four C_3 axes, one coinciding with each of the four Si–F bonds. In addition, there are six mirror planes of symmetry (any pair of F atoms and the central Si atom define a mirror plane, and there are always six ways to choose a pair of objects out of a set of four). Furthermore, there is no centre of symmetry. Thus, the set (E, $4C_3$, 6σ, no i) describes this molecule and corresponds to the T_d point group. SiF_4 has the same symmetry as SO_4^{2-}; if you have problems with this exercise, review Self-test 3.2(b). A complete list of symmetry elements is E, $4C_3$, $3C_2$, $3S_4$, and $6\sigma_d$.

d) HCN: Hydrogen cyanide is linear; it has the infinite order symmetry axes, C_∞, as well as the infinite number of vertical mirror planes. Thus, it belongs to either the $D_{\infty h}$ or the $C_{\infty v}$ point group (see Figure 3.9). Since it does not possess a centre of symmetry, which is a requirement for the $D_{\infty h}$ point group, it belongs to the $C_{\infty v}$ point group.

e) SiFClBrI: This tetrahedral molecule does not possess any element of symmetry other than the identity element, E. Thus, it is asymmetric and belongs to the C_1 point group, the simplest possible point group.

(f) BF_4^-: This anion is also tetrahedral and belongs to the T_d point group as SiF_4 above.

E3.5 Overall, benzene molecule has seven mirror planes: $3\sigma_v$, $3\sigma_d$, and $1\sigma_h$. It has a C_6 axis perpendicular to six C_2 axes and a horizontal mirror plane that contains all the atoms in the ring. Hence, the point group symmetry of the benzene molecule is D_{6h}. The chloro-substituted, $C_6H_3Cl_3$, with chlorines on alternating carbons around the ring (1,3,5-trichlorobenzene), has a D_{3h} symmetry and four mirror planes: $3\sigma_v$ and $1\sigma_h$.

E3.6 **a) A p orbital:** The two lobes of a p orbital are not equivalent because they have different sign, + and –, and therefore cannot be interchanged by potential elements of symmetry. Thus, a p orbital does not possess a mirror plane of symmetry perpendicular to the long axis of the orbital. It does, however, possess an infinite number of mirror planes that pass through both lobes and include the long axis of the orbital. In addition, the long axis is a C_n axis, where n can be any number from 1 to ∞ (in group theory this is referred to as a C_∞ axis).

b) A d_{xy} orbital: The two pairs of + and – lobes of a d_{xy} orbital are interchanged by the centre of symmetry that this orbital possesses. It also possesses three mutually perpendicular C_2 axes, each one coincident with one of the three Cartesian coordinate axes. Furthermore, it possesses three mutually perpendicular mirror planes of symmetry, which are coincident with the xy plane and the two planes that are rotated by 45° about the z-axis from the xz plane and the yz plane.

c) A d_{z^2} orbital: Unlike a p_z orbital, a d_{z^2} orbital has two large lobes with positive sign along its long axis, and a torus of negative sign around the middle (in xy plane). In addition to the symmetry elements possessed by a p orbital (see above), the infinite number of mirror planes that pass through both lobes and include the long axis of the orbital as well as the C_∞ axis, a d_{z^2} orbital also possesses (i) a centre of symmetry, (ii) a mirror plane that is perpendicular to the C_∞ axis, (iii) an infinite number of C_2 axes that pass through the centre of the orbital and are perpendicular to the C_∞ axis, and (iv) an S_∞ axis.

E3.7 **a)** Using the decision tree shown in Figure 3.9, you will find the point group of SO_3^{2-} anion (trigonal pyramidal geometry) to be C_{3v} (it is nonlinear; it only has one proper rotation axis, a C_3 axis; and it has three σ_v mirror planes of symmetry).

b) Inspection of the C_{3v} character table (Resource Section 4) shows that the characters under the column headed by the identity element, E, are 1 and 2. Therefore, the maximum degeneracy possible for molecular orbitals of SO_3^{2-} anion is 2.

c) According to the character table in Resource Section 5, the S atom 3s and $3p_z$ orbitals are each singly degenerate (and belong to the A_1 symmetry type), but the $3p_x$ and $3p_y$ orbitals are doubly degenerate (and belong to the E symmetry type). Thus, the $3p_x$ and $3p_y$ atomic orbitals on sulfur can contribute to molecular orbitals that are twofold degenerate.

E3.8 **a) Point group:** The PF_5 molecule has a trigonal bipyramidal geometry. The decision tree should lead you to D_{3h} point group (the molecule is nonlinear; it has only one high-order proper rotation axis, a C_3 axis; it has three C_2 axes that are perpendicular to the C_3 axis; and it has a σ_h mirror plane of symmetry).

b) Degenerate MOs: Inspection of the D_{3h} character table (Resource Section 4) reveals that the characters under the *E* column are 1 and 2, so the maximum degeneracy possible for a molecule with this symmetry is 2.

c) Which p orbitals have the maximum degeneracy: The P atom $3p_x$ and $3p_y$ atomic orbitals, which are doubly degenerate and are of the E′ symmetry type (i.e., they have E′ symmetry), can contribute to molecular orbitals that are twofold degenerate. In fact, if they contribute to molecular orbitals at all, they *must* contribute to twofold degenerate ones.

E3.9 PF_5 is a non-linear molecule, thus the expected number of vibrational modes is $3N - 6 = 3 \times 6 - 6 = 12$. To determine the symmetries of vibrations, we have to consider the reducible representation of D_{3h}. Since we have six atoms with three displacement directions each, there are a total of $6 \times 3 = 18$ displacements in this molecule. These displacements remain unchanged after identity operation; hence the character for *E* is 18. You can follow this procedure for all symmetry elements in D_{3h} (as well as consult Example 3.14) to construct the following reducible representation:

D_{3h}	E	$2C_3$	$3C_2$	σ_h	$2S_3$	$3\sigma_v$
Γ_{3N}	18	0	−2	4	−2	4

Γ_{3N} reduces to: $2A_1' + A_2' + 4E' + 3A_2'' + 2E''$; subtracting Γ_{trans} ($E' + A_2''$) and Γ_{rot} ($A_2' + E''$), we obtain Γ_{vib}: $2A_1' + 3E' + 2A_2'' + E''$. Thus, we expect six Raman bands: $2A_1' + 3E' + E''$ (note that A_2'' is inactive in Raman because symmetry type does not contain the same symmetry as the function $x^2 + y^2 + z^2$).

E3.10 **a) In the plane of the nuclei:** SO_3 has a trigonal-planar geometry. The number of vibrational modes is then calculated using $3N - 6$ formula. Since SO_3 has four atoms, the number of vibrational modes is $3 \times 4 - 6 = 6$. If you consider the C_3 axis to be the *z*-axis, then each of the four atoms has two independent displacements in the *xy* plane, namely along the *x* and along the *y* axis. The product (4 atoms) × (2 displacement modes/atom) gives eight displacement modes, not all of which are vibrations. There are two translation modes in the plane of the nuclei, one each along the *x*- and *y*-axes. There is also one rotational mode around the *z*-axis. Therefore, if you subtract these three nonvibrational displacement modes from the total of 8 displacement modes, you arrive at a total of five vibrational modes in the plane of the nuclei.

D_{3h}	E	$2C_3$	$3C_2$	σ_h	$2S_3$	$3\sigma_v$
Γ_{3N}	12	0	−2	4	−2	2

Γ_{3N} reduces to: $A_1' + A_2' + 3E' + 2A_2'' + E''$.

Subtracting Γ_{trans} ($E' + A_2''$) and Γ_{rot} ($A_2' + E''$), we obtain Γ_{vib}: $A_1' + 2E' + A_2''$.

In the plane: A_1' is a symmetric stretch; $2E'$ are modes consisting of (i) mainly asymmetric stretching and (ii) deformation.

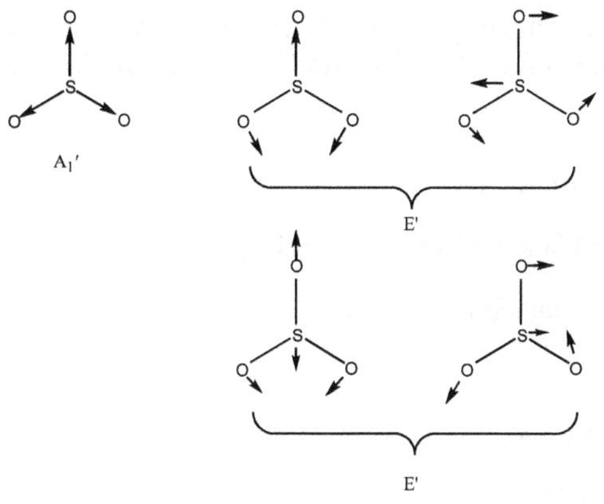

b) Perpendicular to the molecular plane: You can use your answer to part (a), above, to answer this question. Since there are four atoms in the molecule, there are 3(4) – 6 = 6 vibrational modes. You discovered that there are five vibrational modes in the plane of the nuclei for SO_3, so there must be only one vibrational mode perpendicular to the molecular plane. A_2'' is the deformation in and out of the plane.

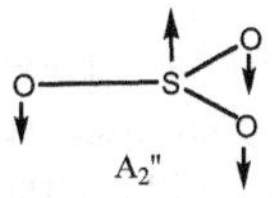

E3.11 **a) SF_6:** Since sulfur hexafluoride has a centre of symmetry, the exclusion rule applies. Therefore, none of the vibrations of this molecule can be *both* IR and Raman active. A quick glance at the O_h character table in Resource Section 4 confirms that the functions *x*, *y*, and *z* (required for IR activity) have the T_{1u} symmetry type and that all the binary product functions such as x^2, *xy*, etc. (required for Raman activity) have different symmetry types.

O_h	E	$8C_3$	$6C_2$	$6C_4$	$6C_2$	i	$6S_4$	$8S_6$	$3\sigma_h$	$6\sigma_d$
Γ_{3N}	21	0	−1	3	−3	−3	−1	0	5	3

Γ_{3N} reduces to: $A_{1g} + E_g + T_{1g} + T_{2g} + 3T_{1u} + T_{2u}$.

Then subtracting Γ_{trans} (T_{1u}) and Γ_{rot} (T_{1g}), we obtain Γ_{vib} : $A_{1g} + E_g + T_{2g} + 2T_{1u} + T_{2u}$.

None of these modes is both IR and Raman active as there is a centre of inversion.

b) BF_3: Boron trifluoride does not have a centre of symmetry. Therefore, it is possible that some vibrations are both IR and Raman active. You should consult the D_{3h} character table in Resource Section 4. Notice that the pairs of functions (*x*, *y*) and ($x^2 - y^2$, *xy*) have the E′ symmetry type. Therefore, any E′ symmetry vibration will be observed as a band in both IR and Raman spectra.

D_{3h}	E	$2C_3$	$3C_2$	σ_h	$2S_3$	$3\sigma_v$
Γ_{3N}	12	0	−2	4	−2	2

Γ_{3N} reduces to: $A_1' + A_2' + 3E' + 2A_2'' + E''$.

Subtracting Γ_{trans} ($E' + A_2''$) and Γ_{rot} ($A_2' + E''$), we obtain Γ_{vib}: A_1' (Raman) + $2E'$ (IR and Raman) + A_2'' (IR).

The E′ modes are active in both IR and Raman.

E3.12 The SALCs for methane have the same forms as the expressions for the AOs as SALCs of sp^3 hybrid orbitals.

$$s = (\tfrac{1}{2})(\varphi_1 + \varphi_2 + \varphi_3 + \varphi_3)\ (= A_1)$$

$$p_x = (\tfrac{1}{2})(\varphi_1 - \varphi_2 + \varphi_3 - \varphi_3)\ (= T_2)$$

$$p_y = (\tfrac{1}{2})(\varphi_1 - \varphi_2 - \varphi_3 + \varphi_3)\ (= T_2)$$

$$p_z = (\tfrac{1}{2})(\varphi_1 + \varphi_2 - \varphi_3 - \varphi_3)\ (= T_2)$$

The 2s atomic orbitals have symmetry species A_1 and the 2p atomic orbitals have symmetry species T_2. Therefore, the MOs would be constructed from SALCs with H1s and 2s and 2p atomic orbitals on C.

E3.13 **a) BF_3:**

$(^1/_{\sqrt{3}})(\varphi_1 + \varphi_2 + \varphi_3)$ (= A_1')

$(^1/_{\sqrt{6}})(2\varphi_1 - \varphi_2 - \varphi_3)$ and $(^1/_{\sqrt{2}})(\varphi_2 - \varphi_3)$ (= E')

Note that $(^1/_{\sqrt{2}})(\varphi_2 - \varphi_3)$ is obtained by combining the other two E′-type SALCs, that is, $(2\varphi_2 - \varphi_3 - \varphi_1) - (2\varphi_3 - \varphi_1 - \varphi_2)$.

b) PF_5: (axial F atoms are $\varphi_4 + \varphi_5$)

$(^1/_{\sqrt{2}})(\varphi_4 + \varphi_5)$ (= A_1')

$(^1/_{\sqrt{2}})(\varphi_4 - \varphi_5)$ (= A_2'')

$(^1/_{\sqrt{3}})(\varphi_1 + \varphi_2 + \varphi_3)$ (= A_1')

$(^1/_{\sqrt{6}})(2\varphi_1 - \varphi_2 - \varphi_3)$ and $(^1/_{\sqrt{2}})(\varphi_2 - \varphi_3)$ (= E')

E3.14 The molecular orbital energy diagram for ammonia is shown in Figure 3.30. The interpretation given in the text was that the $2a_1$ molecular orbital is almost nonbonding, so the electron configuration $1a_1^2 1e^4 2a_1^2$ results in only three bonds ((2 + 4)/2 = 3). Since there are three N–H bonds, the average N–H bond order is 1 (3/3 = 1).

E3.15 The molecular orbital energy diagram for sulfur hexafluoride is shown in Figure 3.32. The HOMO of SF_6 is the nonbonding *e* set of MOs. These are pure F atom symmetry-adapted orbitals, and they do not have any S atom character whatsoever. On the other hand, the antibonding $2t_1$ orbitals, a set of LUMO orbitals, have both a sulfur and a fluorine character. Since sulfur is less electronegative than fluorine, its valence orbitals lie at higher energy than the valence orbitals of fluorine (from which the *t* symmetry-adapted combinations were formed). Thus, the $2t_1$ orbitals lie closer in energy to the S atomic orbitals and hence they contain more S character.

Guidelines for Selected Tutorial Problems

T3.1 Use VSEPR theory to deduce the molecular geometry of IF_3O_2, then analyse all possible arrangement of F and O atoms around the central I atom that cannot be superimposed on each other (you should get three isomers). Follow the decision tree to determine the point group of each isomer.

T3.3 The cation NH_4^+ has a tetrahedral geometry and hence belongs to T_d point group. Looking at the character table we can see that we have to consider the degeneracy; for example, T_2 has (*x, y, z*) in the last column. This means that the vibration is going to be active along all three axes and degeneracy should be expected.

T3.5 From VSEPR, $AsCl_5$ should be trigonal pyramidal with symmetry D_{3h}. We first obtain the representation Γ_{3N}:

D_{3h}	E	$2C_3$	$3C_2$	σ_h	$2S_3$	$3\sigma_v$
Γ_{3N}	18	0	−2	4	−2	4

Γ_{3N} reduces to: $2A_1' + A_2' + 4E' + 3A_2'' + 2E''$; subtracting Γ_{trans} $(E' + A_2'')$ and Γ_{rot} $(A_2' + E'')$, we obtain Γ_{vib}: $2A_1' + 3E' + 2A_2'' + E''$. Thus, we expect six Raman bands: $2A_1' + 3E' + E''$ (note that A_2'' is inactive in Raman because symmetry type does not contain the same symmetry as the function $x^2 + y^2 + z^2$).

T3.7 The easiest way to approach this tutorial problem is to sketch the Cl orbitals in tetrahedral arrangement around Co^{2+} as shown below (note that all p orbitals face the central metal cation with the same phase. Then, start applying the symmetry elements and see how each set of orbitals is behaving. The sketch shows the location of one of the four three-fold rotation axes, and rotation for 120° round this axis leaves all orbitals unchanged.

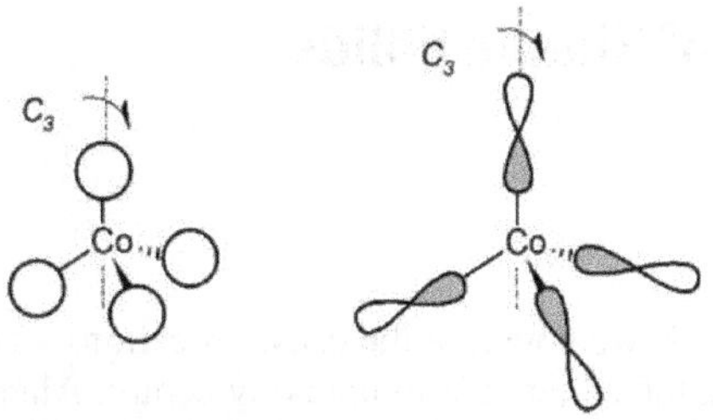

Keep in mind that both the orbital shape and the phases of individual lobes must match to claim that orbitals are unchanged. To show that indeed both sets of Cl orbitals transform in identical manner, all symmetry elements for the T_d point group must be checked.

For bonding, consider the symmetry requirements for orbital bonding and the possibility of p orbitals contribution to Co–Cl bonding.

T3.9 Take a 3d transition metal as a central atom surrounded by six monoatomic ligands. First, assign the metal's 3d, 4s and 4p valence orbitals to a proper symmetry species using the first table in Resource Section 5.

Now you have to construct SALCs that would have the same symmetry species as some metal orbitals. For the σ bonding use simple s atomic orbital. Your SALCs should look like the ones provided in the later part of Resource Section 5 under O_h point group.

For π bonding use p orbitals. Your final result is also given in Resource Section 5 for O_h point group.

Chapter 4 The Structures of Simple Solids

Self-Tests

S4.1 By examining Figures 4.7 and 4.32, we note that the caesium cations sit on a primitive cubic unit cell (lattice type P) with chloride anion occupying the cubic hole in the body centre. Alternatively, one can view the structure as P-type lattice of chloride anions with caesium cation in cubic hole. Keep in mind that caesium chloride does not have a body centred cubic lattice although it might appear so at a first glance. The body centred lattice has all points identical, whereas in CsCl lattice the ion at the body centre is different from those at the corners.

S4.2 The 3D structure of SiS_2 is shown below:

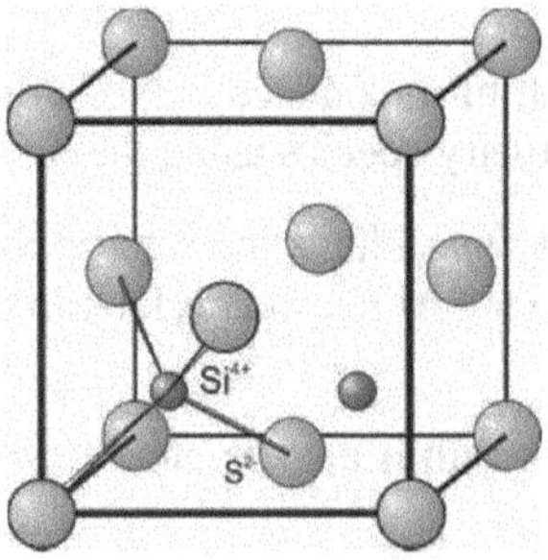

S4.3 **a)** Figures 4.3 and 4.23 show the primitive cubic unit cell. Each unit cell contains one sphere (equivalent to 8 × 1/8 spheres on the vertices of the cell in contact along the edges). The volume of a sphere with radius r is $4/3\pi r^3$, whereas the volume of the cubic unit cell is $a^3 = (2r)^3$ (where a is a length of the edge of the unit cell). Thus, the fraction filled is $\frac{^3/_4\pi r^3}{(2r)^3} = 0.52$: 52% of the volume in the primitive cubic unit cell is filled.

b) Figures 4.8 and 4.29(a, the structure of iron) show the body-centred unit cell. Each unit cell contains two spheres (one in the middle of the cell and an equivalent of one sphere at eight corners of a cube, see part (i) above). The volume of a sphere is $4/3\pi r^3$ (where r is the sphere radius), thus the total volume is $2 \times 4/3\pi r^3 = 8/3\ \pi r^3$. Now, we have to find a way to represent the length of the unit cell, a, in terms of r. In a body-centred cubic unit cell the spheres are in contact along the space diagonal of the cube (labelled D on the figure below). Thus

$$D = r + 2r + r = 4r \qquad (1)$$

based on the same arguments as in Example 4.3.

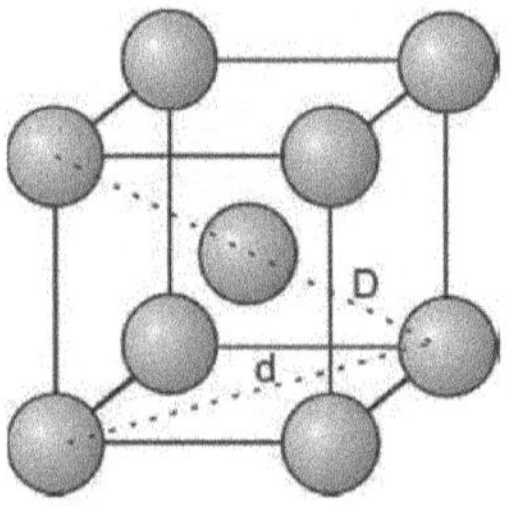

If we apply Pythagoras' theorem to this case, we obtain

$$D^2 = d^2 + a^2, \qquad (2)$$

where d is the length of a face diagonal of the cube and a the length of the cube's edge. As we have also seen in Example 3.3, the face diagonal of the cube can be calculated from $d^2 = 2a^2$. Substituting this result in equation (1), we have:

$$D = 2a^2 + a^2 = 3a^2. \qquad (3)$$

Finally, we can square the result in (1) and equate this to (3) to obtain the relationship between r and a in a body-centred unit cell as:

$$16r^2 = 3a^2, \text{ and } a = \frac{4r}{\sqrt{3}}.$$

From here the volume of the cell is $a^3 = \frac{64r^3}{(\sqrt{3})^3}$. Thus, the fraction of space occupied by identical spheres in this unit cell is:

$$\frac{\frac{8}{3}\pi r^3}{\frac{64r^3}{(\sqrt{3})^3}} = \frac{8\pi(\sqrt{3})^3}{3\times 64} = 0.68 \text{ or } 68\%.$$

From these results and Example 4.3, it becomes clear that close-packed cubic lattice has the best space economy (best packing, least empty space), followed by the body-centred lattice, whereas the simple cubic packing has the lowest space economy with the highest fraction of unoccupied space.

S4.4 See Figure 4.19b for important distances and geometrical relations. Note that the distance S–T = $r + r_{hole}$, by definition. Therefore, you must express S–T in terms of r. Note also that S–T is the hypotenuse of the right triangle STX, with sides SX, XT, and TS. Point X is at the midpoint of line SS, and because SS = $2r$, so SX = r. The angle θ is 54.74°, one-half of the tetrahedral angle S–T–S (109.48°). Therefore, $\sin 54.74° = r/(r + r_{hole})$, and $r_{hole} = 0.225r$. This is the same as $r_{hole} = ((3/2)^{1/2} - 1)r$.

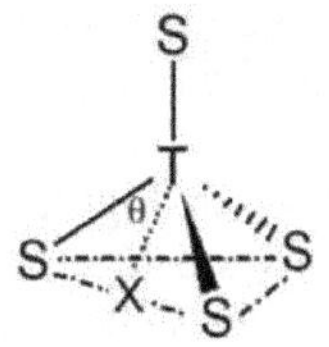

S4.5 The position of tetrahedral holes in a ccp unit cell is shown in Figure 4.20(b). From the figure, we see that each unit cell contains eight tetrahedral holes, each inside the unit cell. As outlined in Example 4.5, the ccp unit cell has four identical spheres. Thus, the spheres-to-tetrahedral holes ratio in this cell is 4 : 8 or 1 : 2.

S4.6 The ccp unit cell contains four silver atoms that weigh together (4 mol Ag × 107.87 g/mol)/(6.022 × 10^{23} Ag/mol)g or 7.165 × 10^{-22} g. The volume of the unit cell is a^3, where a is the length of the edge of the unit cell. The density equals mass divided by volume or:

$$10.5 \text{ g cm}^{-3} = 7.165 \times 10^{-22} \text{ g}/a^3 \text{ with } a \text{ in cm}$$

$$a = 4.09 \times 10^{-8} \text{ cm or } 409 \text{ pm}$$

S4.7 For the bcc structure, $a = 4r/\sqrt{3}$. This relation can be derived simply by considering the right triangle formed from the body diagonal, face diagonal, and edge of the bcc unit cell (see also self-test 4.3(a)). Using the Pythagorean theorem, we have $(4r)^2 = a^2 + (a\sqrt{2})^2$. Solving for a in terms of r, we have $a = 4r/\sqrt{3}$. From Example 4.7, we know that the metallic radius of Po is 174 pm. Therefore, using the derived relation we can calculate a = 401 pm.

S4.8 The lattice type of this Fe/Cr alloy is that of body-centred packing which can be deduced by comparing the unit cells shown in Figure 4.29(a) and (c). The stoichiometry is 1.5 Cr [4 × (1/8) corners + 1 × (1) body] atoms and 0.5 Fe [4 × (1/8) corners] atoms per unit cell. This would give Cr-to-Fe ratio of 1.5 : 0.5 which when multiplied by 2 provides whole numbers required for the formula 3 : 1. The formula for the alloy is therefore $FeCr_3$.

S4.9 For close-packed structures, N close-packed ions lead to N octahedral sites. Therefore, if we assume three close-packed anions (A), then there should also be three octahedral holes. Since only two thirds are filled with cations, there are two cations (X) in two octahedral sites and the stoichiometry for the solid is X_2A_3.

S4.10 Following the lead from Example 4.10, we'll first use Table 4.4 to find possible structures for each compound, and then consult Resource Section 1 to compare ionic sizes and suggest a possible structure.

a) PrO_2 is an AX_2 compound. Looking at Table 4.4 we see that it can have one of the two commonly adopted AX_2 structures: either fluorite (CaF_2) or rutile (TiO_2). To decide which one, we have to compare the ionic radii. A glance at the periodic table of elements can provide a shortcut: Pr is in the f block just like U (UO_2 is in the list of compounds with fluorite-type structure) so it is a good idea to first compare ionic radii of Pr^{4+} and U^{4+}. From Resource Section 1 we find $r(Pr^{4+}) = 96$ pm and $r(U^{4+}) = 100$ pm. The two values are very close bringing us to the conclusion that PrO_2 very likely has the same structure as UO_2, i.e. fluorite-type. Note that we compared the radii for coordination number 8. This is because the cation's coordination number in fluorite-type structure is eight (see Figure 4.38 in your textbook and accompanying description in the text).

b) CrO_2 can have again either fluorite or rutile structure type. Cr is in the d block like most metallic elements under "Rutile" line in Table 4.4. Thus, it is reasonable to compare Cr^{4+} radius with the radii of these cations. The coordination of cation in TiO_2 is close to octahedral with coordination number 6 (more strictly, the coordination is 4 + 2, see Figure 4.39 and the explanation in the text). For Cr^{4+} $r = 55$ pm and for Mn^{4+} $r = 53$ pm; the two values are very close so it is reasonable to suggest that CrO_2 likely has rutile-type structure.

c) The $CrTaO_4$ analysis follows the same reasoning as for $LiNiO_2$ found in the discussion of rock salt type of structure and in Example 4.10(d). The general formula can be written as $(AX_2)_2$, with A = $Cr_{½}Ta_{½}$ and X = O, so the compound again can have either fluorite or rutile structure. We have already seen that Cr^{4+} is very similar in size to Mn^{4+}, now we can confirm that Ta^{4+} is similar in size to W^{4+}; 68 pm vs. 66 pm, respectively. Thus, $CrTaO_4$ likely has rutile-type structure as well.

d) AcOF can be viewed as an AB_2 compound and hence have either fluorite- or rutile-type structure. The ionic radius of Ac^{3+} (81 pm) is close to that of U^{4+} (86 pm) and we can expect fluorite-type structure for AcOF.

e) The case of Li_2TiF_6 is similar to the one in Example 4.10(c). This compound is composed of Li^+ cations and $[TiF_6]^{2-}$ anions and can be rewritten as having the general formula A_2X. Looking at the Table 4.4, we see that only antifluorite structure has this formula. Thus, the likely structure is antifluorite.

S4.11 The perovskite, $CaTiO_3$, structure is shown in Figure 4.42. The large calcium cations occupy site A whereas smaller Ti^{4+} cations site B. The O^{2-} anions are located at site X. Thus, we can see two different coordination environments of O^{2-}: one with respect to Ca^{2+} (A site) and the other with respect to Ti^{4+} (B site). Ti^{4+} is located on the corners of the cube whereas O^{2-} anions are located on the middle of every edge. Thus, there are two Ti^{4+} surrounding each O^{2-}, which are oxygen's nearest neighbours. To see how many Ca^{2+} surround each O^{2-}, we have to add three unit cells horizontally, each sharing a common edge. Then we can see that there are four Ca^{2+} cations as next nearest neighbours. Thus, the coordination of O^{2-} is two Ti^{4+} and four Ca^{2+} and longer distances, this is sometimes written as CN 2+4.

S4.12 La^{3+} would have to go on the larger A site, and to maintain charge balance it would be paired with the smaller In^{3+} and thus the composition would be $LaInO_3$.

S4.13 Follow the procedure outlined in the example. From Resource Section 1 we find the following radii for coordination number 6: $r(Ca^{2+}) = 100$ pm, $r(Bk^{4+}) = 63$ pm and $r(O^{2-}) = 140$ pm. From here we can calculate the ratios:

- For CaO: 100 pm/140 pm = 0.714, within the 0.414–0.732 range for the AB structure type, and the most plausible structure is NaCl structure.

- For BkO_2: 63 pm/140 pm = 0.450, is also within the 0.414–0.732 range but this time we have to look under AB_2 structure-type column; thus, we can predict TiO_2 structure type. For this oxide, however, this is only a predicted structure—BkO_2 actually has a fluorite-type structure. Thus, this example shows a limitation of this method. Even if we were to use ionic radii for correct coordination numbers (8 for Bk^{4+} and 4 for O^{2-} in CaF_2 structure type), the γ value would still fall within the TiO_2 structure range.

S4.14 You should proceed as in the example, calculating the total enthalpy change for the Born–Haber cycle and setting it equal to $\Delta_L H$. In this case, it is important to recognize that two Br^- ions are required, so the enthalpy changes for (i) vaporization of Br_2(l) and (ii) breaking the Br–Br bond in Br_2(g) are used without dividing by 2, as was done for Cl_2 in the case of KCl described in Example 4.14. The Born–Haber cycle for $MgBr_2$ is shown below, with all of the enthalpy changes given in kJ mol^{-1}. These enthalpy changes are not to scale. The lattice enthalpy is equal to 2421 kJ mol^{-1}. Note that $MgBr_2$ is a stable compound despite the enormous enthalpy of ionization of magnesium.

This is because the very large lattice enthalpy more than compensates for this positive enthalpy term. Note the standard convention used; lattice enthalpies are positive enthalpy changes.

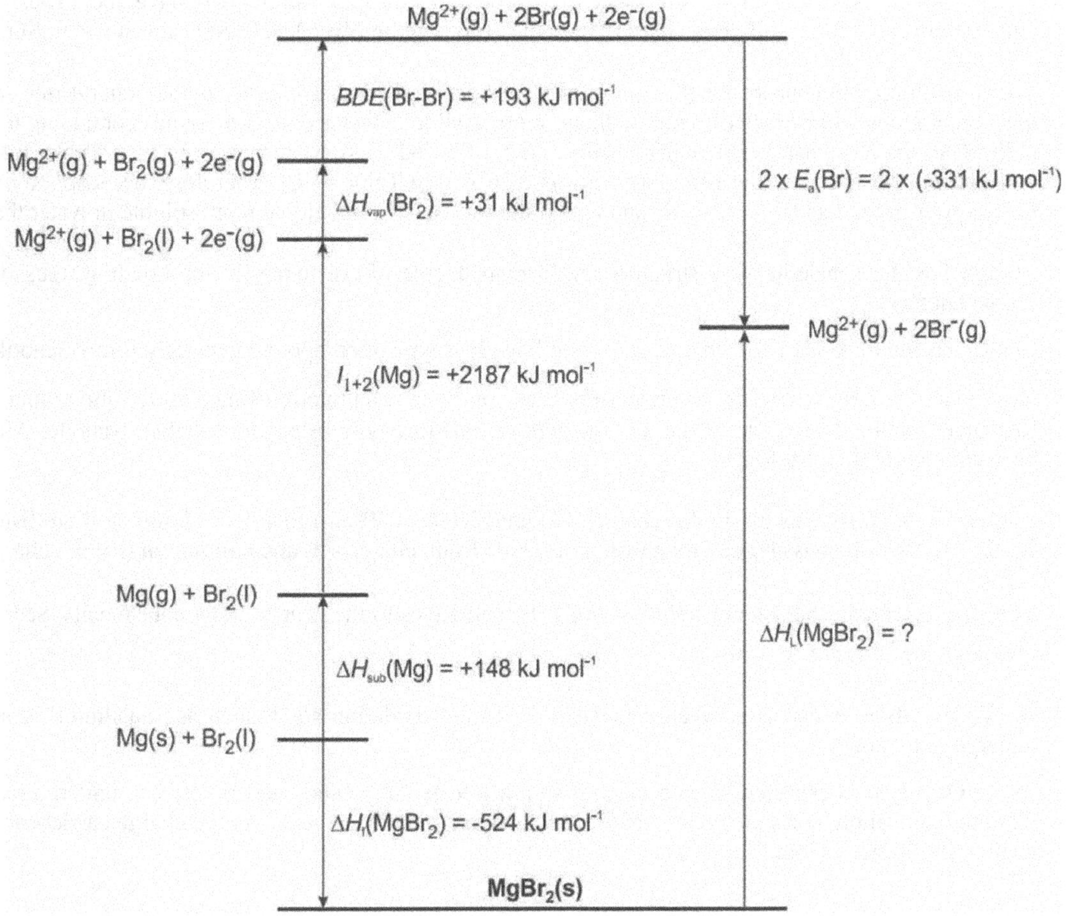

The Born–Haber cycle for $MgBr_2$

S4.15 This compound is unlikely to exist owing to a large positive value for the heat of formation for $CsCl_2$ that is mainly due to the large second ionization energy for Cs (I_1 = 375 kJ mol^{-1} vs. I_2 = 2420 kJ mol^{-1}; the I_2 alone for Cs is higher than the sum of I_1 and I_2 for Mg; see Self-Test 4.14.). The compound is predicted to be unstable with respect to its elements mainly because the large ionization enthalpy to form Cs^{2+} is not compensated by the lattice enthalpy.

S4.16 Closely following Example 4.16, using $r(Li^+)$ = 76 pm and provided thermochemical radius for N_2^{2-}:

$$\Delta H_L = \frac{3\times|(+1)(-2)|}{(76+173)\,\text{pm}}\times(1.08\times10^5)\,\text{kJ mol}^{-1} = 2602\,\text{kJ mol}^{-1}$$

S4.17 The enthalpy change for the reaction

$$MSO_4(s) \rightarrow MO(s) + SO_3(g)$$

includes several terms, including the lattice enthalpy for MSO_4, the lattice enthalpy for MO, and the enthalpy change for removing O^{2-} from SO_4^{2-}. The last of these remains constant as you change M^{2+} from Mg^{2+} to Ba^{2+}, but the lattice enthalpies change considerably. The lattice enthalpies for $MgSO_4$ and MgO are both larger than those for $BaSO_4$ and BaO, simply because Mg^{2+} is a smaller cation than Ba^{2+}. However, the *difference* between the lattice enthalpies for $MgSO_4$ and $BaSO_4$ is a smaller number than the difference between the lattice enthalpies for MgO and BaO (the larger the anion, the less changing the size of the cation affects ΔH_L). Thus, going from $MgSO_4$ to MgO is thermodynamically more favourable than going from $BaSO_4$ to BaO because the *change* in ΔH_L is greater for the former than for the latter. Therefore, magnesium sulfate will have the lowest decomposition temperature and barium sulfate the highest, and the order will be $MgSO_4 < CaSO_4 < SrSO_4 < BaSO_4$.

S4.18 The most important concept for this question from Section 4.15(c) *Solubility* is the general rule that compounds that contain ions with widely different radii are more soluble in water than compounds containing ions with similar radii. The six-coordinate radii of Na^+ and K^+ are 1.02 and 1.38 Å, respectively (see Table 1.4), whereas the thermochemical radius of the perchlorate ion is 2.36 Å (see Table 4.10). Therefore, because the radii of Na^+ and ClO_4^- differ more than the radii of K^+ and ClO_4^-, the salt $NaClO_4$ should be more soluble in water than $KClO_4$.

S4.19 **a)** HgS has the sphalerite type structure and a high degree of covalency in the bonding, thus it would favour Frenkel defects.

b) CsF has the rock-salt structure and ionic bonding. This type of compound generally forms Schottky defects.

Note that you can determine possible structures for both compounds using radius ratio γ and radii given in Resource Section 1. You can make a judgement on bonding type using the Ketelaar triangle. All of these have been covered in this chapter.

S4.20 We need to identify ions of similar charge (+4) and size (r = 40 pm) to silicon. Ionic radii are listed in Resource Section 1. Two obvious choices are phosphorus (r = 31 pm, charge +3) and aluminium (r = 53, charge +3).

S4.21 The $d_{x^2-y^2}$ and d_{z^2} have lobes pointing along the cell edges to the nearest neighbour metals. See Figure 1.15 for review of the shape of 3d orbitals.

S4.22 **(a) V_2O_5:** *n*-type is expected when a metal is in a high oxidation state, such as vanadium(V), and is likely to undergo reduction.

(b) CoO: *p*-type is expected when a metal is in a lower oxidation state and is likely to undergo oxidation. Recall that upon oxidation, holes are created in the conduction band of the metal and the charge carriers are now positive, leading the classification.

Exercises

E4.1 Consulting Table 4.1, for the monoclinic crystal system we have the following unit cell parameters: $a \neq b \neq c$ and $\alpha = 90°$, $\beta \neq 90°$, $\gamma = 90°$. See Figure 4.2 for a three-dimensional structure of this type of unit cell. Note that angle β is between a and c edges. The projection along b direction means that the b length is perpendicular to the paper. Hence the projection is a rhombus (Figure E4.1-A). The origin of the unit cell is marked with 0, while the dark circles at the corners represent lattice points. The angle β is also shown. In Figure E4.1-B we have replaced the lattice points with atoms (grey spheres). To show that the space is indeed filled, we translate the projection along a and c (Figure E4.1-C).

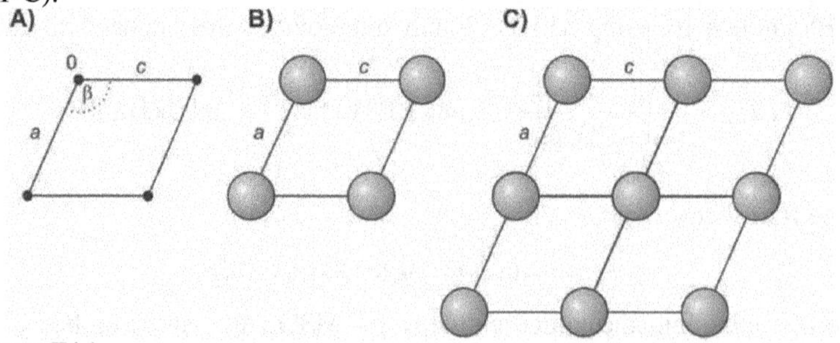

Figure E4.1

E4.2 The drawings are shown on the Figure E4.2 below. Figure E4.2-A represents only the outline of the tetragonal unit cell without lattice points. A face-centred tetragonal unit cell is shown on E4.2-B whereas a body-centred lattice unit cell is shown on Figure E4.2-C. Two face-centred unit cells next to each other (along one *a* direction) are shown in Figure E4.2-D. The two face-centred unit cells are now outlined with dashed lines whereas the new, body-centred cell is outlined in solid lines. We can see that we can form the top of this new unit cell if we connect the lattice points in the middle of top faces of face-centred unit cells with the corner lattice points shared between two adjacent face-centred unit cells. The same applies to the bottom face of the new cell. On the left is shown the new body-centred lattice. The length of *c* dimension remains the same; the length of new *a* direction (marked with X), however must be determined.

X can be determined easily if we look at the projection of two adjacent face-centred unit cells along *c* (Figure E4.2-E). Again, the dashed lines indicate the edges of two adjacent, face-centred unit cells, whereas the solid lines outline the new, body-centred unit cell. The fractional coordinates of selected, important lattice points have been indicated as well as original length *a*. You can see that the common lattice point (located at the centre of the two shared faces) is now at the centre of the new body-centred unit cell. We can see that X is half of the face diagonal of the old unit cell. Using Pythagoras's theorem, we have:

$$\mathsf{X} = \frac{d}{2} = \frac{\sqrt{a^2 + a^2}}{2} = \frac{a\sqrt{2}}{2} = \frac{a}{\sqrt{2}}.$$

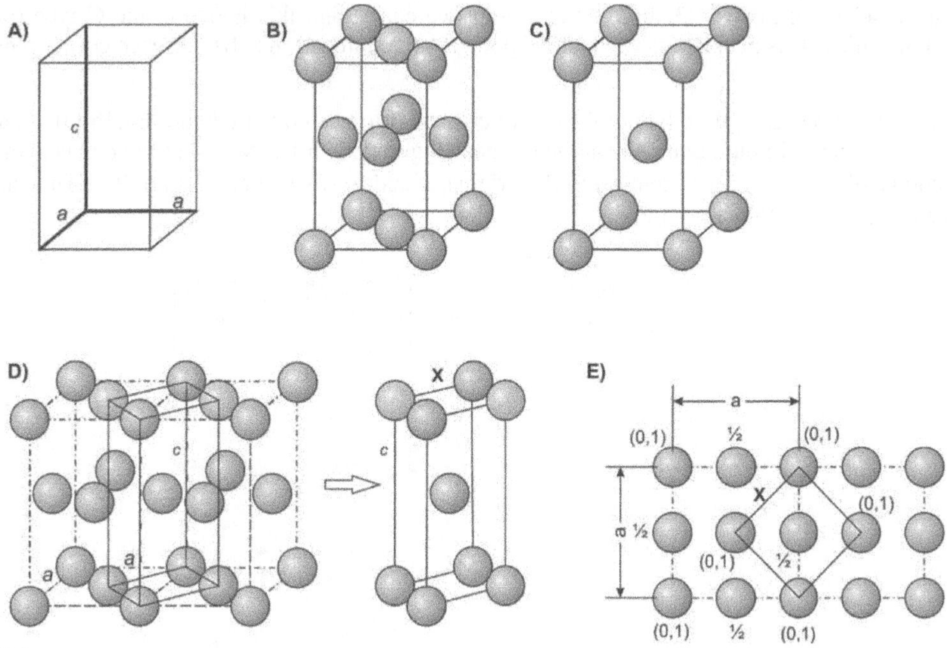

Figure E4.2

E4.3 **a)** First start with an outline of a cubic unit cell with defined edges and origin (Figure E4.3a-A). Then add the three points along the edges: (½, 0, 0), (0, ½, 0) and (0, 0, ½) (Figure E4.3a-B, three dark circles). Since the cell is cubic, the three directions must be the same, thus there must be a point in the middle of each edge. These are shown now on Figure E4.3a-C. Finally, the point (½, ½, ½) is added in Figure E4.3a-D. If you had trouble with this part, have a look at Figures 4.3 and 4.4 of your textbook. To make the lattice type more obvious we can translate this unit cell to get two of them sharing one face, Figure E4.3a-E. It is easy to see now that we have a face-centred unit cell and cubic close-packing (Figure E4.3a-F). Note that in final figure our original two unit cells have been indicated with dashed lines while the new, face-centred unit cell with solid lines. Figure E4.3a-E also shows that the new unit cell has the same parameters as the original one.

The number of lattice points in face-centred unit cell is four (see Section 4.1(a) of your textbook if necessary).

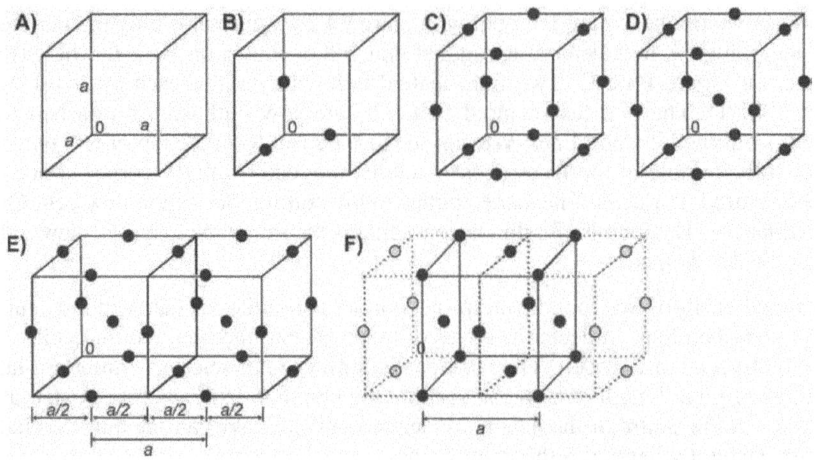

Figure E4.3a

b) The approach is similar to that in part a). The general parameters of an orthorhombic unit cell can be found in Table 4.1 and the outline in Figure 4.2. The starting point is a simple outline of an orthorhombic unit cell. Figure E4.3b-A shows such an outline with indicated cell origin and cell dimensions. Figure E4.3b-B shows lattice points that correspond to assigned point (0, 0, 0). Note that this point is translated from the origin to each corner. The (½, ½, 0) point is added in Figure E3.4b-C. At this point we can assign this lattice to the C type (or base-centred) orthorhombic lattice. This packing has two lattice points in the unit cell: 8 × 1/8 for each corner plus 2 × ½ for two faces.

We go one step further, and place two of these unit cells next to each other (Figure E4.3b-D). Analysing this set-up, we see that a new, simpler, unit cell can be defined (Figure E4.3b-E). Note that the *c* axis is the same but two new directions, *a*' and *b*', are not identical to the original *a* and *b*. Also, angle γ is not 90°. This unit cell has only one lattice point: 8 × 1/8 for each corner.

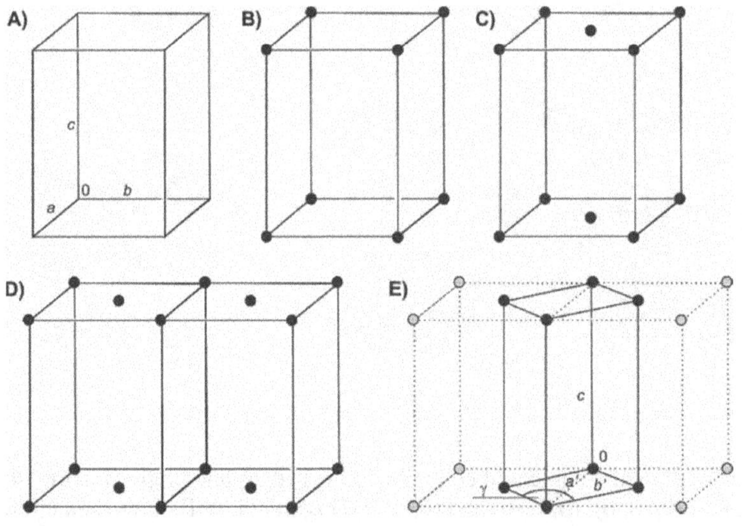

Figure E4.3b

E4.4 If you had problems with this exercise, you should revise Example and Self-Test 4.1. The final graphical solution is provided on Figure E4.4. Remember that the numbers in parentheses correspond to the displacement in three directions. In the top right corner is outlined a projection of a cubic unit cell along one direction (height) with origin indicated with (0, 0, 0). We first find the location of Ti atom by translating along the cell's length for ½ (solid arrow), then for ½ along the cell's width (dashed line), and finally for ½ of cell's height (number ½). That is where we place Ti atom.

Next, we locate oxygen atoms. We start first with O atom at (½, ½, 0). First, we translate along the cell's length for ½ (solid arrow), and then for ½ along the cell's width (dashed line). There is no translation along the cell's height because the last number is 0. Since the cell is cubic, the equivalent ½ length, ½ width displacement will repeat exactly at one full height of the unit cell. Thus, we have two oxygen atoms—one at 0 height and the other at full (1) height, which is indicated by (0, 1) fractional coordinates next to a white circle representing O atom. Similarly, we obtain the positions of O atoms at (0, ½, ½) and (½, 0, ½), each giving two O atoms. Finally, Ba atom is exactly at the origin (0, 0, 0). Since the cell is cubic, we must have Ba atom at every corner of the unit cell with height-fractional coordinates of either 0 or 1 (as indicated next to the first Ba atom represented by a dark circle).

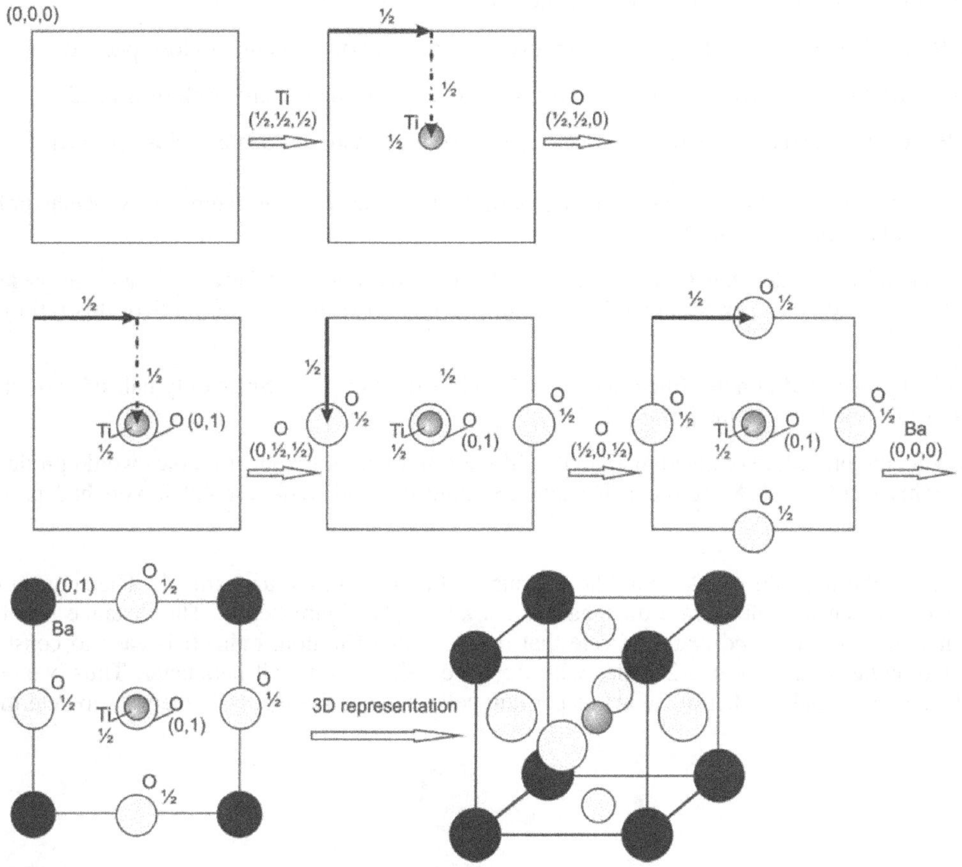

Figure E4.4

The analysis of 3D structure and the last projection with Ba ions added reveals that this compound has a perovskite-type structure (Section 4.9b) with a small cation (Ti^{4+}) in octahedral coordination environment with coordination number six, whereas large cation (Ba^{2+}) lies on sites with coordination number 12.

E4.5 **a) CBACBA…:** Any ordering scheme of planes is close-packed if no two adjacent planes have the same position (i.e., if no two planes are *in register*). When two planes are in register, the packing looks like the figure to the right (the spheres in the upper layer lie exactly on top of the spheres in a layer below), whereas the packing in a close-packed structure allows the atoms of one layer to fit more efficiently into the spaces *between* the atoms in an adjacent layer, like the figure to the left. Notice that the empty spaces between the atoms in the figure to the left are much smaller than in the figure to the right. The efficient packing exhibited by close-packed structures is why, for a given type of atom, close-packed structures are denser than any other possible structure. In the case of a CBACBA… structure, no two adjacent planes are in register, so the ordering scheme is close-packed.

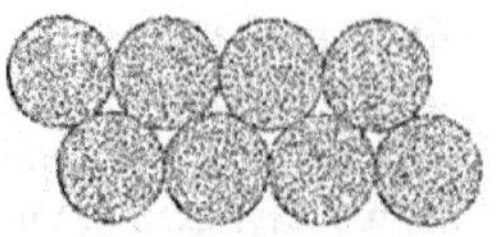

Close-packed structure (not in register, corresponding to AB layers)

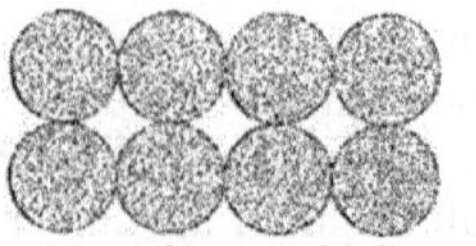

Non-close-packed structure (in register, corresponding to AA layers)

b) ABAC...: Once again, no two adjacent planes are in register, so the ordering scheme is close-packed.

c) ABBA...: The packing of planes using this sequence will put two B planes next to each other as well as two A planes next to each other, so the ordering scheme is not close-packed.

d) ABCBC...: No two adjacent planes are in register, so the ordering scheme is close-packed.

e) ABABC...: No two adjacent planes are in register, so the ordering scheme is close-packed.

f) ABCBCA...: No two adjacent planes are in register, so the ordering scheme is close-packed.

E4.6 Remember that in any close-packed structure with N close-packed ions there are N octahedral holes and $2N$ tetrahedral holes (see Section 4.3).

a) If we had all tetrahedral holes filled, the M-to-X ratio would be 2 : 1 because in a close-packed structure we have $2N$ tetrahedral holes per N anions X. Since only a quarter of the holes are filled, this ratio is ½ : 1 and the formula is MX_2.

b) If we had all octahedral holes filled, the M-to-X ratio would be 1 : 1. Since only half of these holes contain M, the ratio is ½ : 1 or 1 : 2 and the formula is MX_2.

c) Following the procedure outlined in a) above, filling a sixth of all tetrahedral holes would produce 1/3 : 1 M-to-X ratio. This results in MX_3 formula. Review Example 4.9 and Self-Test 4.9 if you had problems with this exercise.

E4.7 The sketch of the unit cell should look like the one on Figure 4.5 in your textbook. The shortest copper atom to copper atom distance is along face diagonal, as shown on the figure below. The distance between two closest copper neighbours is denoted $2r$ to indicate that it equals two Cu atom radii. It is easy to construct a 45-45-90 isosceles triangle formed by two $2r$ sides with the third being a unit cell parameter. Thus, we determine first a from given density and number of Cu atoms per unit cell, then use Pythagoras's theorem to determine $2r$ and then r.

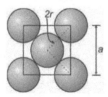

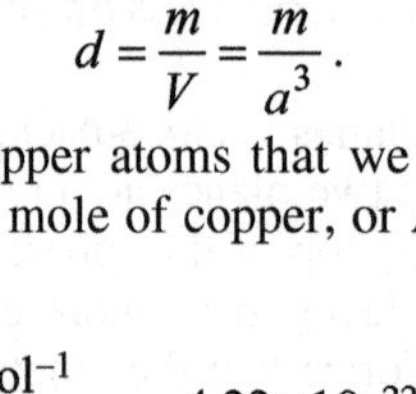

Since Cu has fcc structure, there are four Cu atoms per unit cell (see Section 4.1(a) of your textbook if necessary). The density is a mass-to-volume ratio; fcc unit cell is a cube so the volume is a^3:

$$d = \frac{m}{V} = \frac{m}{a^3}.$$

We have to calculate the mass of four copper atoms that we find in one unit cell volume. The atomic mass of copper is 63.55 g mol^{-1} telling us that one mole of copper, or Avogadro's number of copper atoms, weighs 63.55 g. Thus, our four Cu atoms weigh:

$$m = 4 \times \frac{63.55\,\text{g mol}^{-1}}{6.022 \times 10^{23}\,\text{mol}^{-1}} = 4.22 \times 10^{-22}\,\text{g} = 4.22 \times 10^{-25}\,\text{kg}.$$

Now we can calculate a:

$$a^3 = \frac{m}{d} = \frac{4.22 \times 10^{-25}\,\text{kg}}{8960\,\text{kg}\,\text{m}^{-3}} = 4.71 \times 10^{-29}\,\text{m}^3 \Rightarrow a = \sqrt[3]{4.71 \times 10^{-29}\,\text{m}^3} = 3.61 \times 10^{-10}\,\text{m} = 361\,\text{pm}.$$

Based on the above sketch and applying Pythagoras's theorem, we have:

$$(2r)^2 + (2r)^2 = a^2$$
$$8r^2 = a^2$$
$$r = \sqrt{\frac{a^2}{8}} = \sqrt{\frac{(361\text{pm})^2}{8}} = 128\text{pm}.$$

E4.8 **a)** Within the fcc lattice of fullerides shown in Figure 4.16, there are the equivalent of four close-packed molecules (1/8(8) corners + ½(6) faces). This lattice contains the equivalent of four octahedral holes and eight tetrahedral holes as shown in Figure 3.18. If the potassium cations occupy all of these holes, then we have 12 K^+ ions for every four C_{60} anions and the formula is K_3C_{60} (simplified from $(K_{12}(C_{60})_4)$. See Figure 24.69 for further visualization of the structure. Review Example 4.9 and Self-Test 4.9 if you had problems with this exercise.

b) If this fulleride has a body-centered array of C_{60} anions then its unit cell has two C_{60} anions per unit cell. Any atom or ion located on the edge of a unit cell is shared among four unit cells sharing that edge. Each edge contains two potassium cations, with one fourth inside one unit cell: 2 × ¼ = ½. Cube has twelve edges, so in total one unit cell has 12 × ½ = 6 potassium cations. Finally we have potassium-to-C_{60} ratio of 6 : 2 and the formula is $K_3(C_{60})$.

E4.9 The AAA… layering has hexagonal layers that in the structure lie exactly on top of each other. Inspection of two such layers AA (see Figure E4.9 below) easily reveals the trigonal prismatic holes. A top and side view of the hole is given at the far right of Figure E4.9. Since each S atom in this packing contributes only one-sixth to the hole, and the trigonal prism has six corners, we have *N* trigonal prismatic holes per *N* sulfur atoms. Since the formula is MoS_2, every second hole in the structure contains Mo atom.

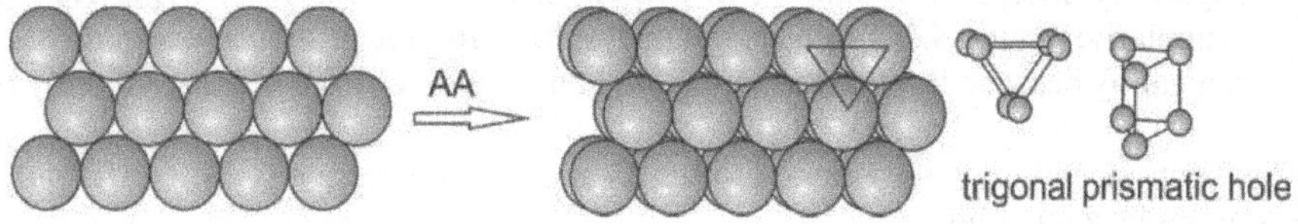

Figure E4.9

E4.10 Tungsten has a body-centred unit cell. Two such adjacent unit cells are shown on Figure E4.10. The site in the middle of the original unit cell is indicated by an open circle. Looking at the arrangement of tungsten atoms around it, we see that the approximate coordination number (CN) of that site is 6—four atoms at the corner defining the shared face plus an atom in the centre of each unit cell. (The true coordination is 2 + 4 as the tungsten atoms in the cell centres are closer to the open circle than those on the original corners). If we fill each of these holes with C atoms we would get the stoichiometry of W_2C_3: two tungsten atoms per body-centred unit cell and 1 C atom on each six faces of the original cube (remembering that the atoms on the faces of the unit cell are shared between two cells; see Example 4.8 and Self-Test 4.8).

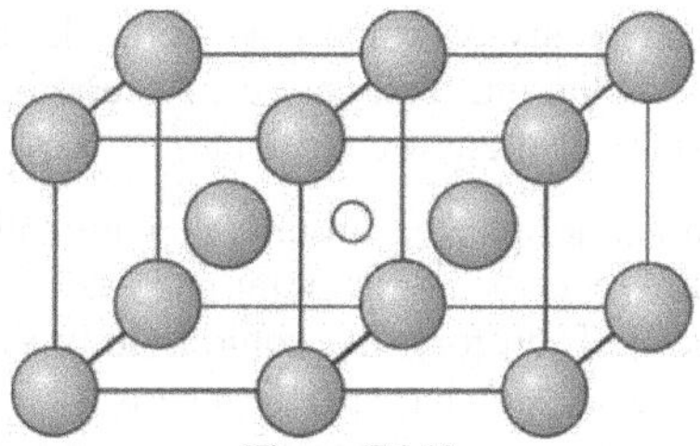

Figure E4.10

The shortest W-C distance is from the face of the cube to the centre and is thus half the cell parameter at (3.60/2 = 1.80 Å (180 pm).

The second shortest W–C distance is found in the plane defined by the four W atoms within the shared cube face. Applying Pythagoras's theorem and using the given cell parameter we have (see exercise 4.7 for some details on the distances):

$$d^2 + d^2 = a^2$$

$$2d^2 = a^2$$

$$d = \sqrt{\frac{a^2}{2}} = \sqrt{\frac{(360\,\text{pm})^2}{2}} = 254\,\text{pm}.$$

E4.11 The bcc unit cell contains two atoms that weigh together (2 mol Li × 6.941 g/mol)/(6.022 × 10^{23} Li/mol) g or 2.305 × 10^{-26} kg. The volume of the unit cell is a^3, where a is the length of the edge of the unit cell. The density equals mass divided by volume or:

$$535\ \text{kg m}^{-3} = 2.305 \times 10^{-26}\ \text{kg}/a^3 \text{ with } a \text{ in m}$$

$$a = 3.506 \times 10^{-10}\ \text{m or } 351\ \text{pm}.$$

E4.12 The composition can be determined by counting atoms in the unit cell shown in Figure 4.75. Six face copper atoms times one-half gives three Cu atoms per unit cell and eight corner gold atoms times one-eighth gives one Au per unit cell with an overall composition for the alloy of Cu_3Au. The unit cell considering just the Au atoms is primitive cubic. The mass % of Au in this alloy is:

$$(\text{mass \% Au}) = m(\text{Au})/m(\text{Cu}_3\text{Au}) = 196.97\ \text{g}/(3 \times 633.55\ \text{g} + 196.97\ \text{g}) \times 100\% = 50.82\%$$

Pure gold is 24 carat (100% gold). This alloy, which contains about 50% by mass gold, would therefore be 12 carat.

E4.13 The electronegativity difference is 0.86 for Sr and Ga and the mean is 1.38. Using Ketelaar's triangle (see Figure 2.28 and particularly Figure 4.27) and calculated values for the y and x axes, respectively, we find that the compound is an alloy because it falls in the metallic bond region of Figure 2.28 and alloy region of Figure 4.27.

E4.14 **a) LiI:** The electronegativities for Li and I are 0.98 and 2.66 respectively. The difference is 1.68 and the mean is 1.82. LiI is an ionic compound.

b) $BeBr_2$: The electronegativities for Be and Br are 1.57 and 2.96 respectively. The difference is 1.39 and the mean is 2.26. $BeBr_2$ is also an ionic compound.

c) SnS: The electronegativities for Sn and S are 1.96 and 2.58 respectively. The difference is 0.62 and the mean is 2.27. SnS is a covalent compound.

d) RbSn: The electronegativities for Rb and Sn are 0.82 and 1.96 respectively. The difference is 1.14 and the mean is 1.39. RbSn is a Zintl phase.

E4.15 **(a) Coordination numbers.** The rock-salt polymorph of RbCl is based on a ccp array of Cl^- ions in which the Rb^+ ions occupy all the octahedral holes. An octahedron has six vertices, so the Rb^+ ions are six-coordinate. Since RbCl is a 1 : 1 salt, the Cl^- ions must be six-coordinate as well. The caesium-chloride polymorph is based on a cubic array of Cl^- ions with Rb^+ ions at the unit cell centres. A cube has eight vertices, so the Rb^+ ions are eight-coordinate, and therefore the Cl^- ions are also eight-coordinate.

(b) Larger Rb^+ radius. If more anions are packed around a given cation, the hole that the cation sits in will be larger. You saw an example of this when the radii of tetrahedral ($0.225r$) and octahedral holes ($0.414r$) were compared (r is the radius of the anion). Therefore, the cubic hole in the caesium-chloride structure is larger than the octahedral hole in the rock-salt structure. A larger hole means a longer distance between the cation and anion, and hence a larger apparent radius of the cation. Therefore, the apparent radius of rubidium is larger when RbCl has the caesium-chloride structure and smaller when RbCl has the rock-salt structure.

E4.16 The unit cell for this structure is shown in Figure 4.30. Each unit cell is surrounded by six equivalent unit cells with which it shares a face. Since each of these unit cells contains a Cs^+ ion at its centre, each Cs^+ ion has six second-nearest neighbours that are Cs^+ ions, one in the centre of each of the six neighbouring unit cells.

The next nearest Cl^- anions lie on the far side of the six adjoining unit cells and there are four of these anions in each of these adjoining unit cells so the total number of third nearest neighbours is 6 × 4 = 24.

E4.17 The ReO_3 unit cell is shown in Figure E4.17a). We can easily see that the coordination number of O atoms is two: each O atom located on the edges is surrounded by two Re atoms. The coordination number of Re atoms is six: we can see this if we extend environment around one Re atom into the neighbouring unit cells (Figure E4.17b). Finally, if we insert one cation in the middle of ReO_3 unit cell (white circle on Figure E4.17c), we generate a perovskite-type structure.

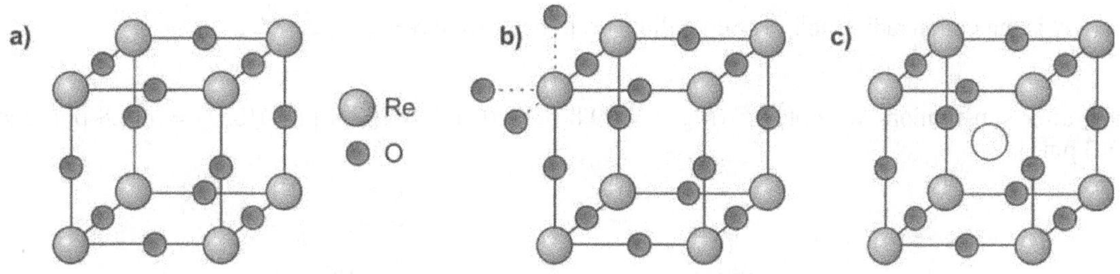

Figure E4.17

E4.18 The number of hydroxide anions would remain the same: three. However, since the hydroxide has a charge of 1– and oxide a charge of 2–, the accompanying metal cation has to have a 3+ charge. The general formula of the hydroxide with ReO_3 type structure would be $M(OH)_3$ and possible metal cations, which need to be fairly large for this structure type to form, include La^{3+} and In^{3+}.

E4.19 In the perovskite structure, cations in site A have coordination number 12, whereas those in site B have coordination number six. See Figure 4.42 for more information on the perovskite structure.

E4.20 Keep in mind that if a close-packed structure is made of N anions X, then the lattice has N octahedral and $2N$ tetrahedral holes.

(a) If all octahedral holes were filled, the stoichiometry would be MX; with half the octahedral holes filled, the stoichiometry becomes MX_2.

(b) If all tetrahedral holes were filled, the stoichiometry would be M_2X; with one quarter it becomes MX_2.

(c) If all octahedral holes were filled, the stoichiometry would be MX; with two-thirds filled the ratio M : X is 2/3 : 1 or 2 : 3 giving M_2X_3 as final formula.

The coordination of M is given by the hole it fills, thus in (a) it is six and in (b) it is four. For the anion X the number of neighbours is one-half of the M coordination with a stoichiometry of MX_2 (as each M must have twice as many M-X bonds as each X to produce this stoichiometry)—therefore three in (a) and two in (b).

E4.21 The close-packed array of A and X in the perovskite structure has $N = 4$ and so there would be 4(= N) octahedral holes per AX_3 unit. Therefore, to preserve the stoichiometry, one quarter of the octahedral holes have to be filled with B cations (these are the quarter that have just X anion at the octahedron vertices).

E4.22 **a)** $r(U^{4+}) = 100$ pm; $r(O^{2-}) = 138$ pm; $\gamma = (100 \text{ pm})/(138 \text{ pm}) = 0.724$; TiO_2-type (in fact, this compound has CaF_2-type structure; this example shows some limitations of the radius ratio rule).

b) $r(Fr^{+}) = 196$ pm; $r(I^{-}) = 220$ pm; $\gamma = (196 \text{ pm})/(220 \text{ pm}) = 0.891$; CsCl-type.

c) $r(Be^{2+}) = 45$ pm; $r(S^{2-}) = 184$ pm; $\gamma = (45 \text{ pm})/(184 \text{ pm}) = 0.245$; ZnS-type.

d) $r(In^{3+}) = 80$ pm; $r(N^{3-}) = 146$ pm; $\gamma = (80 \text{ pm})/(146 \text{ pm}) = 0.548$; NaCl- or NiAs-type.

E4.23 The unit cell of rock-salt structure is shown on Figure 4.30. We have to remember that the cations fill the holes in close-packed structure of Se^{2-} anion. The geometrical approach to calculating the cationic radii is shown in Figure E4.23. Both sketches show only one face of a rock-salt unit cell. The one on the left is for MgSe, showing Se^{2-} anions in contact and smaller Mg^{2+} filling the holes. The sketch on the right shows the other three cases. It is clear that we need Se^{2-} radius first to calculate radii of cations. From MgSe structure:

$$d^2 = (4r)^2 = a^2 + a^2 \quad \text{and} \quad 16r^2 = 2a^2 \Rightarrow r = \frac{a}{2\sqrt{2}} = \frac{545 \text{ pm}}{2\sqrt{2}} = 192.7 \text{ pm.}$$

Where a is the length of unit cell, r is radius of Se^{2-} anion and d is the length of face diagonal of the unit cell. From shown measurements we see that in each case the length of the unit cell, a, is given by:

$$a = 2r + 2r^+$$

where r^+ is the cation radius and r is Se^{2-} radius. From here, we have: $r^+ = \dfrac{a - 2r}{2}$.

Thus, after substitutions we obtain: $r(Mg^{2+})$ = 79.8 pm, $r(Ca^{2+})$ = 102.8 pm, $r(Sr^{2+})$ = 118.8 pm, and $r(Ba^{2+})$ = 138.3 pm.

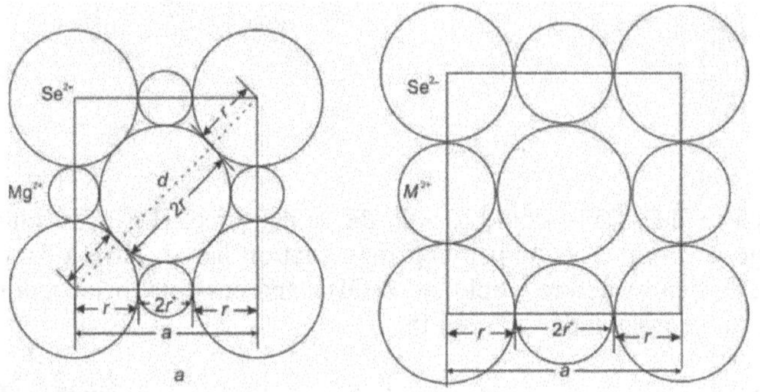

Figure E4.23

E4.24 In order to use the structure map, we first have to calculate the differences in electronegativity, $\Delta\chi$, and average principal quantum number for each MX compound from the problem and then look at the map:

$\Delta\chi$ (LiF) = 3.98 – 0.98 = 3; average n is 2, the point corresponding to LiF belongs in the (6,6) region

$\Delta\chi$ (RbBr) = 2.96 – 0.82 = 2.14; average n is (5 + 4)/2 = 4.5; also in (6,6) region

$\Delta\chi$ (SrS) = 2.58 – 0.95 = 1.63; average n is (5 + 3)/2 = 4; also in (6,6) region, and

$\Delta\chi$ (BeO) = 3.44 – 1.57 = 1.87; average n is 2; also in (6,6) region.

The only discrepancy here is that for BeO (predicted (6,6), actual (4,4)). The reason for this discrepancy is the fact that Be^{2+} with its small size and relatively high charge and electronegativity forms compounds with significant covalent character resulting in higher directionality of chemical bonds.

E4.25 Each of the complex ions in this exercise ($[PtCl_6]^{2-}$, $[Ni(H_2O)_6]^{2+}$, $[SiF_6]^{2-}$, PF_6^-) is highly symmetric and their shape can be approximated with a sphere (each of these structures is octahedral, thus around each a sphere can be constructed with a radius equal to that of half height of octahedron or approximately element–element bond). Thus, for (a) we could predict that K_2PtCl_6 has an antifluorite structure with large anions $[PtCl_6]^{2-}$ forming a ccp array and small K^+ cations occupying each tetrahedral hole. For (b) we have a large cation $[Ni(H_2O)_6]^{2+}$ and a large anion $[SiF_6]^{2-}$ in the stoichiometric ratio 1 : 1 so could predict a CsCl structure type. (c) CN^- is a similar sized anion (the CN^- ionic radius has been estimated as 192 pm) compared to Cs^+ cationic radius of ~180 pm and with a 1 : 1 stoichiometry would predict 8 : 8 coordination and CsCN does indeed adopt a CsCl structure type with CN^- replacing Cl^-. In (d) with the much larger PF_6^- anions (thermochemical radius 242 pm) the ion size mismatch leads to 6 : 6 coordination and adoption of NaCl structure type with alternating Cs^+ and PF_6^- anion.

E4.26 Calcite has a rhombohedral unit cell, whereas NaCl has a cubic one. Both rhombohedral and cubic unit cells have all three dimensions equal ($a = b = c$) but they differ in the angles, with cubic unit cell having all angles at 90° and rhombohedral all angles equal but different from 90°. If you look at the arrangement of cations and anions in the calcite structure, you'll see that when all angles are made 90° the two structures are identical. Calcite's rhombohedral unit cell can be obtained if we take cubic NaCl unit cell and pull its opposite corners across the body diagonal apart.

E4.27 The structure on Figure 4.77 is actually the structure of wurtzite (ZnS, see Figure 4.35 in your textbook). The stoichiometry is therefore AB (p = 1, q = 1). The number of formula units in the unit cell is: 8 Bs on the corners each contributing an eighth to the composition of the cell plus one full B inside the unit cell totals two Bs; 4 As on the unit cell edges each contributing a fourth to the composition of the unit cell plus one full A inside the unit cell totals two As; hence the composition is $(AB)_2$. The coordination number and geometry can be extrapolated from wurtzite structure to be four and tetrahedral (for both A and B). It can be deduced from the unit cell projection as well: consider the A cation inside the unit cell at 1/8 position. It has four neighbours–one B anion at ½ and three B anions at position 1 forming a triangle around A. This gives a tetrahedral geometry. Similarly, we can analyse B anion inside the cell at ½ as being surrounded by four A cations: one at 1/8 and three at 5/8 again forming a triangle. The structure that corresponds to the formula A_pB_{2q} can be derived by removing every second A from the structure of AB.

E4.28 The most important term will be the lattice enthalpy as the compound contains di- and trivalent ions, (ΔH°_L is proportional to $z_A \times z_B$). Also, the bond dissociation energy and third electron gain enthalpy for nitrogen (N_2) will be large. See Section 4.11 for more details.

E4.29 The terms in the Born–Haber cycle to consider are (X stands for F or Cl):

$Ag(s) \rightarrow Ag(g)$, $\Delta_{sub}H(Ag)$

$Ag(g) \rightarrow Ag^{2+}(g)$, $I_1 + I_2$

$X_2(g) \rightarrow 2X(g)$, $BDE(X_2)$

$2X(g) + 2e^- \rightarrow 2X^-(g)$, $2 \times E_a(X)$

$Ag^{2+}(g) + 2X^-(g) \rightarrow AgX_2(s)$, $\Delta H_L(AgX_2)$

Looking at the steps involved, we can expect significant energy demand for the Ag(g) ionization step ($I_1 + I_2$) for both compounds. The bond dissociation energy (BDE) for Cl_2 is higher than that for F_2 adding in energy requirement for $AgCl_2$. The electron affinity, a process that releases energy, is more negative for F(g) than for Cl(g). Also, it is expected that AgF_2 lattice enthalpy will be much higher than that of $AgCl_2$ on the account of smaller F^- size as ΔH°_L is proportional to $1/d$. This combined suggests that $AgCl_2$ is less stable (higher energy demand in gas phase ion formation, but lower energy gain in lattice formation from the gas phase ions) than AgF_2.

E4.30 Lattice enthalpy can be estimated using the Born–Mayer equation (Equation 4.2):

$$\Delta H^\circ_L = \frac{N_A|z_A z_B|e^2}{4\pi\varepsilon_0 d}\left(1-\frac{d^*}{d}\right)A.$$

Considering that all compounds have rock-salt structure, the Madelung constant (A) is the same for all. In this case ΔH_L is simply directly proportional to the product of ion charges ($z_A \times z_B$) and inversely proportional to the distance between cation and anion (d). We are not given the d values for the compounds, but we can make a very good estimate based on the expected relative sizes of cations and anions involved (we'll have to recall the atomic properties covered in Chapter 2).

Based on this overview, we can expect the highest lattice enthalpy for AlP because this compound has the highest charges (3+ and 3-). Next come SrO and NiO. Here, the charges are the same, 2+ and 2-, thus the higher lattice enthalpy would have the compound with smaller d value. Both are oxides, so the difference boils down to the difference in cation sizes. Since Sr is in period 5 and Ni is in period 4, we can expect Ni^{2+} to be smaller than Sr^{2+} (recall that atomic radii increase as we go down the periodic table of the elements). Thus, NiO has higher lattice enthalpy than SrO.

We are left with a group of three halides LiF, RbCl and CsI. All three have the same charges so again the difference is in ionic radii (strictly in d). The cation sizes increase in order $Li^+ < Rb^+ < Cs^+$. The anion sizes increase in order $F^- < Cl^- < I^-$. Consequently, the lattice enthalpies increase in order CsI < RbCl < LiF.

Summing all together, we have the following order of increasing lattice enthalpies:

CsI < RbCl < LiF < SrO < NiO < AlP.

E4.31 The lattice enthalpy for MgO will be approximately equal to *four* times the NaCl value or 3144 kJ mol^{-1} because the charges on both the anion and the cation are doubled in the Born–Mayer equation. For AlN the charges are tripled and the lattice enthalpy will be close to *nine* times the NaCl value or 7074 kJ mol^{-1}.

E4.32 **a)** The Born–Mayer equation is Equation 4.2:

$$\Delta H^{\circ}_L = \frac{N_A|z_A z_B|e^2}{4\pi\varepsilon_0 d}\left(1-\frac{d^*}{d}\right)A.$$

All needed values are known except for d, which has to be calculated after the radius for K^{2+} has been extrapolated. Looking at Table 1.4 and/or Resource Section 1 we see that Ca^{2+} has a radius of 112 pm in coordination eight (i.e., its coordination number in CaF_2). If KF_2 has the same structure, then K^{2+} radius should be similar. From general atomic trends, we know that potassium atom is larger than Ca atom meaning that K^{2+} is larger than Ca^{2+} but *smaller* than K^+. Thus, we can approximate $r(K^{2+})$ as an average between $r(K^+)$ and $r(Ca^{2+})$ for coordination number eight: (151 + 112) pm/2 = 131.5 pm. From the table we also need radius for F^- with coordination number four, 131 pm (the coordination number of F^- in CaF_2 structure). Then we have:

$$\Delta H^{\circ}_L = \frac{N_A|z_A z_B|e^2}{4\pi\varepsilon_0 d}\left(1-\frac{d^*}{d}\right)A$$

$$= \frac{6.022\times10^{23}\text{mol}\times|2\times(-1)|\times(1.602\times10^{-19}\text{C})^2}{4\times3.14\times(8.854\times10^{-12}\ \text{C}^2\ \text{J}^{-1}\ \text{m}^{-1})\times(131.5+131)\times10^{-12}\ \text{m}}\left(1-\frac{34.5\ \text{pm}}{131.5\ \text{pm}+131\ \text{pm}}\right)\times2.159$$

$$=1.99\times10^{6}\ \text{J mol}^{-1} = 1990\ \text{kJ mol}^{-1}.$$

b) What prevents the formation of KF_2 is the massive second ionization energy for K, 3051 kJ mol^{-1}, which is significantly higher than the lattice energy for this compound.

E4.33 To use the Born–Mayer equation, we first must estimate the radius of Ca^+ and choose the appropriate Madelung constant for the equation.

Looking at the trends in atomic properties, we can see that K atom is bigger than Ca and consequently that Ca^+ is smaller than K^+ but bigger than Ca^{2+}. Thus, the ionic radius for Ca^+ can be approximated as in between that of K^+ and Ca^{2+}. Since KCl has an NaCl-type structure, we can safely assume that CaCl would have the same, so we are using cationic radii for coordination six: (138 + 100) pm/2 = 119 pm. We have:

$$\Delta H^{0}_{L} = \frac{N_A|z_A z_B|e^2}{4\pi\varepsilon_0 d}\left(1-\frac{d^*}{d}\right)A$$

$$= \frac{6.022\times10^{23}\text{mol}\times|1\times(-1)|\times(1.602\times10^{-19}\text{C})^2}{4\times3.14\times(8.854\times10^{-12}\ \text{C}^2\ \text{J}^{-1}\ \text{m}^{-1})\times(119+181)\times10^{-12}\ \text{m}}\left(1-\frac{34.5\ \text{pm}}{119\ \text{pm}+181\ \text{pm}}\right)\times1.748$$

$$= 7.17\times10^{5}\,\text{J mol}^{-1} = 717\,\text{kJ mol}^{-1}.$$

The values for ionic radii (Ca^{2+}, K^+ and Cl^-) have been taken from Table 1.4 and the value for the Madelung constant from Table 4.8.

We can now use this lattice energy to construct the Born–Haber cycle for CaCl and show that it is indeed an exothermic compound. The cycle is shown below with all values taken from Table 1.5 in your textbook except the Ca sublimation energy given in the exercise text:

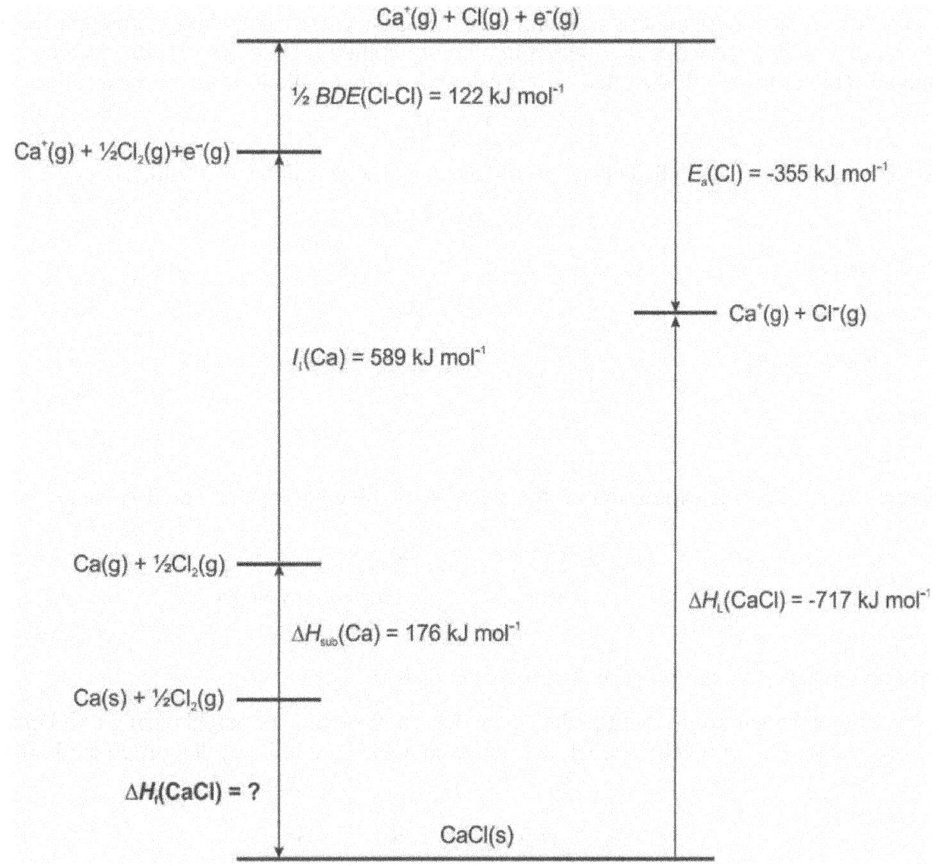

From the cycle, $\Delta_f H$(CaCl) = +176 kJ mol^{-1} + 589 kJ mol^{-1} + 122 kJ mol^{-1} – 355 kJ mol^{-1} – 717 kJ mol^{-1} = –185 kJ mol^{-1}, showing that CaCl is an exothermic salt.

Although CaCl has a favourable (negative) $\Delta_f H$, this compound does not exist and would convert to metallic Ca and $CaCl_2$ (2CaCl $\rightarrow$ Ca + $CaCl_2$). This reaction is very favourable because $CaCl_2$ has a higher lattice enthalpy (higher cation charge). The higher lattice enthalpy combined with gain of I_1 for Ca (Ca^+(g) + e^- $\rightarrow$ Ca(g)) are sufficient to compensate for I_2 (Ca^+(g) $\rightarrow$ Ca^{2+}(g) + e^-) required to form $CaCl_2$.

E4.34 The Madelung constant is in a sense a geometrical factor that depends on the position of ions within a unit cell. The electrostatic potential in which ions reside is given by:

$$V = \frac{(z_+e)(z_-e)}{4\pi\varepsilon_o d}$$

where in this case d is the distance between the species we are considering.
The unit cell of CsCl is shown in Figure E4.34a. Two important parameters are shown as well: space diagonal, D, and face diagonal, d. We can conveniently start by looking at the Cl^- ion at the middle of the unit cell (the darker sphere).

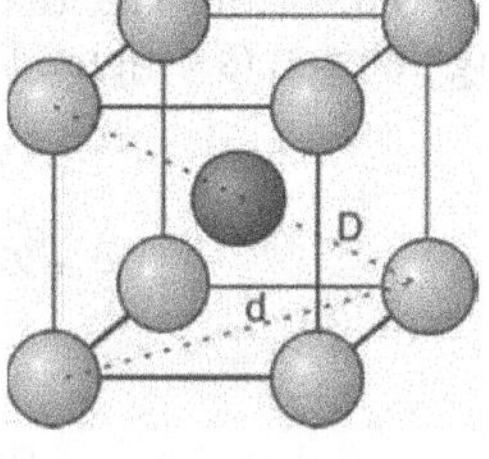

Figure E3.34a

There is only one Cl^- and z_- in the above equation equals 1. Its closest neighbours are eight Cs^+ ions ($z_+ = 8$). To calculate the Cs^+–Cl^- distance we have to resort to a bit of geometry. Note that this distance equals one-half of the space diagonal. The length of this diagonal can be calculated using Pythagoras's theorem as:

$$D^2 = d^2 + a^2$$

where a is the length of the unit cell. The length of face diagonal, d, can be also similarly calculated as:

$$d^2 = a^2 + a^2.$$

After substitution:

$$D^2 = 3a^2$$

and Cs^+–Cl^- distance, d, is:

$$d = \frac{\sqrt{3}a}{2}.$$

Substituting this value into the expression for the electrostatic potential, we get the first value:

$$V_1 = \frac{(z_+e)(z_-e)}{4\pi\varepsilon_o d} = -\frac{8e^2}{4\pi\varepsilon_0 \frac{\sqrt{3}a}{2}} = -\frac{4e^2}{\sqrt{3}\pi\varepsilon_0 a}.$$

The potential has a negative sign because the force is attractive.

The next closest neighbours to the central Cl^- are six Cl^- anions from six neighbouring unit cells: the one sharing the faces with our starting unit cell. The Cl^-–Cl^- distance equals the unit cell length a (see Figure E4.34b). Thus, we have:

$$V_2 = \frac{(z_+e)(z_-e)}{4\pi\varepsilon_o d} = \frac{6e^2}{4\pi\varepsilon_0 a} = \frac{3e^2}{2\pi\varepsilon_0 a}.$$

This potential is positive because the interaction is repulsive.

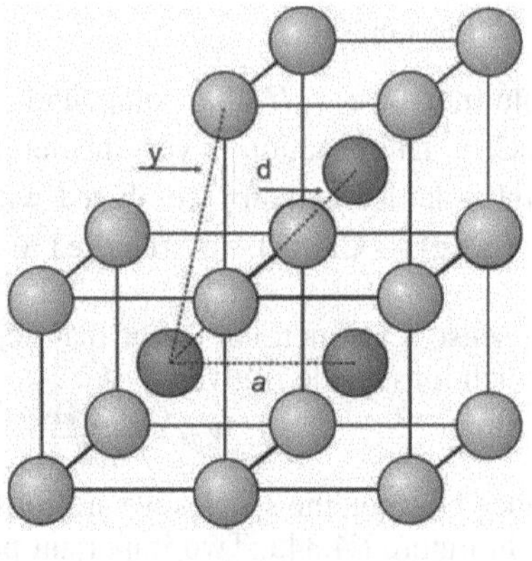

Figure E3.34b

The next neighbour is another Cl^- this time at full face diagonal distance. These are located in 12 unit cells that share an edge with our central, starting unit cell (see figure E4.34b). We have:

$$d^2 = 2a^2$$

$$d = \sqrt{2}a.$$

The potential for this interaction is:

$$V_3 = \frac{(z_+e)(z_-e)}{4\pi\varepsilon_o d} = \frac{12e^2}{4\pi\varepsilon_0\sqrt{2}a} = \frac{3e^2}{\sqrt{2}\pi\varepsilon_0 a}.$$

The fourth and final interaction to consider is between Cl^- and 24 Cs^+ cations found in the neighbouring unit cells that have a common face with our central one. One such Cs^+ is shown in Figure E4.34b at the distance marked y. This distance can be calculated:

$$y^2 = \left(\frac{3a}{2}\right)^2 + \left(\frac{d}{2}\right)^2 = \frac{9a^2}{4} + \frac{2a^2}{4} = \frac{11a^2}{4} \Rightarrow y = \frac{\sqrt{11}a}{2}.$$

And the potential is:

$$V_4 = \frac{(z_+e)(z_-e)}{4\pi\varepsilon_o d} = -\frac{24e^2}{4\pi\varepsilon_0 \frac{\sqrt{11}a}{2}} = -\frac{12e^2}{\sqrt{11}\pi\varepsilon_0 a}.$$

The sign is negative again, because the interaction is attractive.

We can sum the potentials now:

$$V_1 + V_2 + V_3 + V_4 = -\frac{4e^2}{\sqrt{3}\pi\varepsilon_0 a} + \frac{3e^2}{2\pi\varepsilon_0 a} + \frac{3e^2}{\sqrt{2}\pi\varepsilon_0 a} - \frac{12e^2}{\sqrt{11}\pi\varepsilon_0 a} = \frac{e^2}{\pi\varepsilon_0 a}\left(-\frac{4}{\sqrt{3}} + \frac{3}{2} + \frac{3}{\sqrt{2}} - \frac{12}{\sqrt{11}}\right).$$

E4.35 **a)** The Born–Mayer equation is based on electrostatic interaction between cations and anions in the lattice. Thus, it works very well for the ionic structures like LiCl. However, it does not provide good estimates for the salts that have significant covalent character in the bonding. One such case is AgCl.

b) A good example of M^{2+} pair is Ca^{2+} and Hg^{2+} (one main group metal and one late transition metal, just like in the case of M^+ cations in part (a)).

E4.36 **a) BkO_2?** Using $r(Bk^+) = 97$ pm and $r(O^{2-}) = 128$ pm and Equation 4.4 we have:

$$\Delta H_L = [(3 \times 4 \times 2)/(97 \text{ pm} + 128 \text{ pm})][(1 - 34.5 \text{ pm}/(97 \text{ pm} + 128 \text{ pm})] (1.21\times 10^5 \text{ kJ pm mol}^{-1})$$

$$= 0.0179 \times 0.845 \times 1.21 \times 10^5 \text{ kJ mol}^{-1} = 10906 \text{ kJ mol}^{-1}$$

b) K_2SiF_6? Using $r(K^+) = 152$ pm and $r([SiF_6]^{2-}) = 194$ pm and Equation 4.4 we have:

$$\Delta H_L = [(3 \times 1 \times 2)/(152 \text{ pm} + 194 \text{ pm})] \times 0.9 \times 1.21 \times 10^5 \text{ kJ pm mol}^{-1} = 1888 \text{ kJ mol}^{-1}$$

c) $LiClO_4$? Using $r(Li^+) = 90$ pm and $r([ClO_4]^-) = 236$ pm and Equation 4.4 we have:

$$\Delta H_L = [(2 \times 1 \times 1)/(236 \text{ pm} + 90 \text{ pm})] \times 0.895 \times 1.21 \times 10^5 \text{ kJ pm mol}^{-1} = 664 \text{ kJ mol}^{-1}$$

E4.37 **a) $BaSeO_4$ or $CaSeO_4$?** In general, difference in size of the ions favours solubility in water. The thermochemical radius of selenate ion is 240 pm (see Table 4.10), whereas the six-coordinate radii of Ba^{2+} and Ca^{2+} are 135 pm and 100 pm, respectively. Since the difference in size between Ca^{2+} and SeO_4^{2-} is greater than the difference in size between Sr^{2+} and SeO_4^{2-}, $CaSeO_4$ is predicted to be more soluble in water than $BaSeO_4$.

(b) NaF or $NaBF_4$? This exercise can be answered without referring to tables in the text. The ions Na^+ and F^- are isoelectronic, so clearly Na^+ is smaller than F^-. It should also be obvious that the radius of BF_4^- is larger than the radius of F^-. Therefore, the difference in size between Na^+ and BF_4^- is greater than the difference in size between Na^+ and F^-; $NaBF_4$ is more soluble in water than NaF.

E4.38 Of the three halides listed, CsI has the lowest lattice enthalpy and highest solubility. This compound is fragile (breaks easily) and is damaged by moisture (in fact it is very hygroscopic). Crafting the CsI optics and maintaining it in a working order inside the instrument is very challenging.

E4.39 To quantitatively precipitate SeO_4^{2-} ion from an aqueous solution, we must find a cation that is of approximately the same size as SeO_4^{2-}. Table 3.10 lists the thermodynamic radius for this anion as 240 pm. From Resource Section 1, Ba^{2+} emerges as a good candidate. The 2+ charge can help to increase the lattice enthalpy, whereas the relatively large size of this cation reduces the cost of hydration enthalpy. Both factors decrease the solubility of $BaSeO_4$.

The thermochemical radius of phosphate ion is 238 pm. Thus, smaller cations like Na^+ and K^+ would give soluble phosphates, whereas larger cations of heavier Group 2 elements (i.e., Ba^{2+}) would likely produce insoluble phosphates.

E4.40 **a)** $MgCO_3$ will have a lower decomposition temperature than $CaCO_3$. This is because Mg^{2+} is a smaller cation and as such is not as good as Ca^{2+} in stabilizing CO_3^{2-} anion. Also, because Mg^{2+} is smaller, MgO (one of the products) has higher lattice enthalpy than CaO, which also decreases the decomposition temperature of the carbonate.

b) CsI_3 is expected to have a lower decomposition temperature than $N(CH_3)_4I_3$. Cs^+ is smaller than $N(CH_3)_4^+$ as such it is more likely to stabilize the product of the decomposition (CsI) than CsI_3.

E4.41 Of the three anions listed, $[ICl_4]^-$ is the largest. Thus, it can be preferentially stabilized in the presence of a large, 1+ cation such as Cs^+. Also, a good choice would be tetraalkylammonium cations, for example tetraethylammonium.

E4.42 **a)** Ca_3N_2 has ionic bonding and, considering stoichiometry of 3 : 2, both cation and anion should have relatively high coordination numbers. Thus, this compound is likely to exhibit Schottky defects owing to its structure type and bonding.

b) HgS assumes hexagonal close-packed structure with low coordination numbers and partial covalent bonding character. This compound is likely to exhibit Frenkel defects.

E4.43 In sapphire, the blue colour is due to the electron transfer between Fe^{2+} and Ti^{4+} substituting for Al^{3+} in adjacent octahedral sites in Al_2O_3 structure. Considering that beryl also contains Al^{3+} in octahedral sites, it is plausible that the blue colour of aquamarine is due to the same two dopants replacing Al^{3+} (see Table B3.1).

E4.44 As noted in Section 4.17(a), nonstoichiometry is common in the solid-state compounds of d-, f-, and heavier p-block elements. We would therefore expect vanadium carbide and manganese sulfide to exhibit nonstoichiometry (two d-block metal compounds) but not aluminium oxide (a light p-block metal oxide).

E4.45 The formation of defects is normally endothermic because as the lattice is disrupted the enthalpy of the solid rises. However, the term $-T\Delta S$ becomes more negative as defects are formed because they introduce more disorder into the lattice and the entropy rises. If we are not at absolute zero, the Gibbs energy will have a minimum at a nonzero concentration of defects and their formation will be spontaneous. As temperature is raised, this minimum in G shifts to higher defect concentrations as shown in Figure 4.52b in your textbook, so solids have a greater number of defects as temperatures approach their melting points. Increase in pressure would result in fewer defects in the solid. This is because at higher pressures, higher coordination numbers (tighter packing) are preferred and vacancies (present with both Schottky and Frenkel defects) become less energetically favourable. Increase in the pressure also reduces the spacing between the ions in the structure increasing the energy associated with ion removal and displacement within the structure, thereby increasing the free energy of defect formation.

E4.46 As noted in Section 4.17(a) nonstoichiometry is common for the solid state compounds of d-, f-, and heavier p-block elements. All the metals forming oxides are either d-block (Zn and Fe) or f-block (U) metals. However, $Zn_{1+x}O$ would be expected to show nonstoichiometry over very small range because the only significant and sufficiently stable oxidation state for Zn is +2. Thus, in this case as x increases, some of Zn(II) should be reduced to Zn(I), a very unlikely process. The next one would be $Fe_{1-x}O$ because Fe has two adjacent oxidation states Fe(II) and Fe(III). Finally, that leaves UO_{2+x} as the oxide with possible wide range of nonstoichiometry, considering that U, apart from U(IV), has a relatively stable U(VI) state.

E4.47 **a)** Ga is in Group 13 whereas Ge is in Group 14 of the periodic table of elements. Since Ga has one less valence electron than Ge, this would be a p-type semiconductor.

b) As is in Group 15 whereas Si is in Group 14 of the periodic table. Since As has one more valence electron than Si, this would be an n-type semiconductor.

c) $In_{0.49}As_{0.51}$ shows nonstoichiometry: it has more of (formally) As^{3-} than In^{3+} and should be also an n-type semiconductor.

E4.48 Low oxidation number d-metal oxides can lose electrons through a process equivalent to the oxidation of the metal atoms, with the result that holes appear in the predominately metal band. The positive charge carriers result in their p-type semiconductor classification. NiO is an example of this p-type semiconduction. Early transitional metal oxides with low oxidation number such as TiO and VO have metallic properties owing to the extended overlap of the d orbitals of the cations.

E4.49 A semiconductor is a substance with an electrical conductivity that decreases with increasing temperature. It has a small, measurable band gap. A semimetal is a solid whose band structure has a zero density of states and no measurable band gap as shown in Figure 4.70. Graphite is a classic example of a semimetal with conduction in the plane parallel to the sheets of carbon atoms.

E4.50 Ag_2S and CuBr (low-oxidation-number metal chalcogenide and halide) would be a p-type, and VO_2 (high-oxidation-number transition metal oxide) would be an n-type.

E4.51 In KC_8 potassium donates the electrons to the upper band which was originally empty in the graphite structure. In C_8Br bromine removes some electrons from the filled lower band in the graphite structure. In either case the net result is a partially filled band (formed either by addition of electrons to the originally empty band or by removal of electrons from the initially filled band) and both KC_8 and C_8Br should have metallic properties.

Answers to Selected Tutorial Problems

T4.2 Considering that fcc unit cell contains four atoms and given dimension of gold's unit cell, it is possible to calculate the number of Au atoms in the nanoparticle as follows.

The unit cell length of 2000 nm represents 5 unit cells so the nanoparticle cube contains 5 × 5 × 5 = 125 unit cells. A low estimate for the number of atoms would use the number of atoms in a fcc unit cell in an infinite crystal (= 4) giving 4 × 125 = 500 gold atoms; however, unlike an infinite crystal the gold atoms on the faces of the nanocube are not shared with neighbouring unit cells and contribute fully to the number of atoms in the cubic nanoparticle. There are various mathematical ways of tackling this problem but from a crystallographic point of view we can count atoms in the various unit cells that are internal, or on the surface of the nanocube, see the Figure that shows one face of the cube.

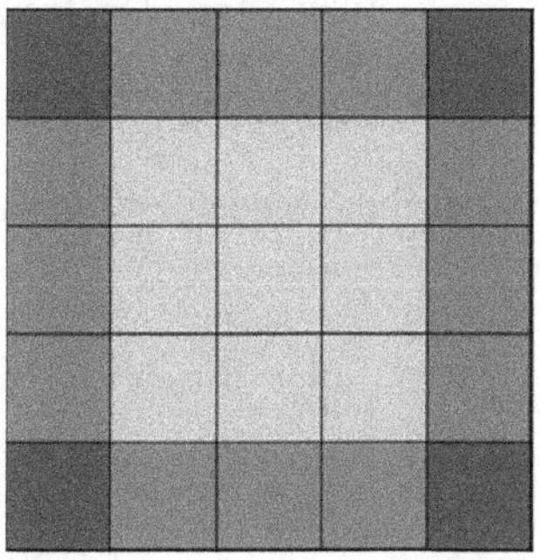

The 5 × 5 × 5 unit will have 3 × 3 × 3 = 27 unit cell units embedded at the centre of the cube and these cells will be equivalent to those within an infinite crystal and therefore have 4 Au atoms per unit cell. The total number of atoms contributing to the nanoparticle is thus 27 × 4 = 108 Au atoms. Consideration of a cube face shows there are nine unit cells that have one external surface, shaded pale grey in the Figure. Counting the number of gold atoms contributing to the gold nanoparticle in this type of cell gives (4 × 1/8) + (5 × ½) + (4 × 1/4) + (1 × 1) = 5; there are 6 faces of the cube giving the total of 6 × 9 × 5 = 270. For the cell shaded mid grey the calculation gives 3(unit cells per edge) × 12(number of edges to nanocube) × 6.25(number of atoms contributed by this type of unit cell) = 225. Finally, for the corner units shaded dark grey the calculation is 8 × 7.875 = 63. The total number of atoms in the nanocube is thus 108 + 270 + 225 + 63 = 666 Au atoms.

For a general nanocube with n unit cells along each edge the number of atoms in the nanocube is given by

$$4m^3 - 6m^2 + 3m.$$

Where m = n + 1, that is the number of atoms along the nanocube edge.

When considering the interaction with light, it is important to consider the wavelengths of the visible light and the size of nanoparticles.

T4.4 The Madelung constant is in a sense a geometrical factor that depends on the position of ions within a unit cell. The electrostatic potential in which Na^+ ions resides is given by:

$$V = \frac{z^2e^2}{4\pi\varepsilon_o d}$$

where, in this case, d is the distance between Na^+ and the neighbouring atoms we are considering. The first neighbours are 6 Cl^- ions at the distance d. Thus, the potential energy of a single Na^+ surrounded by six anions is:

$$V_1 = -\frac{6e^2}{4\pi\varepsilon_o d}.$$

Next, we have 12 Na^+ ions (the ones located on the faces of the cubic unit cell). Using Pythagoras's theorem, we can show that they are at the distance of $d\sqrt{2}$ from the same Na^+ we looked at above (see Figure T4.4 below). This is a repulsive interaction for which we get the following potential V_2:

$$V_2 = +\frac{12e^2}{4\pi\varepsilon_o\sqrt{2}d}.$$

Next to consider are eight Cl^- ions at the middle of each edge of the unit cell. Again, using Pythagoras's theorem we can see that the distance between them and our Na^+ is $d\sqrt{3}$. The potential V_3 is thus:

$$V_3 = -\frac{8e^2}{4\pi\varepsilon_o\sqrt{3}d}.$$

This process can be continued in the same manner until V_6, which would give all six terms in the Madelung series given in this project (Figure T4.4 shows the distances involved in calculation of all V_n).

To find the total potential for Na^+ cation we have to sum V_1, V_2, V_3, etc:

$$\begin{aligned} V_{tot} &= V_1 + V_2 + V_3 + \ldots \\ &= -\frac{6e^2}{4\pi\varepsilon_0 d} + \frac{12e^2}{4\pi\varepsilon_0 d\sqrt{2}} - \frac{8e^2}{4\pi\varepsilon_0 d\sqrt{3}} + \ldots \\ &= -\frac{e^2}{4\pi\varepsilon_0 d}\left(\frac{6}{\sqrt{1}} - \frac{12}{\sqrt{2}} + \frac{8}{\sqrt{3}} - \ldots\right). \end{aligned}$$

The value of the series in the brackets is called the Madelung constant for NaCl type structure.

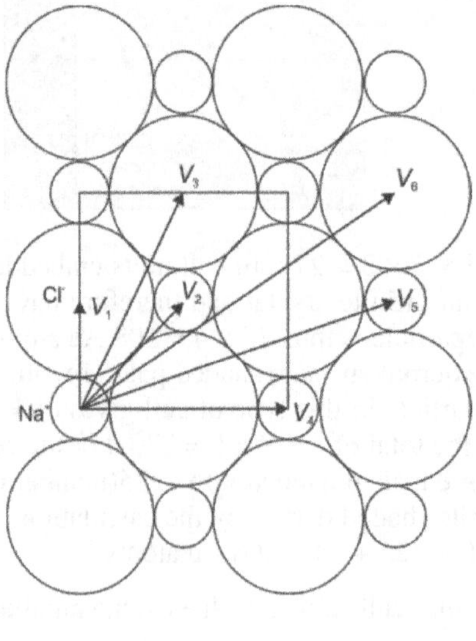

Figure T4.4

T4.8 Note that for the case of MX salts the product of charges is 1 and N_{ion} = 2. The change from pm (in Kapustinskii's equation) to nm (in Bartlett's relationship) is responsible for change in the order of magnitude in the numerator. Also, the volume in Bartlett's relationship is the volume of the formula unit (simple NaCl) and not of the unit cell, and as such is directly proportional to the anion and cation radii in Kapustinskii's equation.

Chapter 5 Acids and Bases

Self-Tests

S5.1 **a) $HNO_3 + H_2O \rightarrow H_3O^+ + NO_3^-$.** The compound HNO_3 transfers a proton *to* water, so it is an acid. The nitrate ion is its conjugate base. In this reaction, H_2O accepts a proton, so it is a base. The hydronium ion, H_3O^+, is its conjugate acid.

b) $CO_3^{2-} + H_2O \rightarrow HCO_3^- + OH^-$. A carbonate ion accepts a proton from water, so it is a base. The hydrogen carbonate, or bicarbonate, ion is its conjugate acid. In this reaction, H_2O donates a proton, so it is an acid. A hydroxide ion is its conjugate base.

c) $NH_3 + H_2S \rightarrow NH_4^+ + HS^-$. Ammonia accepts a proton from hydrogen sulfide, so it is a base. The ammonium ion, NH_4^+, is its conjugate acid. Since hydrogen sulfide donated a proton, it is an acid, while HS^- is its conjugate base.

S5.2 With $K_a < 1$, HF is a weak acid only partially dissociated in water. The dissociation equation is:

$$HF + H_2O \rightleftharpoons H_3O^+ + F^-.$$

Applying what you have learned in general chemistry, this type of problem can best be solved using the initial concentration, change in concentration, and final concentrations according to the table below.

	$HF + H_2O \rightleftharpoons$	H_3O^+ +	F^-
Initial (M)	0.10	0.0	0.0
Change (M)	–x	+x	+x
Equilibrium (M)	0.10–x	x	x

Insert these values into the acidity constant expression for HF.

$$K_a = \frac{[H_3O^+][F^-]}{[HF]} = 3.5\times10^{-4} = \frac{x^2}{0.10-x}.$$

From here we obtain the following quadratic equation:

$$x^2 + 3.5 \times 10^{-4}x - 3.5 \times 10^{-5} = 0.$$

With roots (solutions):

$$x = \frac{-3.5\times10^{-4} \pm \sqrt{(3.5\times10^{-4})^2 - 4\times1\times(3.5\times10^{-5})}}{2\times1}.$$

The two roots are x = 5.7×10^{-3} M and -6.1×10^{-3} M. Obviously, the negative root is physically impossible, so our final result is $[H_3O^+] = 5.7 \times 10^{-3}$ M. Taking the negative log gives the pH of the solution.

$$pH = -\log[H_3O^+] = 2.24.$$

S5.3 Tartaric acid is a diprotic weak acid and must be solved in two equilibrium steps, similar to **S5.2**. The first equilibrium, and ICE table is shown below:

	$H_2C_4O_6$ + $H_2O \rightleftharpoons$	H_3O^+ +	$HC_4O_6^-$
Initial (M)	0.20	0.0	0.0
Change (M)	–x	+x	+x
Equilibrium (M)	0.20–x	x	x

The equilibrium values are substituted in the acidity constant for the first deprotonation step:

$$K_{a1} = 1.0\times10^{-3} = \frac{[H_3O^+][HC_4O_6^-]}{[H_2C_4O_6]} = \frac{x^2}{0.2-x}.$$

Now we must solve the following quadratic equation:

$$x^2 + 1.0 \times 10^{-3}x - 2.0 \times 10^{-4} = 0$$

$$x = \frac{-1.0\times10^{-3} \pm \sqrt{(1.0\times10^{-3})^2 - 4\times1\times(-2.0\times10^{-4})}}{2\times1}$$

$$x = 0.014 \text{ M or } -0.015 \text{ M.}$$

So, our $[H_3O^+] = 0.014$ M which is also equal to $[HC_4O_6^-]$. Now we have to consider the second deprotonation step using the same approach as above and calculate $[C_4O_6^{2-}]$.

	$HC_4O_6^-$ + H_2O ⇌	H_3O^+ +	$C_4O_6^{2-}$
Initial (M)	0.014	0.014	0.0
Change (M)	−x	+x	+x
Equilibrium (M)	0.014–x	0.014+x	x

Insert these values into the equilibrium expression:

$$K_{a2} = 4.6\times10^{-5} = \frac{[H_3O^+][C_4O_6^{2-}]}{[HC_4O_6^-]} = \frac{(0.014+x)x}{(0.014-x)}.$$

Now we can assume that $0.014 \pm x \approx 0.014$, so x is simply equal to 4.6×10^{-5}.

This means our total $[H_3O^+] = 0.014 \text{ M} + 4.6 \times 10^{-5}$, which is essentially 0.014 M meaning the second deprotonation does not affect the pH in this solution. Our pH is:

$$pH = -\log[H_3O^+] = 1.85.$$

S5.4 Since the acid strength of aqua acids increases as the charge density of the central metal cation increases, the strongest acid will have the highest charge and the smallest radius. Thus $[Na(H_2O)_6]^+$, with the lowest charge, will be the weakest of the four aqua acids, and $[Sc(H_2O)_6]^{3+}$, with the highest charge, will be the strongest. The remaining two aqua acids have the same charge, and so the one with the smaller ionic radius, *r*, will have the higher charge density and hence the greater acidity. Since Ni^{2+} is on the right of Mn^{2+} in the periodic table of the elements, it has a smaller radius, and so $[Ni(H_2O)_6]^{2+}$ is more acidic than $[Mn(H_2O)_6]^{2+}$. The order of increasing acidity is $[Na(H_2O)_6]^+ < [Mn(H_2O)_6]^{2+} < [Ni(H_2O)_6]^{2+} < [Sc(H_2O)_6]^{3+}$.

S5.5 **a) H_3PO_4?** Pauling's first rule for predicting the pK_{a1} of an oxoacid is $pK_a \approx 8 - 5p$ (where *p* is the number of oxo groups attached to the central element).
Since $p = 1$ (see the structure of H_3PO_4 below), the predicted value of pK_{a1} for H_3PO_4 is $8 - (5 \times 1) = 3$. The actual value, given in Table 5.3, is 2.1.

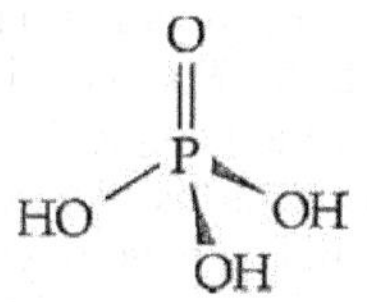

b) $H_2PO_4^-$? Pauling's second rule for predicting the pK_a values of a polyprotic oxoacid is that successive pK_a values for polyprotic acids increase by five units for each successive proton transfer. Since pK_{a1} for H_3PO_4 was predicted to be 3 (see above), the predicted value of pK_a for $H_2PO_4^-$, which is pK_{a2} for H_3PO_4, is $3 + 5 = 8$. The actual value, given in Table 5.3, is 7.4.

c) HPO_4^{2-}? The pK_a for HPO_4^{2-} is the same as pK_{a3} for H_3PO_4, so the predicted value is 3 + (2 × 5) = 13. The actual value, given in Table 5.3, is 12.7.

S5.6 According to Figure 5.6, Ti(IV) is amphoteric. Treatment of an aqueous solution containing Ti(IV) ions with ammonia causes the precipitation of TiO_2, but further treatment with NaOH causes the TiO_2 to redissolve. H_2O_2 would not react with Ti(IV) since Ti(IV) cannot be further oxidized.

S5.7 **a) $FeCl_3 + Cl^- \rightarrow [FeCl_4]^-$?** The acid $FeCl_3$ forms a complex, $[FeCl_4]^-$, with the base Cl^-.

b) $I^- + I_2 \rightarrow I_3^-$? The acid I_2 forms a complex, I_3^-, with the base I^-.

S5.8 If the N atom lone pair of $(H_3Si)_3N$ is delocalized onto the three Si atoms, it cannot exert its normal repulsion influence as predicted by VSEPR rules. The N atom of $(H_3Si)_3N$ is trigonal planar, whereas the N atom of $(H_3C)_3N$ is trigonal pyramidal. The structures of $(H_3Si)_3N$ and $(H_3C)_3N$, excluding the hydrogen atoms are:

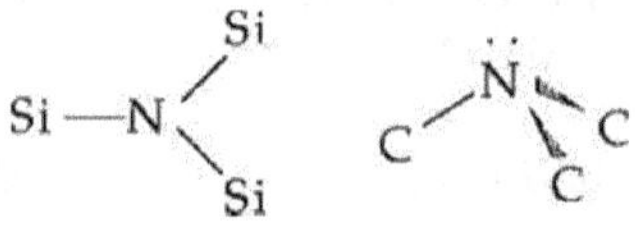

S5.9 The two solvents in Figure 5.12 for which the window covers the range 25 to 27 are dimethylsulfoxide (DMSO) and ammonia.

S5.10 We need to identify the autoionization products of the solvent, BrF_3, and then decide whether the solute increases the concentration of the cation (an acid) or the anion (a base). The autoionization products of BrF_3 are:

$$2BrF_3(l) \rightleftharpoons BrF_2^+(sol) + BrF_4^-(sol)$$

The solute would ionize to give K^+ and BrF_4^-. Since BrF_4^- will increase the amount of the anion, $KBrF_4$ is a base when dissolved in BrF_3.

S5.11 The ether oxygen atom will behave as a Lewis base and donate one of its lone electron pairs forming a dative bond with the boron atom of BF_3. The geometry at the boron atom will go from trigonal planar in BF_3 to tetrahedral in $F_3B–OEt_2$. The structure of the complex is shown below.

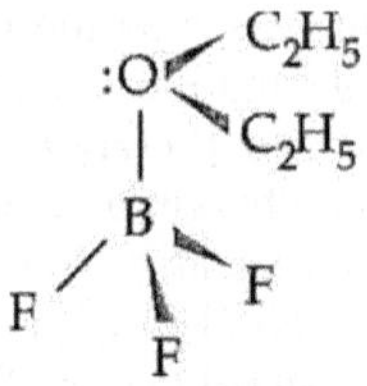

Exercises

E5.1 See the diagram below for the outline of the acid–base properties of main group oxides. The elements that form basic oxides have symbols printed in bold, those forming acidic oxides are underlined, and those forming amphoteric oxides are in italics. Note the diagonal region from upper left to lower right that includes the elements forming amphoteric oxides. The elements As, Sb, and Bi are shown both underlined and in italics. This is because their oxides in highest oxidation state (+5) are acidic, but the oxides in lower oxidation state (+3) are amphoteric. There are no oxides for fluorine. Fluorine forms two compounds with oxygen, O_2F_2 and OF_2; but because F is more electronegative than O these compounds have oxygen in a positive oxidation state (+1 and +2 respectively) and thus are not oxides (oxides have O in a negative oxidation state). For elements 113–117 there is no data.

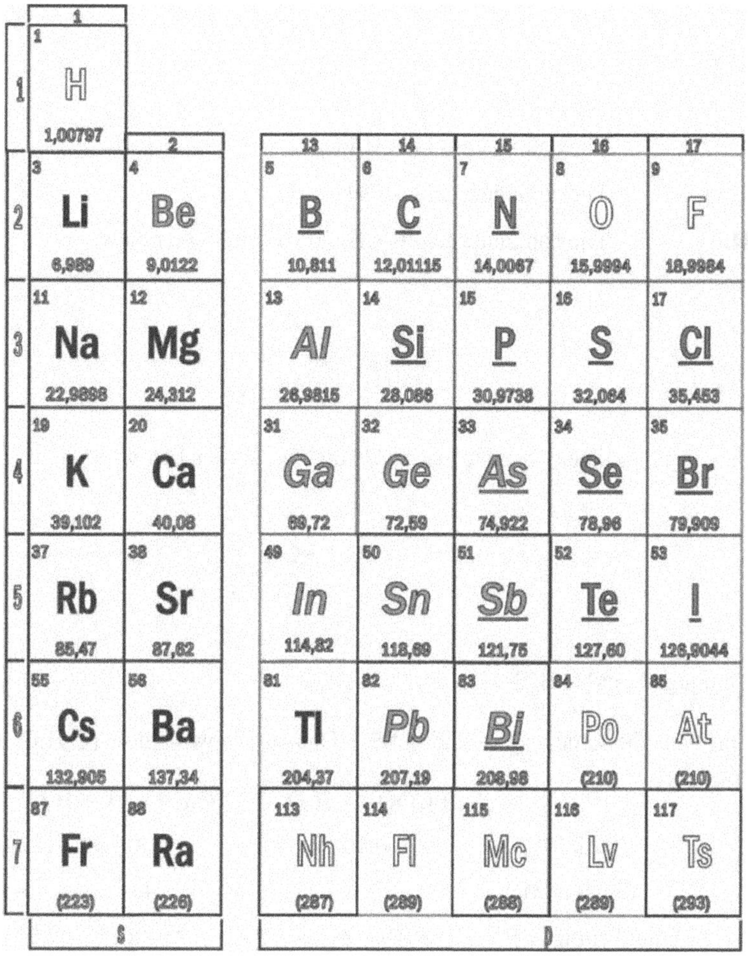

Period	1	2	13	14	15	16	17
1	1 H 1,00797						
2	3 Li 6,989	4 Be 9,0122	5 B 10,811	6 C 12,01115	7 N 14,0067	8 O 15,9994	9 F 18,9984
3	11 Na 22,9898	12 Mg 24,312	13 Al 26,9815	14 Si 28,086	15 P 30,9738	16 S 32,064	17 Cl 35,453
4	19 K 39,102	20 Ca 40,08	31 Ga 69,72	32 Ge 72,59	33 As 74,922	34 Se 78,96	35 Br 79,909
5	37 Rb 85,47	38 Sr 87,62	49 In 114,82	50 Sn 118,69	51 Sb 121,75	52 Te 127,60	53 I 126,9044
6	55 Cs 132,905	56 Ba 137,34	81 Tl 204,37	82 Pb 207,19	83 Bi 208,98	84 Po (210)	85 At (210)
7	87 Fr (223)	88 Ra (226)	113 Nh (287)	114 Fl (289)	115 Mc (288)	116 Lv (289)	117 Ts (293)
	s		p				

E5.2 **a) $[Co(NH_3)_5(OH_2)]^{3+}$:** A conjugate base is a species with one fewer proton than the parent acid. Therefore, the conjugate base in this case is $[Co(NH_3)_5(OH)]^{2+}$, shown below. Note that because H_2O is generally more acidic than NH_3, H_2O will lose one of its protons.

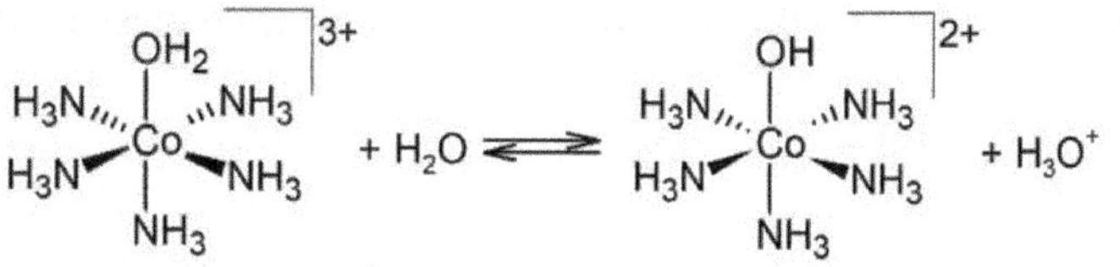

b) HSO_4^-: The conjugate base is SO_4^-.

c) CH_3OH: The conjugate base is CH_3O^-.

d) $H_2PO_4^-$: The conjugate base is HPO_4^{2-}.

e) $Si(OH)_4$: The conjugate base is $SiO(OH)_3^-$.

f) HS^-: The conjugate base is S^{2-}.

E5.3 Remember that a conjugate acid is a species with one more proton than the parent base.

a) C_5H_5N (pyridine): The conjugate acid in this case is the pyridinium ion, $C_5H_6N^+$, shown below.

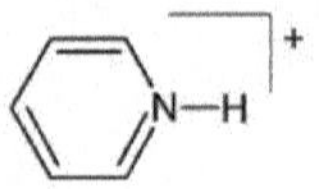

b) HPO_4^{2-}: The conjugate acid is $H_2PO_4^-$.

c) O^{2-}: The conjugate acid is OH^-.

d) CH_3COOH: The conjugate acid is $CH_3C(OH)_2^+$, shown below.

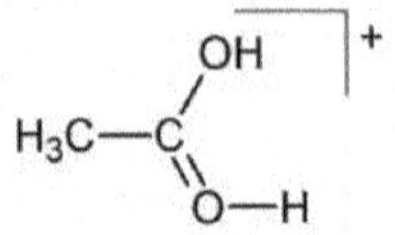

e) $[Co(CO)_4]^-$: The conjugate acid is $HCo(CO)_4$, shown below.

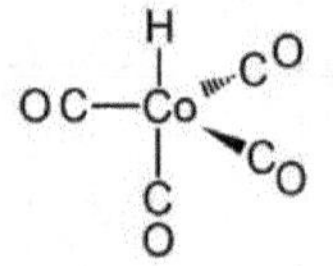

f) CN^-: The conjugate acid is HCN.

E5.4 We solve this problem the same way we did in **S5.2**. Data are shown below (C_3H_7COOH is butanoic acid):

	$C_3H_7COOH + H_2O \rightleftharpoons$	$C_3H_7COO^- +$	H_3O^+
Initial (M)	0.10	0	0
Change (M)	–x	+x	+x
Equilibrium (M)	0.10 – x	x	x

Insert these values into the equilibrium expression:

$$K_a = 1.86\times10^{-5} = \frac{[H_3O^+][C_3H_7COO^-]}{[C_3H_7COOH]} = \frac{x^2}{(0.1-x)}.$$

Now, we can assume that $0.10 - x \approx 0.10$ and then solve for x, which will be our $[H_3O^+]$:

$$1.86\times10^{-5} = \frac{x^2}{0.10}$$

$$x = 1.4 \times 10^{-3} = [H_3O^+].$$

The pH of the solution is: $pH = -\log[H_3O^+] = 2.85$.

E5.5 To find the K_b, use Equation 5.4:

$$K_a \times K_b = K_w$$

$$K_b = K_w/K_a = (1.0 \times 10^{-14})/(1.8 \times 10^{-5}) = 5.6 \times 10^{-10}.$$

E5.6 To find the K_a, use Equation 5.4:

$$K_a \times K_b = K_w$$

$$K_a = K_w/K_b = (1.0 \times 10^{-14})/(1.8 \times 10^{-9}) = 5.6 \times 10^{-6}.$$

E5.7 From Table 5.2, the proton affinities (A_p') of $H_2O(l)$ and $OH^-(aq)$ are 1130 kJ mol^{-1} and 1188 kJ mol^{-1} respectively. Thus, although transfer of H^+ from H_3O^+ to F^- is exothermic by –20 kJ mol^{-1}, the transfer of H^+ from H_2O to F^- is endothermic by 38 kJ mol^{-1}. Therefore, in neutral water, F^- will behave as a base.

E5.8 The structures for chloric acid and chlorous acid are shown below. Chlorine in chloric acid is double-bonded to two oxygen atoms and single-bonded to the oxygen atom of a hydroxide, and has one lone pair, giving it a trigonal pyramidal molecular geometry at chlorine atom. Chlorine in chlorous acid is double-bonded to one oxygen atom and single-bonded to the oxygen atom of a hydroxide, and has two lone electron pairs, thus the molecular geometry is bent.

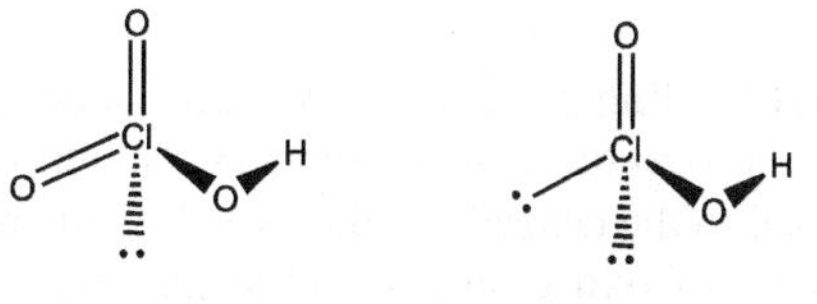

Pauling's first rule for predicting the pK_a of a mononuclear oxoacid is p$K_a \approx 8 - 5p$ (where p is the number of oxo groups attached to the central element). Since $p = 2$ for chloric acid, the predicted pK_a for $HClO_3$ is $8 - (5 \times 2) = -2$. The actual value, given in Table 5.1, is –1. For chlorous acid, $p = 1$, therefore the pK_a for $HClO_2$ is $8 - (5 \times 1) = 3$. The actual value, given in Table 5.1, is 2.

E5.9 **a) CO_3^{2-}, O^{2-}, ClO_4^-, and NO_3^- in water?** You can interpret the term "studied experimentally" to mean that the base in question exists in water (i.e., it is not completely protonated to its conjugate acid) *and* that the base in question can be partially protonated (i.e., it is not so weak that the strongest acid possible in water, H_3O^+, will fail to produce a measurable amount of the conjugate acid). Using these criteria, the base CO_3^{2-} is predicted to be of directly measurable base strength because the equilibrium $CO_3^{2-} + H_2O \rightarrow HCO_3^- + OH^-$ produces measurable amounts of reactants and products. The base O^{2-}, on the other hand, is completely protonated in water to produce OH^-, so the oxide ion is too strong to be studied experimentally in water. The bases ClO_4^- and NO_3^- are conjugate bases of very strong acids, which are completely deprotonated in water. Therefore, since it is not possible to protonate either perchlorate or nitrate ion in water, they are too weak to be studied experimentally.

b) HSO_4^-, NO_3^-, and ClO_4^- in H_2SO_4? The hydrogen sulfate ion, HSO_4^-, is the strongest base possible in liquid sulfuric acid. However, since acids can protonate it, it is not too strong to be studied experimentally. Nitrate ion is a weaker base than HSO_4^-, a consequence of the fact that its conjugate acid, HNO_3, is a stronger acid than H_2SO_4. However, nitrate is not so weak that it cannot be protonated in sulfuric acid, so NO_3^- is of directly measurable base strength in liquid H_2SO_4. On the other hand, ClO_4^-, the conjugate base of one of the strongest known acids, is so weak that it cannot be protonated in sulfuric acid, and hence cannot be studied in sulfuric acid.

E5.10 A comparison of the aqueous pK_a values is necessary to answer this question. These are:

$$HOCN, 4 \qquad H_2NCN, 10.5 \qquad CH_3CN, 20$$

$$H_2O, 14 \qquad NH_3, \text{very large} \qquad CH_4, \text{very large}$$

You know that the values of pK_a are very large for ammonia and methane because these compounds are not normally thought of as acids (i.e., they are extremely weak acids). In all three cases, the cyano-containing compound has a lower pK_a (a *higher* acidity) than the parent compound. In the case of H_2O and HOCN, the latter compound is 10 orders of magnitude more acidic than water (because the difference between pK_a values is 10 pK_a units). The deprotonation equilibrium involves the formation of an anion—the conjugate base of the acid in question. For example:

$$HOCN + H_2O \rightleftharpoons OCN^- + H_3O^+$$

$$H_2O + H_2O \rightleftharpoons OH^- + H_3O^+$$

Since a lower pK_a means a larger K_a, this suggests that the anion OCN^- is better stabilized than OH^-. This occurs because the –CN group is electron withdrawing.

E5.11 The difference in pK_a values can be explained if we look at the structures of the acids in question. (Note: flip through Chapters 15 and 17 to find the structures!)

Structures of H_3PO_4, H_3PO_3, and H_3PO_2

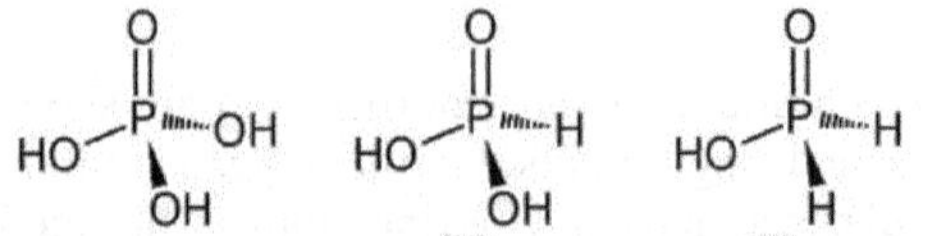

Structures of HOCl, $HClO_2$, and $HClO_3$

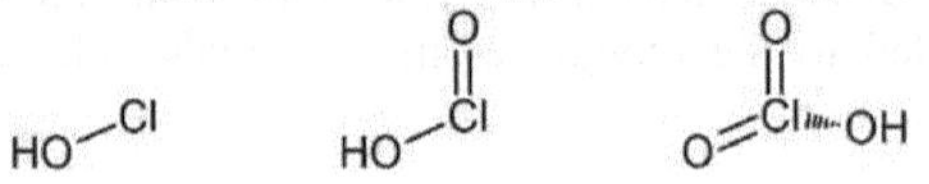

According to the first Pauling rule, p$K_a \approx 8 - 5p$, where p is the number of oxo groups attached to the central element. Looking at the structures of phosphorus containing acids, we can see that all of them actually have only one oxo group attached to the central P atom ($p = 1$). This means that they will all have p$K_a \approx 8 - 5 \times 1 = 3$. On the other hand, the number of oxo groups for chlorine-containing acids is increasing from zero in HOCl to two in $HClO_3$. Thus, the three will have the following approximate pK_a values:

HOCl ($p = 0$): p$K_a \approx 8 - 5 \times 0 = 8$

$HClO_2$ ($p = 1$): p$K_a \approx 8 - 5 \times 1 = 3$

$HClO_3$ ($p = 2$): p$K_a \approx 8 - 5 \times 2 = -2$

E5.12 Consult Example and Self-Test 5.4 as well as Section 5.2 if you had problems with this exercise.

The acidity of aqua acids depends on the charge density on the central cation: the higher the charge density, the higher the acidity. Thus, with the charge of 1+, Na^+ should be the weakest acid. There are three cations with the charge 2+: Mn^{2+}, Ca^{2+}, and Sr^{2+}. Among these three the cation with the largest radius will have the lowest charge density and thus will be the weakest acid. The relative sizes of cations can be determined by looking at their positions in the periodic table: Sr is in period 5 and Sr^{2+} should have the largest radius; Mn is to the right of Ca, and Mn^{2+} should be smaller than Ca^{2+}. Thus, of these three, the weakest acid is Sr^{2+} and the strongest is Mn^{2+}. Finally, using the same arguments for Fe^{3+} and Al^{3+} as for the 2+ cations, we can conclude that Al^{3+} is more acidic in an aqueous solution than Fe^{3+}. The overall order of cations in order of increasing acidity in water is:

$$Na^+ < Sr^{2+} < Ca^{2+} < Mn^{2+} < Fe^{3+} < Al^{3+}.$$

E5.13 According to Pauling's first rule for predicting the pK_a of a mononuclear oxoacid, p$K_a \approx 8 - 5p$ (where p is the number of oxo groups attached to the central element).

pK_a for $HNO_2 \approx 8 - (5 \times 1) = 3$

pK_a for $H_2SO_4 \approx 8 - (5 \times 2) = -2$

pK_a for $HBrO_3 \approx 8 - (5 \times 2) = -2$

pK_a for $HClO_4 \approx 8 - (5 \times 3) = -7$

The lowest pK_a is for $HClO_4$, so this is the strongest acid. Next, we have H_2SO_4 and $HBrO_3$, which have the same pK_a of −2 according to Pauling's rules. (Note that the same conclusion can be reached just by comparing the number of oxo groups in each acid without performing the actual calculations.) Bromine is more electronegative than sulfur; inductively, $HBrO_3$ should be a stronger acid than H_2SO_4. HNO_2 has the highest pK_a and is the weakest acid. Therefore, the order is $HClO_4 > HBrO_3 > H_2SO_4 > HNO_2$.

E5.14 **a) $[Fe(OH_2)_6]^{3+}$ or $[Fe(OH_2)_6]^{2+}$:** The Fe(III) complex, $[Fe(OH_2)_6]^{3+}$, is the stronger acid because Fe^{3+} has a higher charge density than Fe^{2+}—it has both higher charge and smaller radius than Fe^{2+}.

b) $[Al(OH_2)_6]^{3+}$ or $[Ga(OH_2)_6]^{3+}$: In this case, the charges are the same but ionic radii of Al^{3+} and Ga^{3+} are not. Since the ionic radius is smaller for period 3 Al^{3+} than for period 4 Ga^{3+}, the charge density for Al^{3+} is higher than for Ga^{3+} and the aluminium-aqua ion is more acidic.

c) $Si(OH)_4$ or $Ge(OH)_4$: As in part (b) above, the charges on central atoms are the same but their radii are different. The comparison here is also between species containing period 3 and period 4 central atoms in the same group, and the species containing the smaller central atom (smaller *r*), $Si(OH)_4$, is more acidic.

d) $HClO_3$ or $HClO_4$: These two acids are shown below. According to Pauling's first rule for mononuclear oxoacids, the species with more oxo groups has the lower pK_a and is the stronger acid. Thus, $HClO_4$ is a stronger acid than $HClO_3$. Note that the oxidation state of the central chlorine atom in the stronger acid (+7) is higher than in the weaker acid (+5).

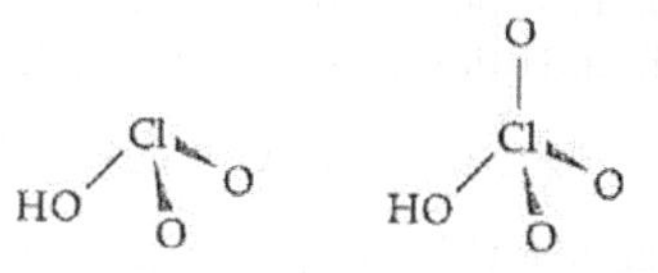

$HClO_3$ $HClO_4$

e) H_2CrO_4 or $HMnO_4$: As in part (d), the oxidation states of these two acids are different—VI for the chromium atom in H_2CrO_4 and VII for the manganese atom in $HMnO_4$. The species with the higher central-atom oxidation state, $HMnO_4$, is the stronger acid. Note that this acid also has more oxo groups—three—than H_2CrO_4, which has two.

f) H_3PO_4 or H_2SO_4: The oxidation state of sulfur in H_2SO_4 is VI, whereas the oxidation state of phosphorus in H_3PO_4 is V. Furthermore, sulfuric acid has two oxo groups attached to the central sulfur atom, whereas phosphoric acid has only one oxo group attached to the central phosphorus atom. Therefore, on both counts (which by now you can see are really manifestations of the same thing) H_2SO_4 is a stronger acid than H_3PO_4.

E5.15 First you pick out the intrinsically acidic oxides among those given (Al_2O_3, B_2O_3, BaO, CO_2, Cl_2O_7, and SO_3), because these will be the *least* basic. The compounds B_2O_3, CO_2, Cl_2O_7, and SO_3 are acidic because the central element for each of them is found in the acidic region of the periodic table (see the s- and p-block diagram in the answer to Exercise 5.1; generally the oxides of non-metallic elements are acidic). The most acidic compound, Cl_2O_7, has the highest central-atom oxidation state, +7, whereas the least acidic, B_2O_3, has the lowest, +3. Of the remaining compounds, Al_2O_3 is amphoteric, which puts it on the borderline between acidic and basic oxides, and BaO is basic. Therefore, a list of these compounds in order of increasing basicity is $Cl_2O_7 < SO_3 < CO_2 < B_2O_3 < Al_2O_3 <$ BaO.

E5.16 The weakest acids, CH_3GeH_3 and NH_3, are easy to pick out of this group (HSO_4^-, H_3O^+, H_4SiO_4, CH_3GeH_3, NH_3, and HSO_3F) because they do not contain any –OH bonds. Ammonia is the weaker acid of the two, primarily because the N-H bond is stronger than the Ge-H bond. Of the remaining species, note that HSO_3F is very similar to H_2SO_4 as far as structure and sulfur oxidation state (VI) are concerned, so it is reasonable to suppose that HSO_3F is a very strong acid, which it is. The anion HSO_4^- is a considerably weaker acid than HSO_3F, for the same reason that it is a considerably weaker acid than H_2SO_4. Since HSO_4^- is not completely deprotonated in water, it is a weaker acid than H_3O^+, which is the strongest possible acidic species in water. Finally, it is difficult to place exactly $Si(OH)_4$ in this group. It is certainly more acidic than NH_3 and CH_3GeH_3, and it turns out to be *less* acidic than HSO_4^-, despite the negative charge of the latter species. Therefore, a list of these species in order of increasing acidity is $NH_3 < CH_3GeH_3 < H_4SiO_4 < HSO_4^- < H_3O^+ < HSO_3F$.

E5.17 Although Na^+ and Ag^+ ions have about the same ionic radius, $Ag^+–OH_2$ bonds are much more covalent than $Na^+–OH_2$ bonds, a common feature of the chemistry of d-block vs. s-block metal ions. The greater covalence of the $Ag^+–OH_2$ bonds has the effect of delocalizing the positive charge of the cation over the whole aqua complex. Therefore, the departing proton is repelled more by the positive charge of $Ag^+(aq)$ than by the positive charge of $Na^+(aq)$, and the Ag^+ ion is the stronger acid.

E5.18 To get to the general rule, the best guide is to analyse one specific case of bridge formation/water elimination, for example

$$2[Al(OH_2)_6]^{3+} + H_2O \rightarrow [(H_2O)_5Al–O–Al(OH_2)_5]^{4+} + 2H_3O^+.$$

The charge per aluminium atom is +3 for the mononuclear species on the left-hand side of the equation, but only +2 for the dinuclear species on the right-hand side. Thus, polycation formation reduces the overall positive charge of the species by +1 per M.

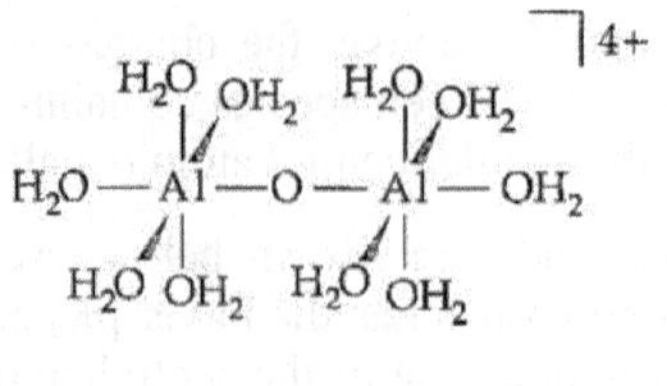

E5.19 **a) H_3PO_4 and Na_2HPO_4:** H_3PO_4 is a polyprotic acid and Na_2HPO_4 is its acidic salt. Considering that for a polyprotic acid, the K_{a1} is the highest, H_3PO_4 will be more acidic and HPO_4^{2-} is going to behave like a base:

$$H_3PO_4(aq) + HPO_4^{2-}(aq) \rightleftharpoons 2H_2PO_4^-(aq)$$

b) CO_2 and $CaCO_3$: Keep in mind that CO_2 in aqueous medium partially reacts with H_2O to produce acid H_2CO_3. The overall equation can be written as follows:

$$CO_2(aq) + CaCO_3(s) + H_2O(l) \rightleftharpoons Ca^{2+}(aq) + 2HCO_3^-$$

E5.20 HF behaves like an acid in anhydrous sulfuric acid, so it must increase the concentration of $H_3SO_4^+$ ions. The equation is:

$$H_2SO_4 + HF \rightleftharpoons H_3SO_4^+ + F^-.$$

HF behaves like a base in liquid ammonia because it decreases the concentration of NH_4^+:

$$NH_4^+ + HF \rightleftharpoons NH_3 + H_2F^+.$$

E5.21 As you go down a family in the periodic chart, the acidity of the homologous hydrogen compounds increases. This is due primarily to the fact that the bond dissociation energy is smaller as you go down a family, due to poorer orbital overlap. Since the H-X bond is weaker for H_2Se, it will release protons more readily than H_2S in each solvent.

E5.22 Lewis acids are electron pair acceptors. Looking at the trend for silicon halides we see that Lewis acidity increases with increasing electronegativity and decreasing size of the halogen element—SiF_4 with the smallest and most electronegative halide is the strongest Lewis acid while SiI_4 with the largest and least electronegative halide is the weakest Lewis acid. Fluorine's high electronegativity results in electron withdrawal from the central Si atom making the Si very nucleophilic. This fact, combined with the small size of F atom (little steric hindrance at Si) makes SiF_4 the strongest Lewis acid of the four.

The order of Lewis acidity for boron halides is inverse to that expected—BI_3 is the strongest and BF_3 is the weakest Lewis acid. This is because additional electronic effects play an important role. BF_3 is stabilized by resonance (see Section 2.2 *Resonance*) with three resonance hybrids having a double B=F bond. This π interaction stabilizes a planar geometry at the B atom, and the B 2p orbital is less available for formation of a Lewis adduct. The stabilization of planar BF_3 can also be explained with a molecular orbital diagram. As we go down the halogen group, the atoms become larger and the π interaction decreases, with Lewis acidity of boron halides increasing in the same order. See also Section 5.7(b) *Group 13 Lewis acids* in the textbook.

E5.23 **a) $SO_3 + H_2O \rightarrow HSO_4^- + H^+$:** The acids in this reaction are the Lewis acids SO_3 and H^+ and the base is the Lewis base OH^-. The complex (or adduct) HSO_4^- is formed by the displacement of the proton from the hydroxide ion by the stronger Lewis acid SO_3. In this way, the water molecule is thought of as a Lewis adduct formed from H^+ and OH^-. Even though this is not explicitly shown in the reaction, the water molecule exhibits Brønsted acidity (not only Lewis basicity). Note that it is easy to tell that this is a displacement reaction instead of just a complex formation reaction because, while there is only one base in the reaction, there are *two* acids. A complex formation reaction only occurs with a single acid and a single base. A double displacement, or metathesis, reaction only occurs with two acids and two bases.

b) $Me[B_{12}]^- + Hg^{2+} \rightarrow [B_{12}] + MeHg^+$: This is a displacement reaction. The Lewis acid Hg^{2+} displaces the Lewis acid $[B_{12}]$ from the Lewis base CH_3^-.

c) $KCl + SnCl_2 \rightarrow K^+ + [SnCl_3]^-$: This is also a displacement reaction. The Lewis acid $SnCl_2$ displaces the Lewis acid K^+ from the Lewis base Cl^-.

d) $AsF_3(g) + SbF_5(g) \rightarrow [AsF_2][SbF_6]$: Even though this reaction produces an ionic substance, it is *not* simply a complex formation reaction. It is a displacement reaction. The very strong Lewis acid SbF_5 (one of the strongest known) displaces the Lewis acid $[AsF_2]^+$ from the Lewis base F^-.

e) EtOH readily dissolves in pyridine: A Lewis acid–base complex formation reaction between EtOH (the acid) and py (the base) produces the adduct EtOH–py, which is held together by a hydrogen bond.

E5.24 **a) Strongest Lewis acid: BF_3, BCl_3, or BBr_3:** The simple argument that more electronegative substituents lead to a stronger Lewis acid does not work in this case (see Section 5.7(b) *Group 13 Lewis acids*). Boron tribromide is observed to be the strongest Lewis acid of these three compounds. The shorter boron–halogen bond distances in BF_3 and BCl_3 than in BBr_3 are believed to lead to stronger halogen-to-boron p–p π bonding. According to this explanation, the acceptor orbital (empty p orbital) on boron is involved in π bonding in BF_3 and BCl_3 to a greater extent than in BBr_3; the acidities of BF_3 and BCl_3 are diminished relative to BBr_3.

$BeCl_2$ or BCl_3: Boron trichloride is expected to be the stronger Lewis acid because the oxidation number of boron in BCl_3 is +3, whereas for the beryllium atom in $BeCl_2$ it is +2. The second reason is related to their structures. The boron atom in BCl_3 is three-coordinate, leaving a vacant site to which a Lewis base can coordinate. On the other hand, $BeCl_2$ is polymeric, each beryllium atom is four-coordinate, and some Be–Cl bonds must be broken before adduct formation can take place:

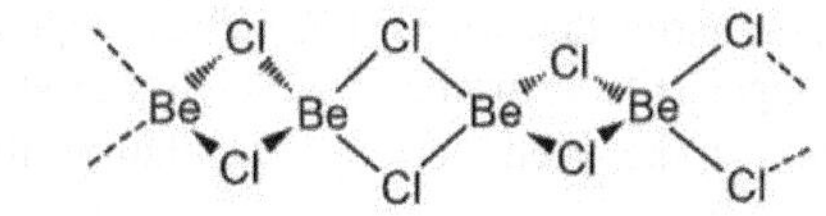

$B(n\text{-}Bu)_3$ or $B(t\text{-}Bu)_3$: The Lewis acid with the unbranched substituents, $B(n\text{-}Bu)_3$, is the stronger of the two because, once the complex is formed, steric repulsions between the substituents and the Lewis base will be less than with the bulky, branched substituents in $B(t\text{-}Bu)_3$.

b) More basic toward BMe_3: NMe_3 or NEt_3: These two bases have nearly equal basicities toward the proton in aqueous solution or in the gas phase. Steric repulsions between the substituents on the bases and the proton are negligible because the proton is very small. However, steric repulsions between the substituents on the bases and *molecular* Lewis acids like BMe_3 are an important factor in complex stability, and the smaller Lewis base NMe_3 is the stronger in this case.

2-Me-py or 4-Me-py: As above, steric factors influence complex formation of two bases that have nearly equal Brønsted basicities. Therefore, 4-Me-py is the stronger base toward BMe_3 because the methyl substituent in this base cannot affect the strength of the B–N bond by steric repulsions with the methyl substituents on the Lewis acid.

E5.25 **a) $R_3P\text{–}BBr_3 + R_3N\text{–}BF_3 \rightleftharpoons R_3P\text{–}BF_3 + R_3N\text{–}BBr_3$:** You know that phosphines are softer bases than amines (see Table 5.4). To determine the position of this equilibrium, you must decide which Lewis acid is softer because the softer acid will preferentially form a complex with a soft base. Boron tribromide is a softer Lewis acid than BF_3, a consequence of the relative hardness and softness of the respective halogen substituents. Therefore, the equilibrium position for this reaction will lie to the left, the side with the soft–soft and hard–hard complexes and the equilibrium constant is less than 1. In general, it is found that soft substituents (or ligands) lead to a softer Lewis acid than for the same central element with harder substituents.

b) $SO_2 + Ph_3P\text{–}HOCMe_3 \rightleftharpoons Ph_3P\text{–}SO_2 + HOCMe_3$: In this reaction, the soft Lewis acid sulfur dioxide displaces the hard acid *t*-butyl alcohol from the soft base triphenylphosphine. The soft–soft complex is favoured, so the equilibrium constant is greater than 1.

c) $CH_3HgI + HCl \rightleftharpoons CH_3HgCl + HI$: Iodide is a softer base than chloride, an example of the general trend that elements later in a group are softer. The soft acid CH_3Hg^+ will form a stronger complex with iodide than with chloride, whereas the hard acid H^+ will prefer chloride, the harder base. Thus, the equilibrium constant is less than 1.

d) $[AgCl_2]^-(aq) + 2CN^-(aq) \rightleftharpoons [Ag(CN)_2]^-(aq) + 2Cl^-(aq)$: Cyanide is a softer and generally stronger base than chloride. Therefore, cyanide will displace the relatively harder base from the soft Lewis acid Ag^+. The equilibrium constant is greater than 1.

E5.26 **a)** The reaction is $CsF + BrF_3 \rightarrow Cs^+ + BrF_4^-$; in this case F^- is a Lewis base whereas BrF_3 is a Lewis acid.

b) The reaction is $ClF_3 + SbF_5 \rightarrow ClF_2^+ + SbF_6^-$; again F^- is a Lewis base, whereas SbF_5 is Lewis acid.

c) The reaction is $B(OH)_3 + H_2O \rightarrow [B(OH)_3(H_2O)] \rightarrow [B(OH)_4]^- + H^+$; the Lewis acid in this reaction is $B(OH)_3$, whereas the Lewis base is H_2O. Note that Lewis acid–base complex $[B(OH)_3(H_2O)]$ behaves as a Brønsted acid in aqueous solution.

d) The reaction is $B_2H_6 + 2PMe_3 \rightarrow 2[BH_3(PMe_3)]$; the base is PMe_3 and the acid is B_2H_6.

E5.27 As you add methyl groups to the nitrogen, the nitrogen becomes more basic due to the inductive effect (electron donating ability) of the methyl groups. So the trend $NH_3 < NH_2(CH_3) < NH(CH_3)_2$ makes sense, but when you put three methyl groups around nitrogen, sterics come into play. Trimethylamine is sterically large enough to fall out of line with the given enthalpies of reaction.

E5.28 **a) Acetone and DMSO:** Since both E_B and C_B are larger for DMSO than for acetone (see Table 5.5), DMSO is the stronger base regardless of how hard or how soft is the Lewis acid. The ambiguity for DMSO is that both the oxygen atom and sulfur atom are potential basic sites.

b) Me_2S and DMSO: Dimethylsulfide has a C_B value that is two-and-a-half times larger than that for DMSO, whereas its E_B value is only one-quarter that for DMSO. Thus, depending on the E_A and C_A values for the Lewis acid, either base could be stronger. For example, DMSO is the stronger base toward BF_3, while SMe_2 is the stronger base toward I_2. This can be predicted by calculating the ΔH of complex formation for all four combinations:

$$\text{DMSO–BF}_3\text{: } \Delta H = -[(20.21)(2.76) + (3.31)(5.83)] = -75.1 \text{ kJ mol}^{-1}$$

$$\text{SMe}_2\text{–BF}_3\text{: } \Delta H = -[(20.21)(0.702) + (3.31)(15.26)] = -64.7 \text{ kJ mol}^{-1}$$

$$\text{DMSO–I}_2\text{: } \Delta H = -[(2.05)(2.76) + (2.05)(5.83)] = -17.6 \text{ kJ mol}^{-1}$$

$$\text{SMe}_2\text{–I}_2\text{: } \Delta H = -[(2.05)(0.702) + (2.05)(15.26)] = -32.7 \text{ kJ mol}^{-1}.$$

E5.29 The metathesis of two hard acids with two hard bases occurs as equilibrium is established between solid, insoluble SiO_2, and soluble H_2SiF_6:

$$SiO_2 + 6HF \rightleftharpoons 2H_2O + H_2SiF_6$$

or

$$SiO_2 + 4HF \rightleftharpoons 2H_2O + SiF_4$$

This is both a Brønsted acid–base reaction and a Lewis acid–base reaction. The Brønsted reaction involves the transfer of protons from HF molecules to O^{2-} ions, whereas the Lewis reaction involves complex formation between Si(IV) centre and F^- ions.

E5.30 The foul odour of damp Al_2S_3 suggests that a volatile compound is formed when this compound comes in contact with water. The only volatile species that could be present, other than odourless water, is H_2S, which has the characteristic odour of rotten eggs. Thus, an equilibrium is established between two hard acids, Al(III) and H^+, and the bases O^{2-} and S^{2-}:

$$Al_2S_3 + 3H_2O \rightleftharpoons Al_2O_3 + 3H_2S$$

E5.31 **a) Favour displacement of Cl^- by I^- from an acid centre:** Since in this case you have no control over the hardness or softness of the acid centre, you must do something else that will favour the acid–iodide complex over the acid-chloride complex. If you choose a solvent that decreases the activity of chloride relative to iodide, you can shift the following equilibrium to the right:

$$\text{acid-Cl}^- + I^- \rightleftharpoons \text{acid-I}^- + Cl^-$$

Such a solvent should interact more strongly with chloride (i.e., form an adduct with chloride) than with iodide. Thus, the ideal solvent properties in this case would be *weak*, *hard*, and *acidic*. It is important that the solvent be a weak acid because otherwise the activity of both halides would be rendered negligible. An example of a suitable solvent is anhydrous HF. Another suitable solvent is H_2O.

b) Favour basicity of R_3As over R_3N: In this case you wish to enhance the basicity of the soft base trialkylarsine relative to the hard base trialkylamine. You can decrease the activity of the amine if the solvent is a hard acid because the solvent–amine complex would then be less prone to dissociate than the solvent–arsine complex. Alcohols such as methanol or ethanol would be suitable.

c) Favour acidity of Ag^+ over Al^{3+}: If you review the answers to parts (a) and (b) of this exercise, a pattern will emerge. In both cases, a hard acid solvent was required to favour the reactivity of a soft base. In this part of the exercise, you want to favour the acidity of a soft acid, so logically a solvent that is a hard base is suitable. Such a solvent will "tie up" (i.e., decrease the activity of) the hard acid Al^{3+} relative to the soft acid Ag^+. An example of a suitable solvent is diethyl ether. Another suitable solvent is H_2O.

d) Promote the reaction $2FeCl_3 + ZnCl_2 \rightarrow Zn^{2+} + 2[FeCl_4]^-$: Since Zn^{2+} is a softer acid than Fe^{3+}, a solvent that promotes this reaction will be a softer base than Cl^-. The solvent will then displace Cl^- from the Lewis acid Zn^{2+}, forming $[Zn(solv)_x]^{2+}$. The solvent must also have an appreciable dielectric constant because ionic species are formed in this reaction. A suitable solvent is acetonitrile, MeCN.

E5.32 The mechanism described in Section 5.7(b) *Group 13 Lewis acids* involves the abstraction of Cl^- from CH_3COCl by $AlCl_3$ to form $AlCl_4^-$ and the Lewis acid CH_3CO^+ (an acylium cation). This cation then attacks benzene to form the acylated aromatic product. An alumina surface, such as the partially dehydroxylated one shown below, would also provide Lewis acidic sites that could abstract Cl^-:

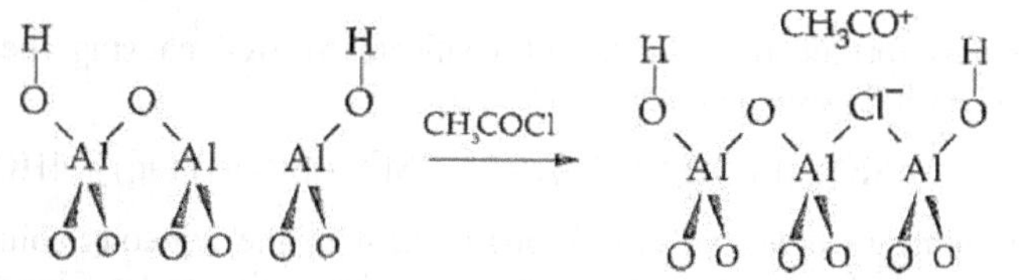

E5.33 Mercury(II), Hg^{2+}, is a soft Lewis acid, and so is found in nature only combined with soft Lewis bases, the most common of which is S^{2-}. Sulfide can readily and permanently abstract Hg^{2+} from its complexes with harder bases in ore-forming geological reaction mixtures. Zinc(II), which exhibits borderline behaviour, is harder and forms stable compounds (i.e., complexes) with hard bases such as O^{2-}, CO_3^{2-}, and silicates, as well as with S^{2-}. The ore that is formed with Zn^{2+} depends on factors including the relative concentrations of the competing bases.

E5.34 **a) CH_3CH_2OH in liquid HF:** Ethanol is a weaker acid than H_2O but a stronger base than H_2O, due to the electron donating property of the $-C_2H_5$ group. The balanced equation is:

$$CH_3CH_2OH + HF \rightleftharpoons CH_3CH_2OH_2^+ + F^-.$$

b) NH_3 in liquid HF: The equation is

$$NH_3 + HF \rightleftharpoons NH_4^+ + F^-.$$

c) C_6H_5COOH in liquid HF: In this case, benzoic acid is a significantly stronger acid than water. Therefore, it will protonate hydrogen fluoride:

$$C_6H_5COOH + HF \rightleftharpoons C_6H_5COO^- + H_2F^+.$$

E5.35 The dissolution of silicates in HF is both a Brønsted acid–base reaction and a Lewis acid–base reaction. The reaction involves proton transfers from HF to the silicate oxygen atoms (a Brønsted reaction), and the formation of complexes such as SiF_6^{2-} from the Lewis acid Si^{4+} and the Lewis base F^- (a Lewis reaction).

E5.36 Since the trivalent lanthanides and actinides (f-block elements) are found as complexes with hard oxygen bases (i.e., silicates) and not with soft bases such as sulfide, they must be hard. Since they are found *exclusively* as silicates, they must be considered very hard, unlike the borderline behaviour of Zn(II) (see Exercise 5.33 as well).

E5.37 From Table 5.5 we find $E = 2.05$ and $C = 2.05$ for I_2 and $E = 8.86$ and $C = 0.09$ for phenol. The Drago-Wayland equation gives:

$$\Delta_f H = -(E_A E_B + C_A C_B) = -20.0\,\text{kJ mol}^{-1}$$

Indicating an exothermic reaction for the formation of an I_2–phenol adduct.

E5.38 As the number of methyl groups bonded to nitrogen atom increases, the basicity of amines increases because methyl groups are good electron donors. Each additional methyl group pushes more electron density onto the nitrogen atom making its lone electron pair more basic. The steric influence of $-CH_3$ groups is relatively small, particularly when considering attachment of the small proton.

When looking at the basic properties of these amines in aqueous solution we have to consider solvation effects as well because they play a crucial role in stability of these cations in water. The Born equation (equation 5.9) shows that solvation energy is inversely proportional to the radius of the cations. This means that the lowest solvation energy is going to be observed for the largest cation of the group, that is $(CH_3)_3NH^+$, and the highest value for the ammonium cation (the smallest of the four).

E5.39 Since $Si(OH)_4$ is a weaker acid than H_2CO_3, H_2CO_3 can replace $Si(OH)_4$ from its salts; that is, it can protonate the conjugate base of $Si(OH)_4$. The following two equilibria in the system $CO_2(g)/CO_2(aq)$ are important for this exercise:

$$CO_2(g) \rightleftharpoons CO_2(aq) \qquad (1)$$

$$CO_2(aq) + H_2O(l) \rightleftharpoons H_2CO_3(aq) \qquad (2)$$

The $H_2CO_3(aq)$ that is formed reacts with solid silicate M_2SiO_4 causing the silicate to dissolve according to equation (3) (one of several possible reactions):

$$M_2SiO_4(s) + 4H_2CO_3(aq) \rightleftharpoons 2M^{2+} + Si(OH)_4(aq) + 4HCO_3^-(aq) \qquad (3)$$

Equilibrium (3) is going to consume $H_2CO_3(aq)$ present in the aqueous solution. This decrease in $H_2CO_3(aq)$ concentration will perturb the equilibrium given by (2). According to Le Chatelier's principle, reaction (2) has to shift to the right—more $CO_2(aq)$ has to react with water to produce H_2CO_3 in order to re-establish the equilibrium conditions. Since reaction (2) has been shifted to the right, so must reaction (1) be shifted to the right as well—the consumption of $CO_2(aq)$ in (2) disturbs the equilibrium (1) and more $CO_2(g)$ has to dissolve in water to produce $CO_2(aq)$. Overall, the more silicate M_2SiO_4 reacts with $H_2CO_3(aq)$, the less $CO_2(g)$ will be present.

E5.40 **a)** The chemical equation is:

$$Fe^{3+}(aq) + 6H_2O(l) \rightleftharpoons Fe(OH)_3(s) + 3H_3O^+ \qquad (1)$$

Where $Fe^{3+}(aq)$ stands for $Fe(H_2O)_6^{3+}$.

b) The molecular weight of $Fe(NO)_3 \cdot 9H_2O$ is 404 g mol^{-1}. We have started with 6.6 kg of this iron salt or 6.6×10^3 g/(404 g mol^{-1}) = 16.3 mol, and when dissolved in 100 dm^3 of water the concentration of Fe^{3+}, prior to precipitation, is (16.3 mol)/(100 L) = 0.163 mol L^{-1}. The equation given, $[Fe^{3+}]/[H_3O^+]^3 = 10^4$ is an equilibrium constant for the inverse of reaction (1), and $[H_3O^+]^3/[Fe^{3+}] = 10^{-4}$ is the equilibrium constant for (1) as written above. If we start from x mol of Fe^{3+} we should get 3x mol of H_3O^+. So, we can set up the equation with equilibrium concentrations as:

$$(3x)^3/(0.163 - x) = 10^{-4}$$

$$27x^3 + 10^{-4}x - 0.163 \times 10^{-4} = 0$$

The only real root of the above equation is 0.0083. This gives $[H_3O^+] = 3x = 3 \times 0.0083$ M = 0.0249 M and pH of $-\log(0.0249) = 1.6$. The equilibrium concentration of $[Fe^{3+}]$ is 0.163 – x = 0.163 – 0.0083 = 0.1547 M. The Fe^{3+}-containing species that have been neglected in this calculation are $Fe(OH)^{2+}$ (or $[Fe(H_2O)_5(OH)]^{2+}$) and $Fe(OH)_2^+$ (or $[Fe(H_2O)_5(OH)]_2^+$).

E5.41 The trend in symmetrical M–O stretching vibration frequencies relates very well with the acidities of these aqua acids. The frequencies increase in the order $Ca^{2+} < Mn^{2+} < Zn^{2+}$, the same order that the M–O bond strength increases. This is a reasonable observation because in the same order the radii of M^{2+} cations are decreasing and charge density is increasing. With increasing charge density, the attraction between M^{2+} cation and partially negative O atom in H_2O is increasing. For the same reason, the acidity of these aqua acids is increasing showing that two trends correlate closely.

E5.42 The solvent acetonitrile, CH_3CN, is a Lewis base because it has a lone electron pair on the nitrogen atom. On the other hand, $AlCl_3$ is a typical Lewis acid of a Group 13 element. This means that the two can react in a Lewis acid–base reaction forming adducts, for example $AlCl_3(NCCH_3)_3$. Considering it is also a polar solvent, CH_3CN can stabilize ions, suggesting that partial ionization of solvated $AlCl_3$ can take place producing the ions $[AlCl_2(NCCH_3)_4]^+$ and Cl^-(sol). It is further possible that solvated Cl^- behaves like a Lewis base toward undissociated $AlCl_3(NCCH_3)_3$, replacing CH_3CN to produce the anion $[AlCl_4(NCCH_3)_2]^-$. These ions are responsible for solution conductivity.

E5.43 Equation (b) is better in explaining the observations described in the exercise. Most notably it contains the Lewis acid–base adduct $[FeCl_2(OPCl_3)_4]^+$ in which $OPCl_3$ is a Lewis base and coordinates via its O atom, consistent with vibrational data. It also has the $Fe_2Cl_6/[FeCl_4]^-$ (i.e., red/yellow) equilibrium. The titration would have to start in either case from a concentrated (red) solution, and the equivalence point at 1 : 1 mole ratio $FeCl_3/Et_4NCl$ can be explained based on the reaction:

$$Fe_2Cl_6 + 2Et_4NCl \rightleftharpoons 2[FeCl_4]^- + 2Et_4N^+.$$

E5.44 The elements mentioned in the exercise are marked on the periodic table shown below: light shaded squares are those that are considered amphoteric in the S^{2-}/HS^- system; dark shaded squares correspond to the elements that are considered less acidic. The borderline between the two is not as clear as is the case with oxides (Figure 5.5). It is possible, however, to note that the boundary has been shifted to the left and toward the bottom of the periodic table. It is in this region that we find soft Lewis acids (large cations with relatively low charge). Thus, this borderline does agree with describing S^{2-} as a softer base than O^{2-}.

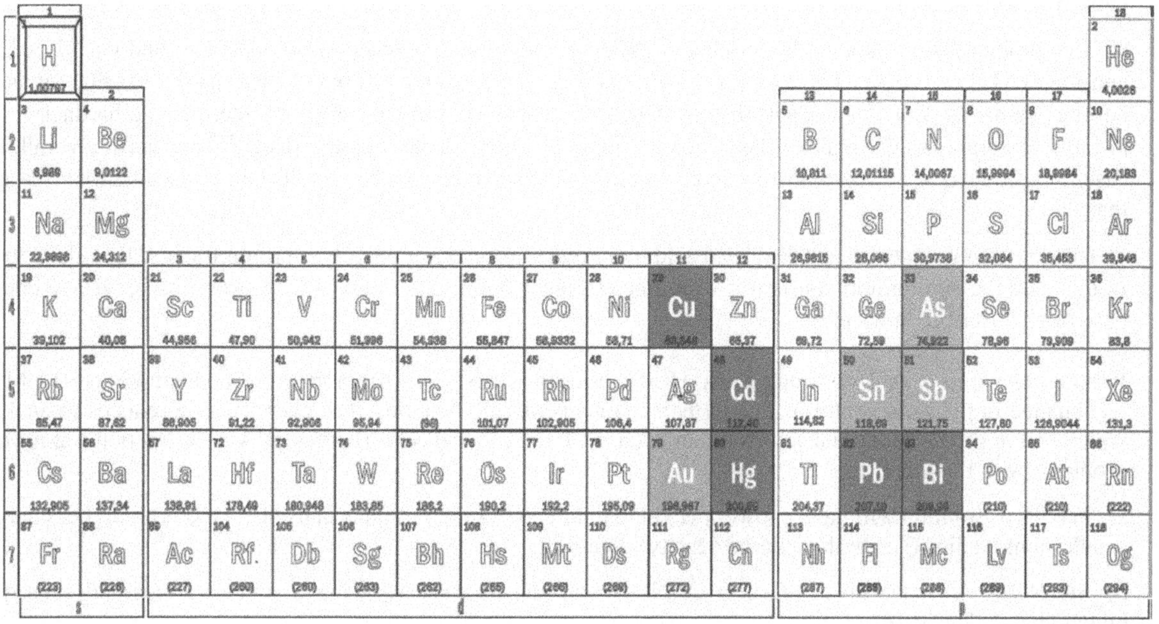

E4.45 The first step in the reaction catalysed by Cl^- ions is formation of $*SO_2Cl^-$ (where *S is the radioactively labelled sulfur atom):

$$*SO_2 + Cl^- \rightleftharpoons *SO_2Cl^-$$

In this reaction SO_2 is a Lewis acid (see Section 5.7(e) *Group 16 Lewis acids*), whereas the catalyst, Cl^-, is a Lewis base. The next step involves reaction between $*SO_2Cl^-$ and $SOCl_2$ in which oxygen and chlorine atoms are exchanged:

$$*SO_2Cl^- + SOCl_2 \rightleftharpoons *SOCl_2 + SO_2Cl^-$$

Finally, newly formed, unlabelled SO_2Cl^- dissociates to release the catalyst and SO_2:

$$SO_2Cl^- \rightleftharpoons SO_2 + Cl^-$$

The first step in the reaction catalysed by $SbCl_5$ is formation of $*SOCl_2{\cdot}SbCl_5$ adduct:

$$*SOCl_2 + SbCl_5 \rightleftharpoons *SOCl_2{\cdot}SbCl_5$$

In this reaction $SbCl_5$ behaves as a Lewis acid (see Section 5.7(d) *Group 15 Lewis acids*), whereas $*SOCl_2$ is a Lewis base. The adduct reacts with SO_2 and the chlorides and oxygen atoms are exchanged:

$$*SOCl_2{\cdot}SbCl_5 \rightleftharpoons SOCl_2{\cdot}SbCl_5 + *SO_2$$

Finally, $SOCl_2{\cdot}SbCl_5$ dissociates to liberate the catalyst:

$$SOCl_2{\cdot}SbCl_5 \rightleftharpoons SOCl_2 + SbCl_5$$

E5.46 The reasons for the difference in stability are steric. Sulfur has the same oxidation state in SO_3 and SF_6, but the differences in molecular geometry are significant. SO_3 has a trigonal planar structure with sulfur atom at the centre. This structure makes the S atom easily accessible for bonding with pyridine. On the other hand, SF_6 has octahedral geometry with sulfur atom "hidden" inside, making it inaccessible for the lone electron pair on pyridine's nitrogen atom. SF_4 has a see-saw molecular geometry in which S atom is rather accessible for bonding with pyridine.

E5.47 **a)** The equilibrium constant for this reaction should be less than 1. This is because the reactants are more stable than the products according to the HSAB theory. According to this theory, hard acids like to bind to hard bases and soft acids like to bind to soft bases (see Section 5.9(a) and Table 5.4). Cd^{2+} is a soft Lewis acid, whereas Ca^{2+} is a hard Lewis acid. That means Cd^{2+} is going to prefer the soft base I^- over the hard base, F^-. The opposite holds for Ca^{2+}—it is going to preferably bind to F^- over I^-. Thus, the reactants are more stable than the products, and the equilibrium constant is lower than 1.

b) This problem is also solved by looking at the HSAB theory. The reaction involves Cu^+ and Cu^{2+} Lewis acids and I^- and Cl^- Lewis bases. Cu^+ as a soft Lewis acid is going to prefer soft base I^- over harder base Cl^-, thus it will be more stable as $[CuI_4]^{3-}$ (product) than as $[CuCl_4]^{3-}$ (reactant). Similarly, Cu^{2+} of intermediate hardness is going to bond preferably to Cl^- making the $[CuCl_4]^{2-}$ (product) more stable than the $[CuI_4]^{2-}$ (reactant). Overall, in this case, the products are expected to be more stable than reactants, and the equilibrium constant should be greater than 1.

c) This reaction is analysed as a Brønsted acid–base reaction because it involves proton transfer from H_2O to NH_2^-. In this case the equilibrium constant is also greater than 1 because we have a very strong base, NH_2^- a conjugate base of NH_3 as acid, in water.

E5.48 In all cases we have salts of strong acids, $HClO_4$ in (a) and HNO_3 in (b), in identical contentrations (0.1 M), thus any change in pH from neutral is due to the aqua acids of metal cations. The exercise then reduces to analysis of which cation gives the more acidic solution. Consult Example and Self-Test 5.4 as well as Section 5.2 if you had problems with this exercise.

a) $Fe(ClO_4)_3$ solution will have a lower pH value than $Fe(ClO_4)_2$ solution because Fe^{3+} has both higher charge and smaller ionic radius (i.e., higher charge density) than Fe^{2+}.

b) $Mg(NO_3)_2$ will have lower pH than $Ca(NO_3)_2$ solution because Mg^{2+} has a smaller radius and consequently higher charge density than Ca^{2+}.

c) $Zn(NO_3)_2$ will have lower pH than $Hg(NO_3)_2$ solution because Zn^{2+} has a smaller radius and consequently higher charge density than Hg^{2+}.

E5.49 In general, cations such as I_2^+ and Se_8^{2+} should behave like good Lewis acids. Therefore, these cations should react (even) with a weakly basic solvent and will be lost. Hence, they can survive and be studied *only* in strongly acidic solvents that do not react with them. The same reasoning applies to anions such as S_4^{2-} and Pb_9^{4-}, which are good Lewis bases and can be stabilized by strongly basic solvents because they would react with acidic ones.

E5.50 Sharp titration end points are obtained when a strong acid is titrated with a strong base, and vice versa. Regardless which strong acid and strong base are used, the overall titration reaction is the same:

$$H^+(aq) + OH^-(aq) \rightleftharpoons H_2O(l)$$

That is, the strongest acid in water is neutralized with the strongest base that can exist in water. In any other titration case, the end point is not easily detected. In the case of titration of a weak base with a strong acid we have to find a solvent other than water in which a weak base is going to behave as a strong one and give us a sharp and easy to detect end point. To achieve this, we have to find solvent more acidic than water, which can be acetic acid.

E5.51 **a)** Both reactions have the same Lewis base, NMe_3, and the difference in thermodynamic properties between two reactions must lie with the Lewis acids. Note that the steric properties of $B(OMe)_3$ and BEt_3 are about the same. However, oxygen atom has two lone electron pairs. These pairs can donate electrons to the empty p orbital on central B atom reducing the Lewis acidity of this compound through p-p π interaction (similar effect is observed with halogens, see Exercise 5.22). This interaction lowers the Lewis acidity of $B(OMe)_3$ which is reflected in the thermodynamic values.

b) Situation is now inversed on comparison to part a). $N(SiH_3)_3$ is weaker Lewis base than NMe_3 because the lone pair on N atom can be delocalized on Si atoms. This delocalization makes lone electron pair less available for the reaction with Lewis acids (see also Self-Test 5.8).

E5.52 For CO_2:

$$CO_2(g) \rightleftharpoons CO_2(aq)$$

$$CO_2(aq) + H_2O(l) \rightleftharpoons H_2CO_3(aq)$$

$$H_2CO_3(aq) + H_2O(l) \rightleftharpoons HCO_3^-(aq) + H_3O^+(aq)$$

For $B(OH)_3$:

$$B(OH)_3(s) + H_2O(l) \rightleftharpoons [B(OH)_3(H_2O)](aq)$$

$$[B(OH)_3(H_2O)](aq) + H_2O(l) \rightleftharpoons [B(OH)_4]^-(aq) + H_3O^+(aq)$$

Guidelines for Selected Tutorial Problems

T5.5 Start from Section 5.16 *Superacids and superbases* which covers "classical" examples of superbases, nitrides, and hydrides of s-block elements. However, there are several groups of superbases that you can look at. For example, amides (diisopropylamide related) are commonly used superbases. More modern are phosphazene bases which are neutral rather bulky compounds containing phosphorus and nitrogen. An interesting topic is a "proton sponge" (or Alder's base) and its derivatives as well as several theoretical and practical approaches used in rational design of superbases.

Chapter 6 Oxidation and Reduction

Self-Tests

S6.1 Following the procedure outlined in Example 6.1, we find the following two half-reactions:

$MnO_4^-(aq) + 8H^+(aq) + 5e^- \rightarrow Mn^{2+}(aq) + 4H_2O(l)$ reduction half-reaction

$Zn(s) \rightarrow Zn^{2+}(aq) + 2e^-$ oxidation half-reaction

The reduction half-reaction requires five electrons whereas the oxidation half-reaction requires two electrons; therefore, we must balance electrons so that an equal number of electrons are required in both reactions. The reduction reaction must be multiplied by 2 and the oxidation reaction by 5 to give 10 electrons in each case:

$2MnO_4^-(aq) + 16H^+(aq) + 10e^- \rightarrow 2Mn^{2+}(aq) + 8H_2O(l)$ reduction half-reaction

$5Zn(s) \rightarrow 5Zn^{2+}(aq) + 10e^-$ oxidation half-reaction

Summing two half-reactions and cancelling $10e^-$ found on both the left and right sides of the equation, we obtain the full reaction:

$$2MnO_4^-(aq) + 5Zn(s) + 16H^+(aq) \rightarrow 5Zn^{2+}(aq) + 2Mn^{2+}(aq) + 8H_2O(l).$$

S6.2 No, copper metal is not expected to react with dilute HCl because the $E°$ for the reaction $Cu^{2+}(aq) + 2e^- \rightarrow Cu(s)$, $E°(Cu^{2+}, Cu) = +0.34$ V is positive. The oxidation of Cu metal in HCl, $Cu(s) + 2H^+(aq) \rightarrow Cu^{2+}(aq) + H_2(g)$, is not favoured thermodynamically because the cell potential for the reaction is negative (–0.34 V), resulting in a positive ΔG.

S6.3 The standard reduction potential for the $Cr_2O_7^{2-}/Cr^{3+}$ couple is +1.38 V, so it can oxidize any couple whose reduction potential is less than +1.38 V. Since the reduction potential for the Fe^{3+}/Fe^{2+} couple (+0.77 V) is less positive than $Cr_2O_7^{2-}/Cr^{3+}$ reduction potential, Fe^{2+} will be oxidized to Fe^{3+} by dichromate. The reduction potential for the Cl_2/Cl^- couple (+1.36 V) is slightly less positive than that of the $Cr_2O_7^{2-}/Cr^{3+}$ couple, so a side reaction with Cl^- should be expected.

S6.4 The potential difference of the cell from Example 6.45 was calculated to be +1.23 V. To calculate the potential difference under non-standard conditions we start with the Nernst equation given by the formula:

$$E_{cell} = E° - \frac{RT}{\nu_e F}\ln Q\text{, where } Q \text{ is the reaction quotient.}$$

For the fuel-cell reaction, $O_2(g) + 2H_2(g) \rightarrow 2H_2O(1)$, the reaction quotient is:

$$Q = \frac{1}{p(O_2)p(H_2)}.$$

Recalling that ν is the number of electrons exchanged, we have:

$$E_{cell} = E° - [(0.059\text{ V})/4][\log(1/(5.0)^2)]$$

$$= +1.23\text{ V} + 0.02\text{ V} = +1.25\text{ V}.$$

Therefore, the new potential difference for the fuel cell is +1.25 V.

S6.5 The complete coupled reaction for oxidation of $SO_2(aq)$ by atmospheric oxygen is given below:

$$2SO_2(aq) + O_2(g) + 2H_2O(l) \rightarrow 2SO_4^{2-}(aq) + 4H^+(aq)$$

Since the standard reduction potential for the SO_4^{2-}/SO_2 couple is +0.16 V and the standard reduction potential for the O_2/H_2O couple is +1.23 V then the potential difference for the equation is $E° = +1.23\text{ V} - 0.16\text{ V} = +1.07$ V.

Since this potential is large and positive, the Gibbs energy of this reaction is negative (reaction is hence spontaneous) and will be driven nearly to completion ($K > 1$). Thus, the expected thermodynamic fate of SO_2 is its conversion to sulfate (or neutral sulfuric acid vapour). This aqueous solution of SO_4^{2-} and H^+ ions precipitates as acid rain, which can have a pH as low as 2 (the pH of rain water that is not contaminated with sulfuric or nitric acid is ~5.6).

S6.6 The disproportionation of Fe^{2+} involves the reduction of one equivalent of Fe^{2+} to Fe^0, a net gain of two equivalents of electrons, and the concomitant oxidation of two equivalents of Fe^{2+} to Fe^{3+}, a net loss of two equivalents of electrons:

$$3\,Fe^{2+}(aq) \rightarrow Fe^0(s) + 2Fe^{3+}(aq)$$

The value of E_{cell} for this reaction can be calculated by subtracting E^o for the Fe^{2+}/Fe couple (–0.44 V) from E^o for the Fe^{3+}/Fe^{2+} couple (+0.77 V),

$$E = -0.44 - 0.77\text{ V} = -1.21\text{ V}.$$

This potential is large *and negative*, resulting in a positive Δ_rG (non-spontaneous reaction), so the disproportionation will *not* occur.

S6.7 Compared to the aqua redox couple of +0.25 V, the given value is higher by +1.01 V and we conclude that bpy binds preferentially to Ru(II).

Using Equation 5.12b from your textbook, we obtain

$$E°(M) - E°(ML) = (0.059V) \times \log(K_{ox}/K_{red})$$

$$E°(Ru^{3+}(aq)) - E°([Ru(bpy)_3]^{3+}) = +0.25\text{ V} - (+1.26\text{ V}) = -1.01\text{ V}$$

$$-1.01\text{ V} = (0.059\text{ V}) \times \log(K_{ox}/K_{red})$$

$$-\frac{1.01\text{ V}}{0.059\text{ V}} = -17.11 \log\frac{K_{ox}}{K_{red}} \Rightarrow 17.11 \log\frac{K_{red}}{K_{ox}}$$

Thus, the binding of three bpy ligands to Ru(III) is about 17 orders of magnitude decreased relative to binding to Ru(II).

S6.8 We need to consider a thermodynamic cycle that links the solubility product under the experimental condition of Cl^- concentration (here simplified, since $[Cl^-] = 1.0\text{ mol dm}^{-3}$). For the thermodynamic cycle $\Delta G = 0$, so we obtain $E_{AgCl/Ag} = RT \ln K_{sp} + E_{Ag+/Ag} = -0.577\text{ V} + 0.80\text{ V} = +0.223\text{ V}$.

S6.9 **(a)** Pu(IV) disproportionates to Pu(III) and Pu(V) in aqueous solution because Pu(IV) is a stronger oxidant in acidic solution compared to Pu(V).

(b) Pu(V) does not disproportionate into Pu(VI) and Pu(IV).

S6.10 See the following figure. This plot was made using the potentials given, $NE° = 0$ V for Tl^0 ($N = 0$), $NE° = -0.34$ V for Tl^+ ($N = 1$), and $NE° = 2.16$ V for Tl^{3+} ($N = 3$). Note that Tl^+ is stable with respect to disproportionation in aqueous acid. Note also that Tl^{3+} is a strong oxidant (i.e., it is very readily reduced) because the slope of the line connecting it with either lower oxidation state is large and positive.

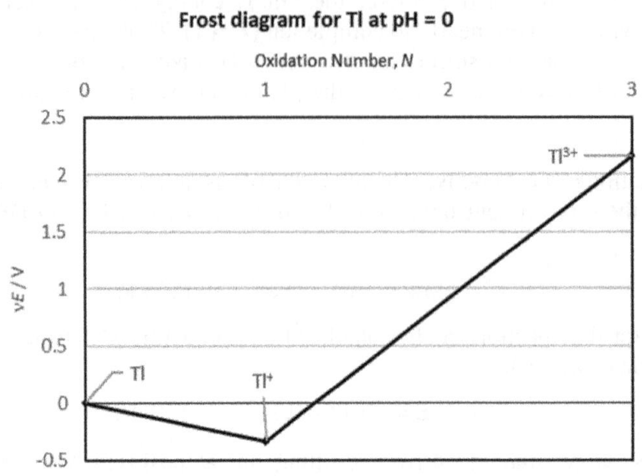

S6.11 When permanganate, MnO_4^-, is used as an oxidant in aqueous solution, the manganese species that will remain is the most stable manganese species under the conditions of the reaction (i.e., acidic or basic). Inspection of the Frost diagram for manganese, shown in Figure 6.10, shows that Mn^{2+}(aq) is the most stable species present because it has the most negative $\Delta_f G$. Therefore, Mn^{2+}(aq) will be the final product of the redox reaction when MnO_4^- is used as an oxidizing agent in aqueous acid.

S6.12 For kinetic reasons, the reduction of nitrate ion usually proceeds to NO, instead of to N_2O or all the way to N_2. That means that our redox couple is NO_3^-/NO and we have to compare the slope of the line connecting the NO_3^- and NO points for both acidic and basic solution. If you compare the Frost diagram for nitrogen in acidic and in basic solutions, you see that the slope for the NO_3^-/NO couple in acidic solution is positive while the slope for basic solution is negative. Therefore, nitrate is a stronger oxidizing agent (i.e., it is more readily reduced) in acidic solution than in basic solution. Even if the reduction of NO_3^- proceeded all the way to N_2, the slope of that line is still less than the slope of the line for the NO_3^-/NO couple in acid solution.

S6.13 According to Figure 6.12, a typical waterlogged soil (rich in organic material but oxygen depleted) has an average pH of about 4 and an average potential of about –0.1 V. If you find this point on the Pourbaix diagram for naturally occurring iron species, shown on Figure 6.11, you see that $Fe(OH)_3$ is not stable and Fe^{2+}(aq) will be the predominant species. In fact, as long as the potential remains at –0.1 V, Fe^{2+} is the predominant species below pH = 8. Above pH = 8, Fe^{2+} is oxidized to Fe_2O_3 at this potential.

S6.14 According to the Ellingham diagram shown in Figure 6.16, at about 2000°C the line (a) for the reducing agent (C/CO) dips below the MgO line, which means that at that temperature the reactions $2Mg(l) + O_2(g) \rightarrow 2MgO(s)$ and $2C(s) + O_2(g) \rightarrow 2CO(g)$ have the same free-energy change. Thus, coupling the two reactions (i.e., subtracting the first from the second) yields the overall reaction $MgO(s) + C(s) \rightarrow Mg(l) + CO(g)$ with $\Delta G = 0$. At 2000°C or above, MgO can be conveniently reduced to Mg by carbon.

Exercises

E6.1 Oxidation numbers for each element in each species are given below the element's symbol:

$2NO(g)$ +		$O_2(g)$	$\rightarrow$	$2NO_2(g)$					
(+2) (–2)		(0)		(+4) (–2)					
$2Mn^{3+}(aq)$	+	$2H_2O$	$\rightarrow$	MnO_2	+	Mn^{2+}	+	$4H^+(aq)$	
(+3)		(+1) (–2)		(+4) (–2)		(+2)		(+1)	
$LiCoO_2(s)$	+	$C(s)$	$\rightarrow$	$LiC(s)$	+	$CoO_2(s)$			
(+1) (+3) (–2)		(0)		(+1)(–1)		(+4) (–2)			
$Ca(s)$	+	$H_2(g)$	$\rightarrow$	$CaH_2(s)$					
(0)		(0)		(+2) (–1)					

E6.2 **a) Oxidation of HCl to Cl_2:** Assume that these transformations are occurring in acidic solution (pH 0). Referring to Resource Section 3, you will find—from the Group 17 Latimer diagrams—that the reduction potential for the Cl_2/Cl^- couple is +1.358 V. To oxidize Cl^- to Cl_2, you need a couple with a reduction potential more positive than 1.358 V, because then the net potential will be positive and ΔG will be negative. Examples include the $S_2O_8^{2-}/SO_4^{2-}$ couple (E° = +1.96 V), the H_2O_2/H_2O couple (E° = +1.763 V), the α–PbO_2/Pb^{2+} couple (E° = +1.468 V), and MnO_4^-/Mn^{2+} couple (E° = +1.51 V). Therefore, since the *oxidized* form of the couple will get *reduced* by *oxidizing* Cl^-, you would want to use either $S_2O_8^{2-}$, H_2O_2, α–PbO_2 or MnO_4^- to oxidize Cl^- to Cl_2.

b) Reducing Cr^{3+}(aq) to Cr^{2+}(aq): In this case, the reduction potential, found from the Latimer diagrams for the Group 6 elements, is –0.424 V. Therefore, to have a net $E^\circ > 0$, you need a couple with a reduction potential more negative than –0.424 V. Examples include Mn^{2+}/Mn (E° = –1.18 V), Zn^{2+}/Zn (E° = –0.7626 V), and NH_3OH^+/N_2 (E° = –1.87 V). Remember that it is the reduced form of the couple that you want to use to reduce Cr^{3+} to Cr^{2+}, so in this case you would choose metallic manganese, metallic zinc or NH_3OH^+.

c) Reducing Ag^+(aq) to Ag(s): The reduction potential is +0.799 V. As above, the reduced form of any couple with a reduction potential less positive than +0.799 V will reduce Ag^+ to silver metal. One possible example comes from the same group: Cu^{2+}/Cu (+0.340 V).

d) Reducing I_2 to I^-: The reduction potential is +0.535 V. As above, the reduced form of any couple with a reduction potential less than +0.535 V will reduce iodine to iodide. One example is S/H_2S with reduction potential +0.144 V.

E6.3 For all species given in the exercise, you must determine whether they can be oxidized by O_2. The standard potential for the reduction $O_2 + 4H^+ + 4e^- \rightarrow 2H_2O$ is +1.23 V. Therefore, only redox couples with a reduction potential less positive than 1.23 V will be driven to completion to the oxidized member of the couple by the reduction of O_2 to H_2O.

a) Cr^{2+}: Since the Cr^{3+}/Cr^{2+} couple has E° = –0.424 V, Cr^{2+} *will* be oxidized to Cr^{3+} by O_2. The balanced equation is:

$$4Cr^{2+}(aq) + O_2(g) + 4H^+(aq) \rightarrow 4Cr^{3+}(aq) + 2H_2O(l) \qquad E^\circ = +1.65\ V$$

b) Fe^{2+}: Since the Fe^{3+}/Fe^{2+} couple has E° = +0.771 V, Fe^{2+} *will* be oxidized to Fe^{3+} by O_2. The balanced equation is:

$$4Fe^{2+}(aq) + O_2(g) + 4H^+(aq) \rightarrow 4Fe^{3+}(aq) + 2H_2O(l) \qquad E^\circ = +0.46\ V$$

c) Cl^-: Both of the following couples have E° values, shown in parentheses, more positive than +1.23 V, so there will be no reaction when acidic Cl^- solution is aerated: ClO_4^-/Cl^- (+1.387 V); Cl_2/Cl^- (+1.358 V). Therefore, Cl^- is stable under these conditions and there is no reaction.

d) HOCl: Since the $HClO_2/HClO$ couple has E° = +1.701 V, the cell potential for oxidation of HOCl by O_2 is –0.47 V (negative, unfavourable). Thus, HClO will not be oxidized to $HClO_2$ by O_2 and there is no reaction.

e) Zn(s): Since the Zn^{2+}/Zn couple has E° = –0.763 V, metallic zinc *will* be oxidized to Zn^{2+} by O_2. The balanced equation is:

$$2Zn(s) + O_2(g) + 4H^+(aq) \rightarrow 2Zn^{2+}(aq) + 2H_2O(l) \qquad E^\circ = +1.99\ V$$

In this case, there is a competing reaction $Zn(s) + 2H^+(aq) \rightarrow Zn^{2+}(aq) + H_2(g)$ (E° = +0.763 V).

E6.4 When a chemical species is dissolved in aerated acidic aqueous solution, you need to think about four possible redox reactions. These are: (i) the species might oxidize water to O_2, (ii) the species might reduce water (hydronium ions) to H_2, (iii) the species might be oxidized by O_2, and (iv) the species might undergo disproportionation.

a) Fe^{2+}: In the case of Fe^{2+}, the two couples of interest are Fe^{3+}/Fe^{2+} (E° = +0.77 V) and Fe^{2+}/Fe (E° = –0.44 V). Consider the four possible reactions (refer to Resource Section 3 for the potentials you need): (i) The O_2/H_2O couple has E° = +1.23 V, so the oxidation of water would only occur if the Fe^{2+}/Fe potential were *greater* than +1.23 V. In other words, the net reaction below is not spontaneous because the net E° is less than zero:

$$
\begin{array}{ll}
2Fe^{2+} + 4e^- \rightarrow 2Fe & E^\circ = -0.44\text{ V} \\
2H_2O \rightarrow O_2 + 4H^+ + 4e^- & E^\circ = -1.23\text{ V} \\
\hline
2Fe^{2+} + 2H_2O \rightarrow 2Fe + O_2 + 4H^+ & E^\circ = -1.67\text{ V}
\end{array}
$$

Therefore, Fe^{2+} will not oxidize water. (ii) The H_2O/H_2 couple has $E^\circ = 0$ V at pH = 0 (by definition), so the reduction of water would only occur if the potential for the oxidation of Fe^{2+} to Fe^{3+} was positive, and it is –0.77 V (note that it is the *reduction* potential for the Fe^{3+}/Fe^{2+} couple that is *positive* 0.77 V). Therefore, Fe^{2+} will not reduce water. (iii) Since the O_2/H_2O has $E^\circ = +1.23$ V, the reduction of O_2 would occur as long as the potential for the oxidation of Fe^{2+} to Fe^{3+} was less negative than –1.23 V. Since it is only –0.77 V (see above), Fe^{2+} will reduce O_2 and in doing so will be oxidized to Fe^{3+}. (iv) The disproportionation of a chemical species will occur if it can act as its own oxidizing agent and reducing agent, which will occur when the potential for reduction minus the potential for oxidation is *positive*. For the Latimer diagram for iron in acidic solution (see Resource Section 3), the difference (–0.44 V) – (+0.771 V) = –1.21 V, so disproportionation will not occur.

b) Ru^{2+}: Consider the Ru^{3+}/Ru^{2+} ($E^\circ = +0.25$ V) and Ru^{2+}/Ru ($E^\circ = +0.80$ V) couples. Since the Ru^{2+}/Ru couple has a potential that is less positive than +1.23 V, Ru^{2+} will not oxidize water. Also, since the Ru^{3+}/Ru^{2+} couple has a positive potential, Ru^{2+} will not reduce water. However, since the potential for the oxidation of Ru^{2+} to Ru^{3+}, –0.25 V, is less negative than –1.23 V, Ru^{2+} will reduce O_2 and in doing so will be oxidized to Ru^{3+}. Finally, since the difference (+0.8 V) – (0.25 V) is positive, Ru^{2+} will disproportionate in aqueous acid to Ru^{3+} and metallic ruthenium. It is not possible to tell from the potentials whether the reduction of O_2 by Ru^{2+} or the disproportionation of Ru^{2+} will be the faster process.

c) $HClO_2$: Consider the $ClO_3^-/HClO_2$ ($E^\circ = +1.181$ V) and $HClO_2/HClO$ ($E^\circ = +1.674$ V) couples. Since the $HClO_2/HClO$ couple has a potential that is more positive than +1.23 V, $HClO_2$ *will* oxidize water. However, since the $ClO_3^-/HClO_2$ couple has a positive potential, $HClO_2$ will not reduce water. Since the potential for the oxidation of $HClO_2$ to ClO_3^-, –1.181 V, is less negative than –1.23 V, $HClO_2$ will reduce O_2 and in doing so will be oxidized to ClO_3^-. Finally, since the difference (+1.674 V) – (1.181 V) is positive, $HClO_2$ will disproportionate in aqueous acid to ClO_3^- and HClO. As above, it is not possible to tell from the potentials whether the reduction of O_2 by $HClO_2$ or the disproportionation of $HClO_2$ will be the faster process.

d) Br_2: Consider the Br_2/Br^- ($E^\circ = +1.065$ V) and $HBrO/Br_2$ ($E^\circ = +1.604$ V) couples. Since the Br_2/Br^- couple has a potential that is less positive than +1.23 V, Br_2 will not oxidize water. Also, since the $HBrO/Br_2$ couple has a positive potential, Br_2 will not reduce water. Since the potential for the oxidation of Br_2 to HBrO, –1.604 V, is more negative than –1.23 V, Br_2 will not reduce O_2. Finally, since the difference (+1.065 V) – (1.604 V) is negative, Br_2 will not disproportionate in aqueous acid to Br^- and HBrO. In fact, the equilibrium constant for the following reaction is 7.2×10^{-9}.

$$Br_2(aq) + H_2O(l) \rightleftharpoons Br^-(aq) + HBrO(aq) + H^+(aq).$$

E6.5 From Section 6.10 *The influence of complexation* we see that the reduction potential of the complex species ML, E°(ML), is related to the reduction potential of the corresponding aqua ion M, E°(M), through the Equation 6.12a:

$$E^\circ(\text{ML}) = E^\circ(\text{M}) - \frac{RT}{\nu_e F}\ln\frac{K^{ox}}{K^{red}}.$$

The equation shows that E°(ML) is influenced by the temperature in two ways: 1) the RT/ν_eF ratio and 2) the values of the equilibrium constants K. The influence of RT/ν_eF is the same for both Ru and Fe complex and E° (ML) values. Thus, the reason must be the equilibrium constant: the change in temperature has the opposite effect on one or both K values for the two complexes. This is reflected in the variation of their respective reduction potentials.

E6.6 Balanced half-reactions and total reaction are provided below:

$$
\begin{array}{ll}
2 \times [MnO_4^-(aq) + 8H^+(aq) + 5e^- \rightarrow Mn^{2+}(aq) + 4H_2O(l)] & \text{reduction half-reaction} \\
5 \times [H_2O(l) + H_2SO_3(aq) \rightarrow HSO_3^-(aq) + 2e^- + 3H^+(aq)] & \text{oxidation half-reaction} \\
\hline
2MnO_4^-(aq) + 5H_2SO_3(aq) + H^+(aq) \rightarrow 2Mn^{2+}(aq) + 5HSO_3^-(aq) + 3H_2O(l) &
\end{array}
$$

The potential decreases as the pH increases and the solution becomes more basic (See Section 6.6 for further clarification).

E6.7 The standard reduction potentials are equal to the potential difference measured against SHE:

$$M^{3+}(aq) + \tfrac{1}{2}H_2(g) \rightarrow M^{2+}(aq) + H^+(aq) \qquad \Delta_rG = -\nu FE^{\circ}.$$

To determine the Δ_rG for the above reaction we have to analyse its thermodynamic cycle. Recall that potential and the Gibbs energy are related through Equation 6.2: $\Delta_rG = -\nu FE^{\circ}$. Recall also that for any process $\Delta_rG = \Delta_rH - T\Delta_rS$. We are going to assume that for each M^{3+}/M^{2+} couple $\Delta_rG \approx \Delta_rH$ because $T\Delta_rS$ is significantly smaller than Δ_rH.

The above redox reaction can be broken into reduction and oxidation reaction:

$$\text{Reduction: } M^{3+}(aq) + e^- \rightarrow M^{2+}(aq) \qquad \Delta_{Red}H$$

$$\text{Oxidation: } \tfrac{1}{2}H_2(g) \rightarrow H^+(aq) + e^- \qquad \Delta_{Ox}H.$$

From here we see that $\Delta_rG \approx \Delta_rH = \Delta_{Red}H + \Delta_{Ox}H$. $\Delta_{Ox}H$ equals +445 kJ mol^{-1} (see Section 6.3 *Trends in standard potentials* if you need to recall how this value is obtained) but we have to calculate $\Delta_{Red}H$ which can be done by analysing the following thermodynamic cycle for reduction reaction:

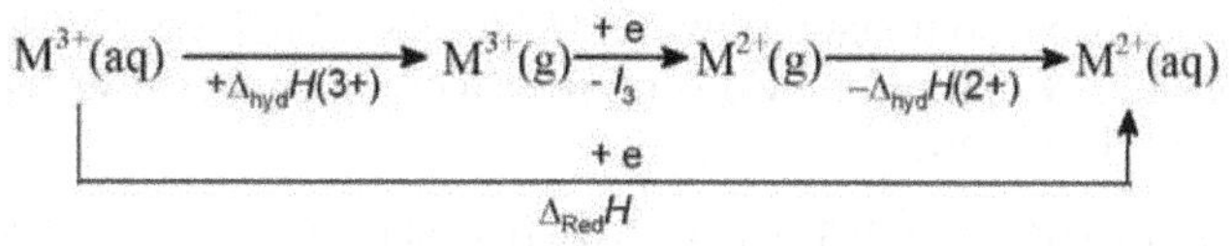

From the cycle, we can see that

$$\Delta_{Red}H = \Delta_{hyd}H(3+) - I_3 - \Delta_{hyd}H(2+).$$

The table below contains the data for all steps of the cycle, estimated ΔrG from the cycle, and final E° values calculated keeping in mind that for these reactions $\nu = 1$:

Redox couple	Ti^{3+}/Ti^{2+}	V^{3+}/V^{2+}	Cr^{3+}/Cr^{2+}	Mn^{3+}/Mn^{2+}	Fe^{3+}/Fe^{2+}	Co^{3+}/Co^{2+}
$\Delta_{hyd}H(3+)$/kJ mol^{-1}	4154	4375	4560	4544	4430	4651
$-I_3$/kJ mol^{-1}	–2652	–2828	–2987	–3247	–2957	–3232
$-\Delta_{hyd}H(2+)$/kJ mol^{-1}	–1882	–1918	–1904	–1841	–1946	–1996
$\Delta_{Red}H$/kJ mol^{-1}	–380	–371	–331	–544	–473	–577
$\Delta_{Ox}H$/kJ mol^{-1}	+445	+445	+445	+445	+445	+445
Δ_rH/kJ mol^{-1}	65	74	114	–89	–28	–122
E° ($\approx -\Delta_rH/F$) / V	**+0.67**	**+0.77**	**+1.18**	**+0.92**	**–0.29**	**–1.26**

Although the relative ordering of E° is correct, the quantitative agreement with experimental data is poor, reflecting the fact that we ignored the contributions of entropy.

E6.8 Recall that the Nernst equation is given by the formula: $E = E^{\circ} - (RT/\nu_eF)(\ln Q)$ where Q is the reaction quotient.

a) The reduction of O_2? For the reduction of oxygen, $O_2(g) + 4H^+(aq) + 4e^- \rightarrow 2H_2O(l)$,

$$Q = \frac{1}{p(O_2)[H^+]^4} \text{ and } E = E^{\circ} - \frac{0.059\text{ V}}{4}\log\frac{1}{p(O_2)[H^+]^4}.$$

The potential in terms of pH would be:

$$E = E^{\circ} - \frac{0.059\text{V}}{4}\log\frac{1}{p(O_2)[H^+]^4} = E^0 - \frac{0.059\text{V}}{4}\left(\log\frac{1}{p(O_2)} + 4\text{pH}\right).$$

Therefore, the potential for O_2 reduction at pH = 7 and $p(O_2)$ = 0.20 bar is:

$$E = +1.23\text{ V} - 0.42\text{ V} = +0.81\text{ V}.$$

b) The reduction of Fe_2O_3(s)? For the reduction of solid iron(III) oxide, $Fe_2O_3(s) + 6H^+(aq) + 6e^- \rightarrow 2Fe(s) + 3H_2O(l)$, we have:

$$Q = 1/[H^+]^6 \text{ and } E = E^o - (RT/\nu_eF)\ln(1/[H^+]^6).$$

The potential in terms of pH is:

$$E = E^\circ - (RT/\nu_eF)\ln(1/[H^+]^6) = E^\circ - (0.059\text{ V}/6) \times \log([H^+]^{-6}) = E^\circ - (0.059\text{ V}/6) \times 6\text{pH} = E^\circ - 0.059\text{ V} \times \text{pH}.$$

E6.9 **a)** The Frost diagram for chlorine in basic solution is shown in Figure 6.18 (blue line). If the points for Cl^- and ClO_4^- are connected by a straight line, Cl_2 lies above it. Therefore, Cl_2 is thermodynamically susceptible to disproportionation to Cl^- and ClO_4^- when it is dissolved in aqueous base. Note that if we would connect Cl^- with other points of the diagram, Cl_2 would always lie above the line. This means that Cl_2 can disproportionate to Cl^- and any other species containing Cl in a positive oxidation state. In practice, however, the disproportionation stops at ClO^- because further oxidation is slow (i.e., a solution of Cl^- and ClO^- is formed when Cl_2 is dissolved in aqueous base).

b) The Frost diagram for chlorine in acidic solution is shown in Figure 6.18 (red line). If the points for Cl^- and any positive oxidation state of chlorine are connected by a straight line, the point for Cl_2 lies below it (if only slightly). Therefore, Cl_2 will not disproportionate. However, E^o for the Cl_2/Cl^- couple, +1.36 V, is more positive than E^o for the O_2/H_2O couple, +1.23 V. Therefore, Cl_2 is (at least thermodynamically) capable of oxidizing water as follows, although the reaction is very slow:

$$Cl_2(aq) + H_2O(l) \rightarrow 2Cl^-(aq) + 2H^+(aq) + \tfrac{1}{2}O_2(g) \qquad E = +0.13\text{ V}$$

c) The point for ClO_3^- in acidic solution on the Frost diagram lies slightly above the single straight line connecting the points for Cl_2 and ClO_4^-. Therefore, since ClO_3^- is thermodynamically unstable with respect to disproportionation in acidic solution (i.e., it *should* disproportionate), the failure of it to exhibit any observable disproportionation must be due to a kinetic barrier.

E6.10 **a)** The only possibility worth considering in this case is disproportionation of N_2O in a basic solution. The Frost diagram for nitrogen in basic solution is shown in 6.6.

Inspection of the diagram shows that N_2O lies above the line connecting N_2 and NO_3^-. Therefore, N_2O is thermodynamically susceptible to disproportionation to N_2 and NO_3^- in basic solution:

$$5N_2O(aq) + 2OH^-(aq) \rightarrow 2NO_3^-(aq) + 4N_2(g) + H_2O(l)$$

However, the redox reactions of nitrogen oxides and oxyanions are generally slow.

b) The overall reaction is:

$$Zn(s) + I_3^-(aq) \rightarrow Zn^{2+}(aq) + 3I^-(aq)$$

Since E° values for the Zn^{2+}/Zn and I_3^-/I^- couples are –0.76 V and +0.54 V, respectively (these potentials are given in Resource Section 3, rounding to two decimal places), E° for the net reaction above is +0.54 V + 0.76 V = +1.30 V. Since the net potential is positive, this is a favourable reaction, and it should be kinetically facile if the zinc metal is finely divided and thus well exposed to the solution.

c) Since E° values for the I_2/I^- and ClO_3^-/ClO_4^- couples are +0.54 V and –1.20 V respectively (see Resource Section 3), the net reaction involving the reduction of I_2 to I^- and the oxidation of ClO_3^- to ClO_4^- will have a net negative potential, E° = +0.54 V + (–1.20 V) = –0.66 V. Therefore, this net reaction will not occur. However, since E° values for the IO_3^-/I_2 and ClO_3^-/Cl^- couples are +1.19 V and +1.47 V respectively, the following net reaction will occur, with a net E° = +0.28 V (note that the potential for IO_3^-/I_2 couple is not given on the Latimer diagram for iodine; however, it can be calculated easily by using Equation 6.15. in Section 6.12(d) *Nonadjacent species*):

$$5[ClO_3^-(aq) + 6H^+(aq) + 6e^- \rightarrow Cl^-(aq) + 3\,H_2O(l)] \quad \text{reduction}$$

$$3[I_2(s) + 6H_2O(l) \rightarrow 2IO_3^-(aq) + 12H^+(aq) + 10e^-] \quad \text{oxidation}$$

$$3I_2(s) + 5ClO_3^-(aq) + 3H_2O(l) \rightarrow 6IO_3^-(aq) + 5Cl^-(aq) + 6H^+(aq)$$

E6.11 The standard potential for the reduction of Ni^{2+} is +0.25 V. The cell potential, E, is given by the Nernst equation as follows:

$$E = E^\circ - (RT/\nu_e F)\ln Q \text{ where } Q = 1/[Ni^{2+}]$$

$[Ni^{2+}]$ can be determined from K_{sp} for $Ni(OH)_2$:

$$K_{sp}[Ni(OH)_2] = [Ni^{2+}][OH^-]^2 \text{ and } [Ni^{2+}] = K_{sp}/[OH^-]^2.$$

Since at pH = 14, $[OH^-]$ = 1 M, it follows that $[Ni^{2+}] = K_{sp}$.

Hence:

$$E = +0.25\text{ V} - \frac{RT}{\nu_e F}\ln\frac{1}{K_{sp}} = +0.025\text{ V} - \frac{0.059\text{ V}}{2}\log\frac{1}{1.5\times10^{-16}} = +0.25\text{ V} - 0.46\text{ V} = -0.21\text{ V}.$$

The electrode potential is $E = -0.21$ V.

E6.12 You can answer these questions by writing the complete, balanced half-reactions for each case. Then, analyse each half-reaction: if either H^+ or OH^- appear on any side of reaction, the potential of the half-reaction is going to depend on the pH of the solution. Once you have determined this, write down the Nernst equation and see how the potential changes with change in acidity of your solution. If you have problems writing the balanced half-reactions, consult Section 6.1 and Example 6.1.

a) $Mn^{2+} \rightarrow MnO_4^-$: In this case the balanced half-reaction is:

$$Mn^{2+}(aq) + 4H_2O(l) \rightarrow MnO_4^-(aq) + 8H^+(aq) + 5e^-$$

Since hydrogen ions are produced, the reduction potential of this oxidation half-reaction depends on pH. We can write the Nernst equation for this reaction as:

$$E = E^0 - \frac{RT}{\nu eF}\ln Q = E^0 - \frac{0.059\text{ V}}{\nu e}\log\frac{[MnO_4^-][H^+]^8}{[Mn^{2+}]}.$$

Thus, if we increase the H^+ concentration, we are going to increase the value of Q and as a result increase the value of its logarithm. Therefore, E is going to become more negative and reaction is going to be less favoured (keep in mind that $\Delta G = -\nu FE$ so more negative E results in more positive, less favourable ΔG.) The opposite happens if we lower the H^+ concentration (make the solution more basic): E is going to become more positive indicating more favourable reaction.

b) $ClO_4^- \rightarrow ClO_3^-$: The balanced half-reaction is:

$$ClO_4^-(aq) + 2H^+(aq) + 2e^- \rightarrow ClO_3^-(aq) + H_2O(l)$$

Again H^+ ions show up in the balanced half-reaction. Thus, this reaction's potential is going to depend on H^+ concentration. Writing the Nernst equation:

$$E = E^\circ - \frac{RT}{\nu_e F}\ln Q = E^\circ - \frac{0.059\text{ V}}{\nu_e}\log\frac{[ClO_3^-]}{[ClO_4^-][H^+]^2}.$$

In this case, increasing the H^+ concentration would decrease Q and its *log* value resulting in E being more positive than E°. Therefore, the reaction would be more favourable under acidic conditions.

c) $H_2O_2 \rightarrow O_2$: The balanced half-reaction is:

$$H_2O_2(aq) \rightarrow O_2(g) + 2H^+(aq) + 2e^-.$$

The potential for this half-reaction is going to depend on the acidity of solution because H^+ ions are present as products. The Nernst equation can be written as:

$$E = E^\circ - \frac{RT}{\nu_e F}\ln Q = E^\circ - \frac{0.059\ \mathrm{V}}{\nu e}\log\frac{p(O_2)[H^+]^2}{[H_2O_2]}.$$

We can see that this is very similar to part a) of this problem: an increase in H^+ concentration is going to be unfavourable for the half-reaction.

d) $I_2 \rightarrow 2I^-$: The balanced half-reaction is:

$$I_2 + 2e^- \longrightarrow 2I^-.$$

Since neither protons nor OH^- are present in this reduction half-reaction, and since I^- is not protonated in aqueous solution (because HI is a very strong acid), this reaction has the same potential in acidic or basic solution, +0.535 V (this can be confirmed by consulting Resource Section 3).

E6.13 The reaction can be broken into two half-reactions as follows:

$$HO_2(aq) \longrightarrow O_2(g) + H^+(aq) + \tfrac{1}{2}e^- \qquad \text{oxidation half-reaction}$$

$$HO_2 + H^+ + \tfrac{1}{2}e^- \longrightarrow H_2O_2 \qquad \text{reduction half-reaction.}$$

The E° for the reaction will be:

$$E^\circ = E_{red} - E_{ox} = +1.150\ \mathrm{V} - (-0.125\ \mathrm{V}) = +1.275\ \mathrm{V}.$$

Looking at the Latimer diagram, we also see that the potential on the right of HO_2 is more positive than the potential on the left of the same species (+1.150 V vs. –0.125 V) informing us immediately that HO_2 has thermodynamic tendency to undergo disproportionation (see Section 6.12(c)).

E6.14 Consult Section 6.12(b) of your textbook for a detailed explanation. The Latimer diagram for chlorine in acidic solution is given in Resource Section 3. To determine the potential for any non-adjacent couple, you must calculate the *weighted average* of the potentials of intervening couples. In general terms, it is:

$$(n_1E^\circ_1 + n_2E^\circ_2 + \ldots + n_nE^\circ_n)/(n_1 + n_2 + \ldots + n_n)$$

and in this specific case it is:

$$[(2)(+1.201\ \mathrm{V}) + (2)(+1.181\ \mathrm{V}) + (2)(+1.674\ \mathrm{V}) + (1)(+1.630\ \mathrm{V})]/(2 + 2 + 2 + 1) = +1.392\ \mathrm{V}.$$

Thus, the standard potential for the ClO_4^-/Cl_2 couple is +1.392 V. The half-reaction for this reduction is:

$$2ClO_4^-(aq) + 16H^+(aq) + 14e^- \rightarrow Cl_2(g) + 8H_2O(l).$$

E6.15 To determine the potential for any couple, you must calculate the *weighted average* of the potentials of intervening couples. In general terms, it is:

$$(n_1E^\circ_1 + n_2E^\circ_2 + \ldots + n_nE^\circ_n) / (n_1 + n_2 + \ldots + n_n)$$

and in this specific case it is:

$$[(2)(+0.16\ \mathrm{V}) + (2)(+0.40\ \mathrm{V}) + (2)(+0.60\ \mathrm{V})] / (2 + 2 + 2) = +0.387\ \mathrm{V}.$$

E6.16 The standard potential for the reduction $MnO_4^-(aq) + 4H^+(aq) + 3e^- \rightarrow MnO_2(s) + 2H_2O(l)$ is +1.69 V. The pH, or $[H^+]$, dependence of the potential E is given by the Nernst equation ($[MnO_4^-] = 1.0\ \mathrm{mol\ dm^{-3}}$) as follows (see also Section 6.6 of your textbook and "A brief illustration" within this section):

$$E = E^\circ - \frac{0.059\ \mathrm{V} \times \nu_{H^+}}{\nu_e}\mathrm{pH}.$$

Note that $\nu_e = 3$ and $\nu_{H+} = 4$ for the reduction of MnO_4^-. Note also that the factor 0.059 V can only be used at 25°C:

$$E = +1.69\text{ V} - \frac{0.059\text{ V}\times 4}{3}\times 9 = +1.69\text{ V} - 0.71\text{ V} = +0.98\text{ V}.$$

In this case $E' = E°$ because the concentration of MnO_4^- is 1.0 mol dm^{-3} which corresponds to the standard conditions for measuring reduction potentials.

E6.17 To calculate an equilibrium constant using thermodynamic data, you can make use of the expressions $\Delta_rG = -RT\ln K$ and $\Delta_rG = -\nu FE$. You can use the given potential data to calculate ΔG for each of the two half-reactions, then you can use Hess's Law to calculate ΔG for the overall process (Δ_rG), and finally calculate K from Δ_rG. The net reaction $Pd^{2+}(aq) + 4\ Cl(aq) \rightarrow [PdCl_4]^{2-}\ (aq)$ is the following sum:

$$Pd^{2+}(aq) + 2e^- \rightarrow Pd(s)$$

$$Pd(s) + 4Cl^-(aq) \rightarrow [PdCl_4]^{2-}(aq) + 2e^-$$

$$Pd^{2+}(aq) + 4Cl^-(aq) \rightarrow [PdCl_4]^{2-}(aq)$$

ΔG for the first reaction is $-\nu FE = -(2)(96.5\text{ kJ mol}^{-1}\text{ V}^{-1})(+0.915\text{ V}) = -176.6\text{ kJ mol}^{-1}$. ΔG for the second reaction is $-(2)(96.5\text{ kJ mol}^{-1}\text{ V}^{-1})(-0.6\text{ V}) = +115.8\text{ kJ mol}^{-1}$. The net Δ_rG is the sum of these two values, -60.8 kJ mol^{-1}. Therefore, assuming $T = 298$ K:

$$K = \exp[(60.8\text{ kJ mol}^{-1})/(8.31\text{ J K}^{-1}\text{ mol}^{-1})(298)] = \exp(24.5) = 4.37\times 10^{10}.$$

Note that the pH of the solution is 0 because $[H^+] = [HCl] = 1.0$ mol dm^{-3} (HCl is a very strong acid that completely dissociates in water).

E6.18 To calculate an equilibrium constant using thermodynamic and electrochemical data, you can make use of the expressions $\Delta_rG = -RT\ln K$ and $\Delta_rG = -\nu FE$. You can use the given potential data to calculate ΔG for each of the two half-reactions, then you can use Hess's Law to calculate Δ_rG for the overall process, and finally calculate K from Δ_rG.

The net reaction $Au^+(aq) + 2CN^-(aq) \rightarrow [Au(CN)_2]^-(aq)$ is the following sum:

$$Au^+(aq) + e^- \rightarrow Au(s)$$

$$Au(s) + 2CN^-(aq) \rightarrow [Au(CN)_2]^-(aq) + e^-$$

$$Au^+(aq) + 2CN^-(aq) \rightarrow [Au(CN)_2]^-(aq)$$

ΔG for the first reaction is $-\nu FE = -(1)(96.5\text{ kJ mol}^{-1}\text{ V}^{-1})(+1.69\text{ V}) = -163\text{ kJ mol}^{-1}$. ΔG for the second reaction is $-(1)(96.5\text{ kJ mol}^{-1}\text{ V}^{-1})(+0.6\text{ V}) = -58\text{ kJ mol}^{-1}$. The net ΔG is the sum of these two values, -221 kJ mol^{-1}. Therefore, assuming $T = 298$ K:

$$K = \exp[(221\text{ kJ mol}^{-1})/(8.31\text{ J K}^{-1}\text{ mol}^{-1})(298)] = \exp(89.2) = 5.7\times 10^{38}.$$

E6.19 Since $edta^{4-}$ forms very stable complexes with $M^{2+}(aq)$ ions of 3d-block elements but *not* with the zerovalent metal atoms, the reduction of a $M(edta)^{2-}$ complex will be more difficult than the reduction of the analogous M^{2+} aqua ion. Since the reductions are more difficult, the reduction potentials become less positive (or more negative, as the case may be). The reduction of the $M(edta)^{2-}$complex includes a decomplexation step, with a positive free energy change. The reduction of $M^{2+}(aq)$ does not require this additional expenditure of free energy.

E6.20 Consult Section 6.13 and Self-Tests 6.10 and 6.11 for the construction and interpretation of the Frost diagrams. The Frost diagram for Hg, constructed based on the provided data, is shown:

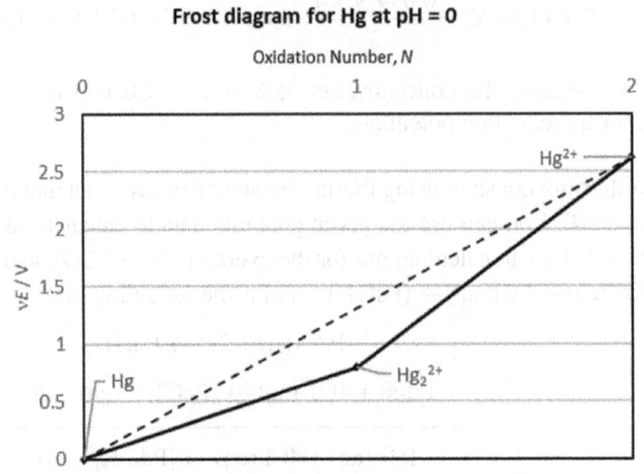

From the diagram, we see that Hg^{2+} and Hg_2^{2+} are both oxidizing agents as they have large positive standard reduction potentials and the line connecting the redox couples is very steep with a positive slope. Thus, these species are likely to undergo reduction encouraging oxidation of the other species in the reaction. Consequently, none of these species are likely to be good reducing agents. Hg_2^{2+} is not likely to undergo disproportionation because the Hg_2^{2+} point lies below the line connecting Hg and Hg^{2+} (the dashed line on the diagram above); rather the opposite is likely to happen: Hg and Hg^{2+} are going to comproportionate and form Hg_2^{2+}.

E6.21 **a) Fe?** According to Figure 6.12, the potential range for surface water at pH 6 is between +0.5 V and +0.6 V, so a value of +0.55 V can be used as the approximate potential of an aerated lake at this pH. Inspection of the Pourbaix diagram for iron (Figure 6.11) shows that at pH 6 and E = +0.55 V, the stable species of iron is $Fe(OH)_3$ precipitate. Therefore, this compound of iron would predominate.

b) Mn? Inspection of the Pourbaix diagram for manganese (Figure 6.13) shows that at pH 6 and E = +0.55 V, the stable species of manganese is solid Mn_2O_3. Therefore, this compound of manganese would predominate.

c) S? At pH 0, the potential for the HSO_4^-/S couple is +0.387 V (this value was calculated using the weighted average of the potentials given in the Latimer diagram for sulfur in Resource Section 3), so the lake will oxidize S_8 all the way to HSO_4^-. At pH 14, the potentials for intervening couples are all negative, so SO_4^{2-} would again predominate. Therefore, HSO_4^- is the predominant sulfur species at pH 6.

E6.22 The following redox reaction describes one of the first steps in iron corrosion:

$$2Fe(s) + O_2(g) + 4H^+(aq) \rightarrow 2Fe^{2+}(aq) + 2H_2O(l).$$

This reaction can be broken in two half-reactions:

$Fe(s) \rightarrow Fe^{2+}(aq) + 2e^-$	$E^\circ = -0.45$ V	oxidation half-reaction
$O_2(g) + 4H^+ + 4e^- \rightarrow 2H_2O$	$E^\circ = +1.23$ V	reduction half-reaction.

From the Section 6.7 *Reactions with water* and Equation 5.10 we know that the reduction potential of the reduction half-reaction depends on pH of the solution as:

$$E = +1.23\text{ V} - (0.059\text{ V} \times \text{pH}).$$

Thus, the potential difference for the overall corrosion reaction can be given as:

$$E_{corr} = +1.23\text{ V} - (0.059\text{ V} \times \text{pH}) - (-0.45\text{ V}) = +1.68\text{ V} - (0.059\text{ V} \times \text{pH}).$$

From the above equation, we see that the air oxidation of iron is favoured in acidic solution: lower pH values would have more positive E_{corr}. Increased levels of carbon dioxide and water generate carbonic acid that lowers the pH of the medium, which makes the corrosion process more favourable.

E6.23 Any species capable of oxidizing either Fe^{2+} or H_2S at pH 6 cannot survive in this environment. According to the Pourbaix diagram for iron (Figure 6.11), the potential for the $Fe(OH)_3/Fe^{2+}$ couple at pH 6 is approximately 0.3 V. Using the Latimer diagrams for sulfur in acid and base (see Resource Section 3), the S/H_2S potential at pH 6 can be calculated as follows:

$$+0.14\ \text{V} - (6/14)[0.14 - (-0.45)] = -0.11\ \text{V}.$$

Any potential higher than this will oxidize hydrogen sulfide to elemental sulfur. Therefore, if H_2S is present, the maximum potential possible is approximately –0.1 V.

E6.24 Any boundary between a soluble species and an insoluble species will change as the concentration of the soluble species changes. For example, consider the line separating $Fe^{2+}(aq)$ from $Fe(OH)_3(s)$ in Figure 6.11. As shown in the text, $E = E^\circ - (0.059\ \text{V}/2)\log([Fe^{2+}]^2/[H^+aq)]^6)$ (see Section 6.14 *Pourbaix diagrams*). Clearly, the potential at a given pH is dependent on the concentration of soluble $Fe^{2+}(aq)$. The boundaries between the two soluble species, $Fe^{2+}(aq)$ and $Fe^{3+}(aq)$, and between the two insoluble species, $Fe(OH)_2(s)$ and $Fe(OH)_3(s)$, will not depend on the choice of $[Fe^{2+}]$.

E6.25 The lines for Al_2O_3 and MgO on the Ellingham diagram (Figure 6.16) represent the change in ΔG° with temperature for the following reactions:

$$\tfrac{4}{3}Al(l) + O_2(g) \rightarrow \tfrac{2}{3}Al_2O_3(s) \text{ and } 2Mg(l) + O_2(g) \rightarrow 2MgO(s).$$

At temperatures below about 1600ºC, the Gibbs energy change for the MgO reaction is more negative than for the Al_2O_3 reaction. This means that under these conditions MgO is more stable with respect to its constituent elements than is Al_2O_3, and that Mg will react with Al_2O_3 to form MgO and Al. However, above about 1600ºC the situation reverses, and Al will react with MgO to reduce it to Mg with the concomitant formation of Al_2O_3. This is a rather high temperature, achievable in an electric arc furnace (compare the extraction of silicon from its oxide, discussed in Section 6.16).

E6.26 **a)** The differences in reduction potential in acidic (pH = 0) and basic (pH = 14) solutions can be explained if we look at the reduction half-reactions. For example, in acidic solutions:

$$H_3PO_4(aq) + 2e^- + 2H^+ \rightarrow H_3PO_3(aq) + H_2O(l),\ E^\circ = -0.276\ \text{V}$$

If we recall Equation 6.8b from the textbook, we can see that reduction potentials that couple transfer of both electrons and protons decrease (become more negative) as the pH is increased (solution becomes more basic). This is exactly what we observe in the case of phosphorus because the same reduction from P(V) to P(III) in basic medium (PO_4^{3-}/HPO_3^{3-} couple) has a more negative reduction potential (–1.12 V).

b) The Frost diagram for phosphorus in acidic (pH = 0, solid line) and basic (pH = 14, dashed line) medium is shown below:

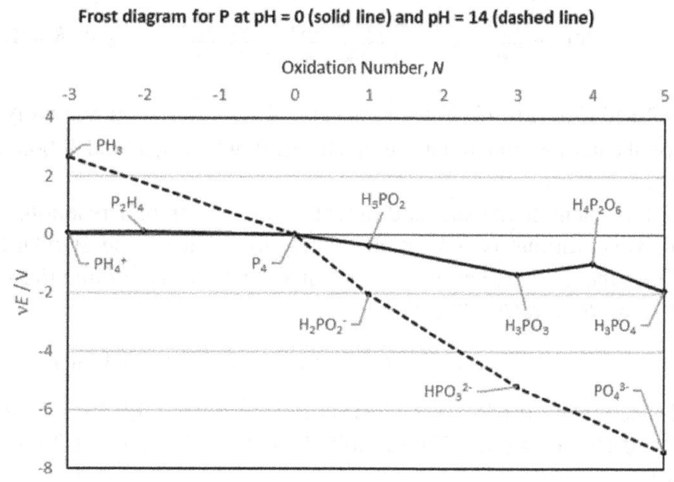

c) If you connect the points for PH_3 and $H_2PO_2^-$ in the diagram (basic solution dashed line), you will note that P_4 lies above the line. This means that in basic solutions P_4 can disproportionate to PH_3 and $H_2PO_2^-$ according to the equation:

$$P_4(aq) + 3OH^- + 3H_2O \rightarrow 3H_2PO_2^- + PH_3.$$

The equilibrium constant can be calculated using equation 6.1, $\Delta_rG° = -RT\ln K$, and Equation 5.2, $\Delta_rG° = -\nu FE°$. From here we have

$$-RT\ln K = -\nu FE° \text{ or } \ln K = (\nu FE°)/RT.$$

$E°$ can be calculated using the data from the Latimer diagram for pH = 14: $E(P_4/PH_3) = -0.89$ V = E_{red} and $E(H_2PO_2^-/P_4) = -2.05$ V = E_{ox}.

$$E° = E_{red} - E_{ox} = -0.89\text{ V} - (-2.05\text{ V}) = +1.16\text{ V}.$$

Thus, with $\nu = 3$ and $T = 25°$, we have

$$\ln K = \frac{\nu FE^0}{RT} = \frac{3 \times 96.48\text{ kC mol}^{-1} \times 1.16\text{ V}}{8.314\text{ J K}^{-1}\text{mol}^{-1} \times 298\text{ K}} = 135.5 \text{ or } K = 7.1 \times 10^{58}.$$

E6.27 Since we are not given the oxidation half-reaction (only two reduction half-reactions) we can assume that the $\Delta_rG°$ and K are calculated using SHE for the oxidation half-reaction. For both reactions, we have to work under basic conditions, so we are going to assume "standard basic solution" of pH = 14 (this is the basic medium for which reduction potentials are reported, hence the assumption—see Resource Section 3 for examples). The reduction potential of SHE depends on pH (Equation 6.9): $E(H^+/H_2) = -0.059\text{ V} \times \text{pH} = -0.059\text{ V} \times 14 = -0.826$ V.

a) For $CrO_4^{2-}(aq)$ we have:

$$E°_{pH=14} = E_{red} - E_{Ox} = -0.11\text{ V} - (-0.826\text{V}) = 0.716\text{ V}.$$

From here we can use Equations 5.2 and 5.5 to calculate $\Delta_rG°$ and K respectively:

$$\Delta_rG° = -\nu FE° = -3 \times 96.48\text{ kC mol}^{-1} \times 0.716\text{ V} = -207.2\text{ kJ mol}^{-1}.$$

$$\ln K = \frac{\nu FE^0}{RT} = \frac{3 \times 96.48\text{ kC mol}^{-1} \times 0.716\text{ V}}{8.314\text{ J K}^{-1}\text{ mol}^{-1} \times 298\text{ K}} = 83.6 \text{ or } K = 2.1 \times 10^{36}.$$

b) For $[Cu(NH_3)_2]^+(aq)$ we have

$$E°_{pH=14} = E_{red} - E_{Ox} = -0.10\text{ V} - (-0.826\text{ V}) = 0.726\text{ V}$$

$$\Delta_rG° = -\nu FE° = -1 \times 96.48\text{ kC mol}^{-1} \times 0.726\text{ V} = -70.0\text{ kJ mol}^{-1}$$

$$\ln K = \frac{\nu FE^0}{RT} = \frac{1 \times 96.48\text{ kC mol}^{-1} \times 0.726\text{ V}}{8.314\text{ J K}^{-1}\text{ mol}^{-1} \times 298\text{ K}} = 28.3 \text{ or } K = 1.9 \times 10^{12}.$$

The values for $\Delta_rG°$ and K in (a) and (b) are noticeably different despite relatively little difference in $E°$ because $E°$ does not depend on the number of exchanged electrons, ν, while $\Delta_rG°$ and K both depend on it.

E6.28 Since we are asked to calculate the standard potential for a given half-reaction, we are using the fact that $\Delta G°$ for SHE under standard conditions is 0 kJ mol^{-1}. Thus, to calculate the standard reduction potential we have to calculate only ΔG for the given reaction and use Equation 5.2 to determine the standard reduction potential. Keep in mind that $\Delta_rG° = -\nu FE° = \Delta_rH° - T\Delta_rS°$.

$$Sc_2O_3(s) + 3H_2O(l) + 6e^- \rightarrow 2Sc(s) + 6OH^-(aq)$$

$\Delta_rH° = [2 \times \Delta_fH°(Sc(s)) + 6 \times \Delta_fH°(OH^-(aq))] - [\Delta_rH°(Sc_2O_3(s) + 3 \times \Delta_rH°(H_2O(l))]$
$= [2 \times 0\text{ kJmol}^{-1} + 6 \times (-230\text{ kJ mol}^{-1})] - [-1908.7\text{ kJ mol}^{-1} + 3 \times (-285.8\text{ kJ mol}^{-1})] = 1386.1$ kJ mol^{-1}

$\Delta_rS° = [2 \times \Delta_mS°(Sc(s)) + 6 \times \Delta_mS°(OH^-(aq))] - [\Delta_mS°(Sc_2O_3(s) + 3 \times \Delta_mS°(H_2O(l))]$
$= [2 \times 34.76\text{ J K}^{-1}\text{ mol}^{-1} + 6 \times (-10.75\text{ kJ mol}^{-1})] - [77.0\text{ J K}^{-1}\text{ mol}^{-1} + 3 \times (69.91\text{ J K}^{-1}\text{ mol}^{-1})]$

$= -0.282\ \text{kJ K}^{-1}\ \text{mol}^{-1}$

$$\Delta_r G° = \Delta_r H° - T\Delta_r S° = 1386.1\ \text{kJ mol}^{-1} - 298\ \text{K} \times (-0.282\ \text{kJ K}^{-1}\ \text{mol}^{-1}) = 1302\ \text{kJ mol}^{-1}$$

$$E° = -\frac{\Delta_r G^o}{\nu F} = -\frac{1302\,\text{kJ mol}^{-1}}{6\times 96.48\,\text{kC mol}^{-1}} = -2.25\ \text{V}.$$

E6.29 The combined Frost diagram for In and Tl is shown below. From the diagram, we can see that thallium is more stable as Tl^+ in aqueous solutions, and that Tl^{3+} can be expected to behave as a strong oxidizing agent (very steep positive slope of the line connecting Tl^+ and Tl^{3+} points). The reverse is observed for indium: In^{3+} is more stable than In^+ in aqueous solutions. In^+ is actually expected to behave as a reducing agent (a negative slope of the line connecting In^+ and In^{3+} points).

Frost diagram for In (solid line) and Tl (dashed line) at pH = 0

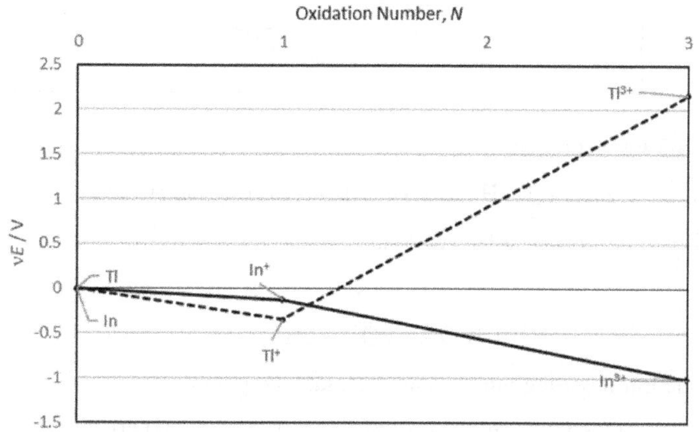

E6.30 (i) Considering that the standard reduction potential of Fe^{3+}/Fe^{2+} couple is more positive than Ru^{3+}/Ru^{2+}, Ru^{2+} should be less stable than Fe^{2+} in acidic aqueous solutions.

(ii) In this case you have to consider several possible outcomes: (a) Fe and Fe^{3+} could react to produce Fe^{2+}, (b) Fe could possibly be oxidized by water, (c) atmospheric oxygen could oxidize Fe, and (d) H^+ could oxidize Fe. If you look at the potential difference of all possible outcomes, you will see that the only expected reaction is oxidation of Fe(s) to Fe^{2+} by H^+ (see also Exercise 6.4 and Self-Test 6.6).

(iii) The reaction is as follows:

$$5Fe^{2+}(aq) + MnO_4^-(aq) + 8H^+(aq) \rightarrow 5Fe^{3+}(aq) + Mn^{2+}(aq) + 4H_2O(l).$$

We are going to use Equation 6.5, $\ln K = \frac{\nu FE^0}{RT}$, with $E° = E°_{red} - E°_{ox} = +1.51\ \text{V} - (+0.77\ \text{V}) = +0.74\ \text{V}$:

$$\ln K = \frac{\nu FE^0}{RT} = \frac{5\times 96.48\ \text{kC mol}^{-1}\times 0.74\ \text{V}}{8.314\ \text{J K}^{-1}\,\text{mol}^{-1}\times 298\ \text{K}} = 144.1 \text{ and } K = 3.7\times 10^{62}.$$

E6.31 i) Following the *weighted average* of the potentials of intervening couples formula:

$$(n_1E°_1 + n_2E°_2 + \ldots + n_nE°_n)/(n_1 + n_2 + \ldots + n_n)$$

for the VO^{2+}/V couple we have:

$$[(1)(+0.34\ \text{V}) + (1)(-0.26\ \text{V}) + (2)(-1.13\ \text{V})] / (1 + 1 + 2) = -2.18\ \text{V}.$$

Balanced chemical equation is:

$$VO^{2+} + 2H^+ + 4e^- \rightarrow V + H_2O.$$

ii) The Frost diagram is shown below:

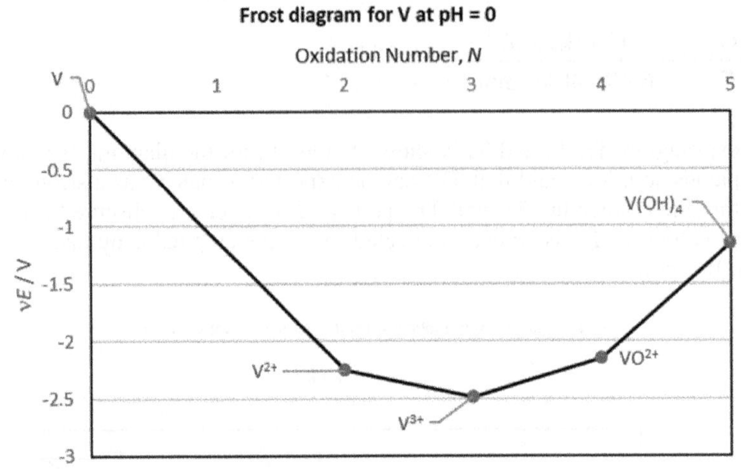

iii) The disproportionation reaction is:

$$2V^{3+}(aq) + H_2O \rightarrow V^{2+}(aq) + VO^{2+}(aq) + 2H^+(aq).$$

The reaction can be broken into two half-reactions:

$$V^{3+} + e^- \rightarrow V^{2+} \qquad \text{reduction reaction}$$

$$V^{3+} + H_2O \rightarrow VO^{2+} + 2H^+ \qquad \text{oxidation reaction}$$

Using the data from the Latimer diagram, we can calculate the potential:

$$E° = E°_{red} - E°_{ox} = (-0.26\ V) - (+0.34\ V) = -0.60\ V.$$

Then we calculate the equilibrium constant using Equation 6.5:

$$\ln K = \frac{\nu FE^0}{RT} = \frac{1 \times 96.48\ kC\ mol^{-1} \times (-0.60\ V)}{8.314\ J\ K^{-1}\ mol^{-1} \times 298\ K} = -23.4 \text{ and}$$

$$K = 4.3 \times 10^{-24}.$$

The value for the equilibrium constant is very small indicating that the disproportionation reaction is actually shifted to the left side (or in other words, the comproportionation reaction is more favourable) telling us that $V^{3+}(aq)$ is stable in acidic solution. The same conclusion can be drawn from the Frost diagram above: $V^{3+}(aq)$ lies below the line connecting V^{2+} and VO^{2+}.

iv) The reduction reaction is:

$$VO^{2+} + 2H^+ + e^- \rightarrow V^{3+} + H_2O.$$

The potential would become more negative with increase in pH. See also Exercise 6.16 and Section 6.6 of your textbook.

E6.32 i) The Frost diagram is shown below:

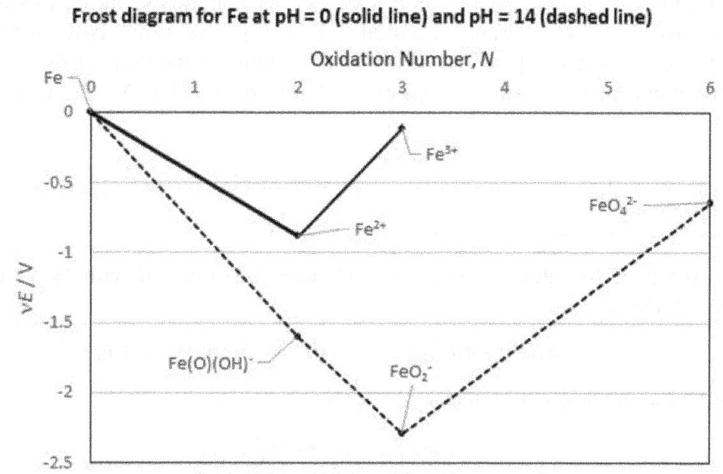

ii) We are dealing again with two nonadjacent species and have to use the weighted average formula:

$$(n_1E°_1 + n_2E°_2 + \ldots + n_nE°_n)/(n_1 + n_2 + \ldots + n_n).$$

For Fe^{3+}/Fe couple we have:

$$[(1)(+0.77\text{ V}) + (2)(-0.44\text{ V})]/(1 + 2) = -0.04\text{ V}.$$

iii) The Fe species present in acidic and alkaline solutions are very different: in acid we have hydrated cations, Fe^{3+}(aq) and Fe^{2+}(aq), while in basic solutions we have oxo-anions, $Fe(O)(OH)^-$ and FeO_2^-. Oxide anion is very good at stabilizing higher oxidation states (in this case +3).

FeO_4^{2-} anion is a very strong oxidizing agent in acidic solutions. Its oxidizing power is sufficient to oxidize water liberating O_2 gas. Due to this reactivity, FeO_4^{2-} is unknown in acidic solutions.

iv) The potential ClO^-/Cl_2 at +0.42 V is not positive enough to oxidize FeO_2^- to FeO_4^{2-} (FeO_4^{2-}/FeO_2^- potential from the Latimer diagram is +0.55 V). But the potential for ClO^-/Cl^- is:

$$E°(ClO^-/Cl^-) = [(1)(+0.42\text{ V}) + (1)(+1.36\text{ V})]/(1 + 1) = +0.89\text{ V}.$$

Thus, ClO^- is going to be reduced to Cl^-. This allows us to set two half-reactions in basic medium:

$$FeO_2^- + 4OH^- \rightarrow FeO_4^{2-} + 2H_2O + 3e^- \qquad \text{oxidation half-reaction}$$

$$ClO^- + H_2O + 2e^- \rightarrow Cl^- + 2OH^- \qquad \text{reduction half-reaction.}$$

To make the number of electrons in two half-reactions the same, we must multiply oxidation half-reaction by 2 and reduction half-reaction by 3:

$$2FeO_2^- + 8OH^- \rightarrow 2FeO_4^{2-} + 4H_2O + 6e^- \qquad \text{oxidation half-reaction}$$

$$3ClO^- + 3H_2O + 6e^- \rightarrow 3Cl^- + 6OH^- \qquad \text{reduction half-reaction.}$$

Now we can sum them and obtain the final reaction:

$$2FeO_2^- + 3ClO^- + 2OH^- \rightarrow 2FeO_4^{2-} + 3Cl^- + H_2O.$$

E6.32 i) Oxidation states of carbon: CO_2, +4; HCHO, 0; CO_2^-, +3; CO, +2; HCO_2^-, +2; CH_3OH, -2; and CH_4, -4.

ii) To construct the Frost diagram, we need to find relevant reduction potentials. A very clear-cut way to do this is by constructing a Latimer diagram first. To start with we order the carbon species in order of decreasing oxidation number (remember that Latimer diagrams are reduction diagrams, the electrons are added from left to right):

$$CO_2 \longrightarrow CO_2^- \longrightarrow CO \longrightarrow HCHO \longrightarrow CH_3OH \longrightarrow CH_4$$

Note that instead of CO you can use HCO_2^- as a species that contains C in +2 oxidation state—the procedure would be the same, just using appropriate potential for HCO_2^- reactivity. Now we start filling the reduction potentials. The second equation given provides us with the reduction potential for the first couple: CO_2/CO_2^-, -1.90 V. The third equation connects CO_2 and nonadjacent CO: CO_2/CO, –0.52 V. We can enter these values into the diagram:

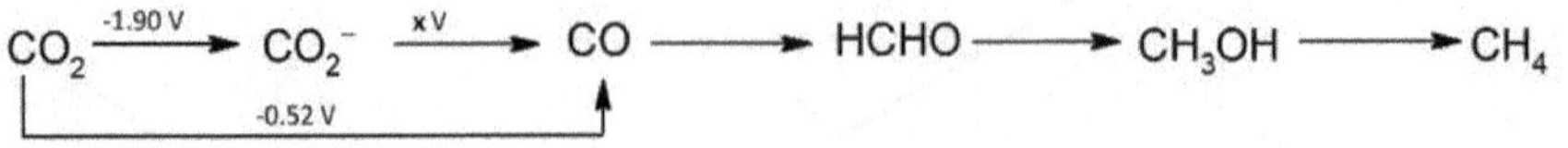

We can use these two values to calculate the CO_2^-/CO potential (marked with 'x V' on the diagram) using the formula for nonadjacent species:

$$(n_1E^\circ{}_1 + n_2E^\circ{}_2 + \ldots + n_nE^\circ{}_n)/(n_1 + n_2 + \ldots + n_n)$$

with a difference being that nonadjacent potential is given.

$$E^\circ_{CO_2/CO} = \frac{n_1 E^\circ_{CO_2/CO_2^-} + n_2 E^\circ_{CO_2^-/CO}}{n_1 + n_2}$$

$$-0.52\ \mathrm{V} = \frac{1\times(-1.90\ \mathrm{V}) + 1\times \mathrm{x}}{1+1}$$

$$\mathrm{x} = E^\circ_{CO_2^-/CO} = +0.86\ \mathrm{V}.$$

We can enter this value in our Latimer diagram:

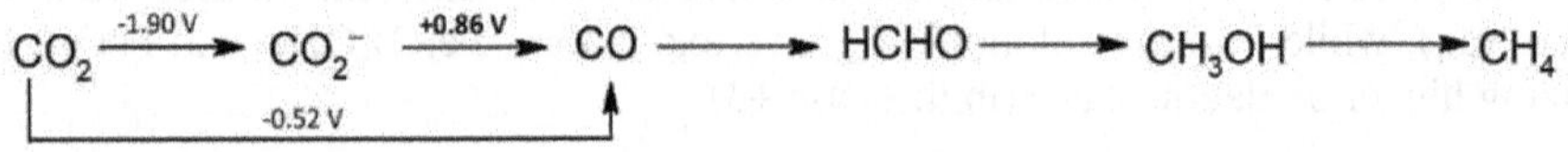

This process is continued until the Latimer diagram is complete:

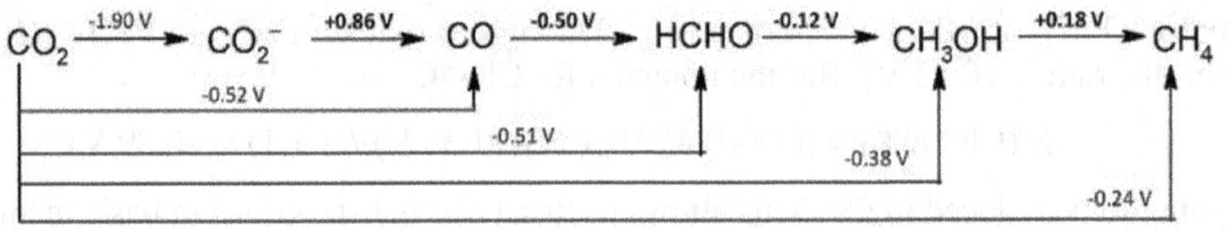

Now the Frost diagram is constructed. Note: if we set HCHO as 0 on the diagram, then moving left from HCHO we have reduction process, but moving right we have oxidation process—important for the signs of potential used for diagram construction.

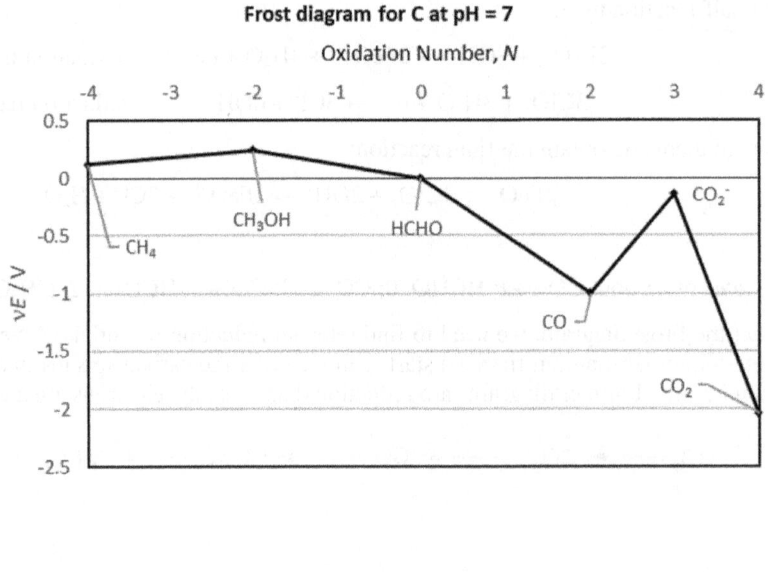

iv) First we have to find a way to make HCO_2^- from CO using given reactions, keeping in mind that the sign of ΔG determines the spontaneity of a reaction. The obvious choice of reaction to consider first is the one where HCO_2^- is the product:

$$CO_2(g) + H^+ + 2e^- \rightarrow HCO_2^-(aq),\ E = -0.43\ V.$$

ΔG for this reaction is (Equation 6.2):

$$\Delta G(CO_2 \rightarrow HCO_2^-) = -\nu FE^0 = -2 \times 96480\ C\ mol^{-1} \times (-0.43\ V) = 83\ kJ\ mol^{-1}.$$

Now we have to find a way to use CO as a starting material. If we invert the third reaction we get:

$$CO(g) + H_2O \rightarrow CO_2(g) + 2H^+ + 2e^-,\ E = +0.52\ V.$$

Note that the reduction potential is now oxidation potential and that the sign changed from minus to plus. ΔG for this reaction is:

$$\Delta G(CO \rightarrow CO_2) = -\nu FE^0 = -2 \times 96480\ C\ mol^{-1} \times (+0.52\ V) = -100\ kJ\ mol^{-1}.$$

Now we can combine the two reactions:

$$CO_2(g) + H^+ + 2e^- \rightarrow HCO_2^-(aq)$$
$$CO(g) + H_2O \rightarrow CO_2(g) + 2H^+ + 2e^-$$
$$\overline{\qquad\qquad\qquad\qquad\qquad\qquad}$$
$$CO(g) + H_2O \rightarrow HCO_2^- + H^+$$

ΔG for this reaction is then:

$$\Delta G(CO \rightarrow HCO_2^-) = \Delta G(CO_2 \rightarrow HCO_2^-) + \Delta G(CO \rightarrow CO_2) = 83\ kJ\ mol^{-1} + (-100\ kJ\ mol^{-1}) = \mathbf{-17\ kJ\ mol^{-1}}.$$

Since $\Delta G(CO \rightarrow HCO_2^-) < 0$, the hydration of CO to give HCO_2^- is a thermodynamically favourable process.

v) Start by inverting hydrogen half-reaction:

$$H_2(g) \rightarrow 2H^+(aq) + 2e^-,\ E = -0.41\ V.$$

To this reaction we can add the third reaction noting the reduction/oxidation half–reactions:

$H_2(g) \rightarrow 2H^+(aq) + 2e^-$	oxidation half-reaction
$CO_2(g) + 2H^+ + 2e^- \rightarrow CO(g) + H_2O$	reduction half-reaction

$$CO_2(g) + H_2(g) \rightarrow CO(g) + H_2O(l)$$

For this reaction:

$$E = -0.52\ V + (-0.41\ V) = -0.93\ V.$$

And the equilibrium constant is:

$$\ln K = \frac{\nu FE}{RT} = \frac{2 \times 96480\ C\,mol^{-1} \times (-0.93\ V)}{8.314\ J\,K^{-1}\,mol^{-1} \times 298\ K} = -72.4 \text{ and } \mathbf{K = 3.5 \times 10^{-32}}.$$

Guidelines for Selected Tutorial Problems

T6.1 The reaction for analytical determination of Fe^{2+} with MnO_4^- is:

$$5Fe^{2+}(aq) + MnO_4^-(aq) + 8H^+(aq) \rightarrow 5Fe^{3+}(aq) + Mn^{2+}(aq) + 4H_2O(l)$$

with $E° = E°_{red} - E°_{ox} = +1.51\ V - (+\ 0.77\ V) = +0.74\ V$. If HCl is present, the following side-reaction can take place:

$$10Cl^-(aq) + 2MnO_4^-(aq) + 16H^+(aq) \rightarrow 5Cl_2(g) + 2Mn^{2+}(aq) + 8H_2O(l)$$

with $E° = E°_{red} - E°_{ox}$ = +1.51 V – (+1.36 V) = +0.15 V. From $E°$ values it is clear that oxidation of Fe^{2+}(aq) is more favourable under the standard conditions than oxidation of Cl^-(aq) because $E°(1) > E°(2)$. However, as reaction (1) is reaching equilibrium its potential difference is changing according to Equation 5.4:

$$E = E° - \frac{RT}{\nu F}\ln\frac{[Fe^{3+}]^5[Mn^{2+}]}{[Fe^{2+}]^5[MnO_4^-][H^+]^8}.$$

As $[Fe^{3+}]$ and $[Mn^{2+}]$ are increasing, so is Q, and E decreases. At one point the two potentials are going to invert, that is, $E(1) < E(2)$ and reaction (2) is going to be more favourable. To avoid the side reaction we have to keep $E(1) > E(2)$. If the phosphate anion stabilizes Fe^{3+}, that means that through complexation we can make the Fe^{3+}/Fe^{2+} potential more negative, meaning that $E(1)$ becomes more positive and reaction (1) more favourable (see Section 6.10 *The influence of complexation*). Addition of Mn^{2+} would influence both reactions but more on reaction (2) than (1) because 2 equivalents of Mn^{2+} are produced in (2) while only one in (1). You can see this effect if you write down reaction quotients for both (1) (given above already) and (2).

T6.2 Keep in mind that the standard reduction potentials can provide not only information about spontaneity of a given reaction in aqueous solutions, but also provide an insight if side reactions are going to take place: Could water be oxidized or reduced? Can a species be oxidized with atmospheric oxygen? How will the reaction direction change with pH etc. It is a good idea to at least mention Pourbaix diagrams and their use in inorganic environmental chemistry and geochemistry (speciation chemical species, their mobility, etc.) as well as point to the problems of corrosion in aqueous media and how reduction potential can be used to both predict the possibility of corrosion and protect from it.

T6.3 If you have problems with this project, review Section 6.3 *Trends in standard potentials* and have a look at Exercises 6.7 and 6.28.

T6.4 Review Section 6.10 *The influence of complexation* to work on this project. You have been given a hint for the value for [Fe(II)(Ent)] stability constant, and you may consult Resource Section 3 for the standard reduction potential for Fe^{3+}/Fe^{2+} couple. Also note the pH conditions for this reduction.

Chapter 7 An Introduction to Coordination Compounds

Self-Tests

S7.1 **a) Diaquadichloridoplatinum(II).** The formula is $[PtCl_2(OH_2)_2]$. Note the order used in naming complexes of the d-block metals (the metal ion is listed, then ligands are listed in alphabetical order).

b) Diammine*tetra*(thiocyanato-κ*N*)chromate(III). The *-ate* suffix added to the name of the metal indicates an overall negative charge of the complex species. Two NH_3 molecules and four NCS^- anions are bonded to the Cr(III) ion. The NCS^- ligands use their N atoms to bond to Cr(III) as indicated by κ*N* after the ligand name. The formula is $[Cr(\underline{N}CS)_4(NH_3)_2]^-$.

c) *Tris*(1,2-diaminoethane)rhodium(III). The metal ion is Rh(III), there are three bidentate (chelating) ligands (see Table 7.1), and the complex is not anionic (-rhodium(III), not -rhodate(III)). The formula is $[Rh(\kappa^2N\text{-}H_2NCH_2CH_2NH_2)]^{3+}$ where κ^2N indicates that the ligand is bound to Rh^{3+} through both N atoms. Since the accepted abbreviation for 1,2-diaminoethane is en, the formula may be written abbreviated as $[Rh(en)_3]^{3+}$.

d) Bromido*penta*carbonylmanganese(I). The manganese has five CO bonded to it along with one bromide. The formula is $[MnBr(CO)_5]$.

e) Chlorido*tris*(triphenylphosphine)rhodium(I). The formula is $[RhCl(PPh_3)_3]$.

S7.2 The hydrate isomers $[Cr(NO_2)(H_2O)_5]NO_2{\cdot}H_2O$ and $[Cr(H_2O)_6](NO_2)_2$ are possible, as are linkage isomers of the NO_2 group. One possible linkage isomer is $[Cr(ONO)(H_2O)_5]NO_2{\cdot}H_2O$.

S7.3 The two square planar isomers of $[PtBrCl(PR_3)_2]$ are shown below. The NMR data indicate that isomer A is the *trans* isomer because the two trialkylphosphine ligands occupy opposite corners of the square plane and are in the identical magnetic environment (i.e., they are NMR equivalent). As a consequence, both P nuclei appear as a singlet in the ^{31}P NMR spectrum. Isomer B is the *cis* isomer. Note that the two phosphine ligands in the *trans* isomer are related by symmetry elements that this C_{2v} molecule possesses, namely the C_2 axis (the Cl-Pt-Br axis) and the σ_v mirror plane that is perpendicular to the molecular plane. Therefore, they exhibit the same chemical shift in the ^{31}P NMR spectrum of this compound. The two phosphine ligands in the *cis* isomer are not related by the σ mirror plane that this C_s molecule possesses. Since they are chemically nonequivalent, they give rise to separate groups of ^{31}P resonances.

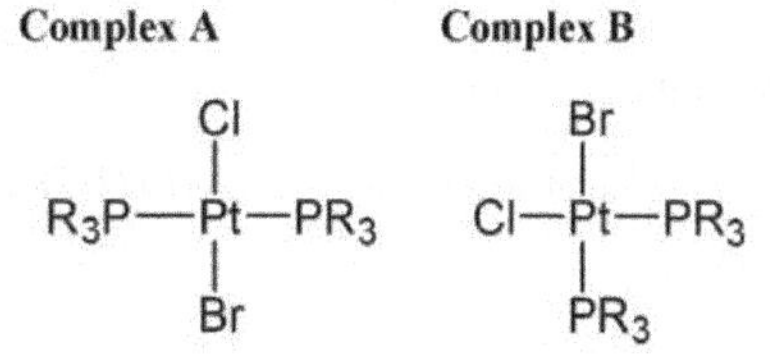

trans-$[PtBrCl(PR_3)]$ *cis*-$[PtBrCl(PR_3)]$

S7.4 The glycinato ligand is an unsymmetrical bidentate ligand (it has a neutral amine nitrogen donor atom and a negatively charged carboxylate oxygen donor atom, see Table 7.1). In the *fac* isomer, the three N atoms (or three O atoms) of glycinato ligand should be adjacent and occupy the corners of one triangular face of the octahedron. In the *mer* isomer, three N atoms should lie in one plane and the O atoms should lie in a perpendicular plane. If you imagine that the complex is a sphere, the three N atoms in the *mer* isomer lie on a *meridian* of the sphere (the largest circle that can be drawn on the surface of the sphere). The two isomers and their non-superimposable mirror images are shown.

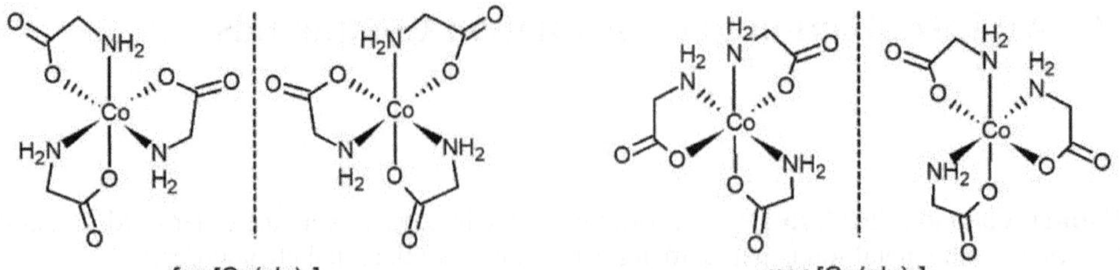

fac-$[Co(gly)_3]$ *mer*-$[Co(gly)_3]$

S7.5 **a) *cis*-$[CrCl_2(ox)_2]^{3-}$:** Drawings of two mirror images of this complex are shown below. They are *not* superimposable and neither molecule possesses a mirror plane or a centre of inversion. Therefore, they represent two enantiomers and this complex is chiral.

b) *trans*-$[CrCl_2(ox)_2]^{3-}$: Drawings of two mirror images of this complex are also shown below. They *are* superimposable, and therefore do not represent two enantiomers but only a single isomer. Note that the complex has at least one mirror plane; for example, one that is defined by two Cl and Cr atoms and bisects the ligand backbones.

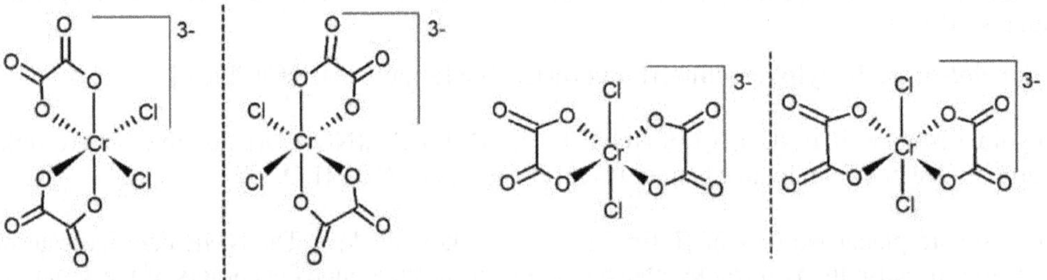

S7.5(a): *cis*-$[CrCl_2(ox)_2]^{3-}$ **S7.5(b)**: *trans*-$[CrCl_2(ox)_2]^{3-}$

c) *cis*-$[RhH(CO)(PR_3)_2]$: This is a complex of Rh(I), which is a d^8 metal ion. Four-coordinate d^8 complexes of period 5 and period 6 metal ions are almost always square planar, and $[RhH(CO)(PR_3)_2]$ is no exception. The bulky PR_3 ligands are *cis* to one another. This compound has C_s symmetry, with the mirror plane coincident with the rhodium atom and the four ligand atoms bound to it. A planar complex cannot be chiral, whether it is square planar, trigonal planar, etc., unless the ligands coordinated to the metal are chiral (see Section 7.11 *Ligand chirality*). In this case none of the ligands is chiral and the complex is achiral.

S7.6 The first substitution step follows the equation:

$$[M(H_2O)_6]^{2+} + L \rightarrow [ML(H_2O)_5]^{2+} + H_2O$$

with $K_{f1} = 1 \times 10^5$.

The second substitution step is:

$$[ML(H_2O)_5]^{2+} + L \rightarrow [ML_2(H_2O)_4]^{2+} + H_2O.$$

We can see that substitution of any of the five remaining waters leads to the desired product. The replacement of the sixth ligand, L, would result in no net reaction—or substitution of five out of six ligands in the reactant leads to the product. Consequently, we can expect K_{f2} to be 5/6 of the K_{f1}:

$$K_{f2} = K_{f1} \times 5/6 = 1 \times 10^5 \times 5/6 = 0.83 \times 10^5.$$

The same analysis can be extended to the third step:

$$[ML_2(H_2O)_4]^{2+} + L \rightarrow [ML_3(H_2O)_3]^{2+} + H_2O.$$

In this case substitution of four out of six ligands leads to the desired product and

$$K_{f3} = K_{f1} \times 4/6 = 1 \times 10^5 \times 4/6 = 0.67 \times 10^5.$$

Following the same procedure, we can find other K values:

$$K_{f4} = K_{f1} \times 3/6 = 1 \times 10^4 \times 3/6 = 0.50 \times 10^5,$$

$$K_{f5} = K_{f1} \times 2/6 = 1 \times 10^5 \times 2/6 = 0.33 \times 10^5, \text{ and}$$

$$K_{f6} = K_{f1} \times 1/6 = 1 \times 10^5 \times 1/6 = 1.7 \times 10^5.$$

From here, the overall formation constant (β_6), is the product of these stepwise formation constants:

$$\beta_6 = 1 \times 10^5 \times 0.83 \times 10^5 \times 0.67 \times 10^5 \times 0.50 \times 10^5 \times 0.33 \times 10^5 \times 0.17 \times 10^5 = 1.6 \times 10^{28}.$$

Exercises

E7.1 **a) $[Ni(CN)_4]^{2-}$:** Tetracyanidonickelate(II), like most d^8 metal complexes with four ligands, has square planar geometry, shown below.

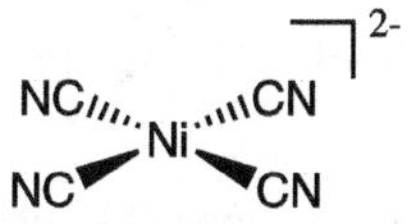

b) $[CoCl_4]^{2-}$: Tetrachloridocobaltate(II) is tetrahedral, like most of the first-row transition metal chlorides.

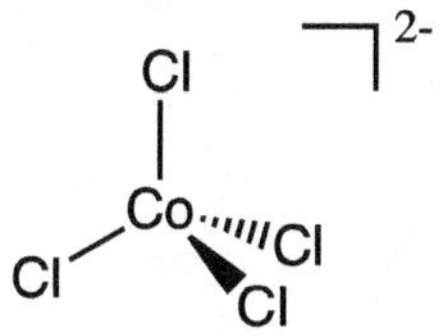

c) $[Mn(NH_3)_6]^{2+}$: Hexaamminemanganese(II) has octahedral geometry, shown below. Coordination number 6 in general can have two geometries: octahedral and trigonal prismatic (see Section 7.4(c) *Six-coordination*). The trigonal prismatic geometry is, however, very rare and we can confidently assign octahedral geometry to a vast majority of complexes with this coordination number.

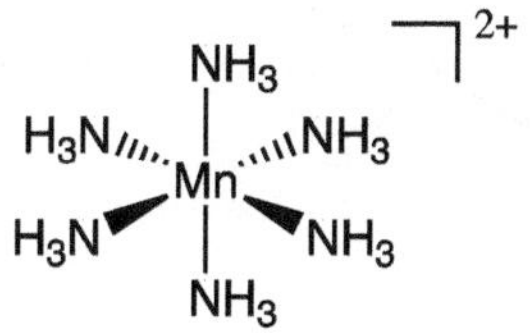

E7.2 The key to writing formulas from a name is keeping the balance of charge between the metal and its ligands. To do this, you need to know if the ligands that directly bond to the metal are neutral or ionic; this will help you determine what the oxidation state is on the metal. The chemical name gives the oxidation state of the metal as a roman numeral; simply balance charge with the ligands that directly bond to the metal and the counter-ions, if necessary.

a) $[CoCl(NH_3)_5]Cl_2$, because it is cobalt(III) and the ammonia ligands are neutral and the chloride bond to the metal is anionic, you know that you need two chlorides as your counter-ion.

b) $[Fe(OH_2)_6](NO_3)_3$, again, because the water ligands are neutral, you need three nitrates as your counter-ion to balance charge.

c) *cis*-$[RuCl_2(en)_2]$ has a divalent metal centre, the chlorides are anionic, and 1,2-diaminoethane ligands are neutral; therefore you do not need a counter-ion to achieve a 2+ charge for the ruthenium atom, rendering the complex neutral.

d) $[Cr(NH_3)_5\text{-}\mu\text{–}OH\text{–}Cr(NH_3)_5]Cl_5$, the Greek symbol μ means you have a bridging ligand between two metal complexes, so you need to balance charge for the entire dimeric complex. The ammonia ligands are neutral, and the bridging hydroxide is an anionic ligand, giving one of the chromium atoms a 1+ charge; thus you need five chloride counter-ions to bring each chromium atom up to a 3+ charge. This gives us an overall 6+ charge for the dimeric complex.

E7.3 **a) *cis*-$[CrCl_2(NH_3)_4]^+$:** *cis*-tetraamminedichloridochromium(III).

b) *trans*-$[Cr(NCS)_4(NH_3)_2]^-$: *trans*-diamminetetra(κ*N*-thiocyanato)chromate(III).

c) $[Co(C_2O_4)(en)_2]^+$: bis(1,2-diaminoethane)ethylenediamineoxalatocobalt(III), which is neither *cis* nor *trans* but does have one optical isomer, shown below.

E7.3(c): $[Co(C_2O_4)(en)_2]^+$

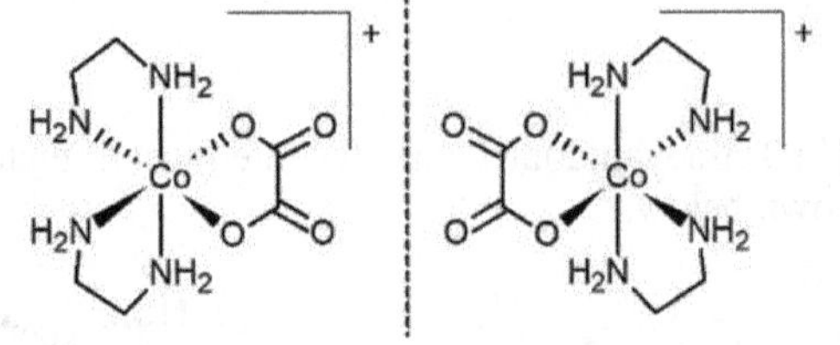

E7.4 **a)** With four-coordinate complexes, the two possible geometries are tetrahedral or square planar, and both are shown below. Most of the first-row divalent transition metal halides are tetrahedral, whereas metals and their ions that have a d^8 electronic configuration tend to be square planar.

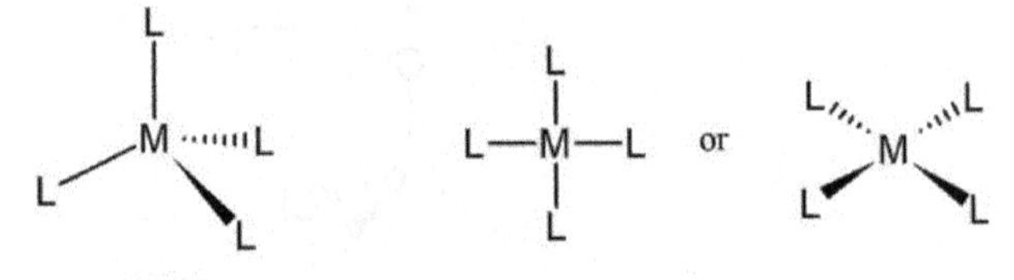

b) For a tetrahedral complex, there are no isomers for MA_2B_2; however, for a square planar complex, there are two isomers, *cis* and *trans*, shown below.

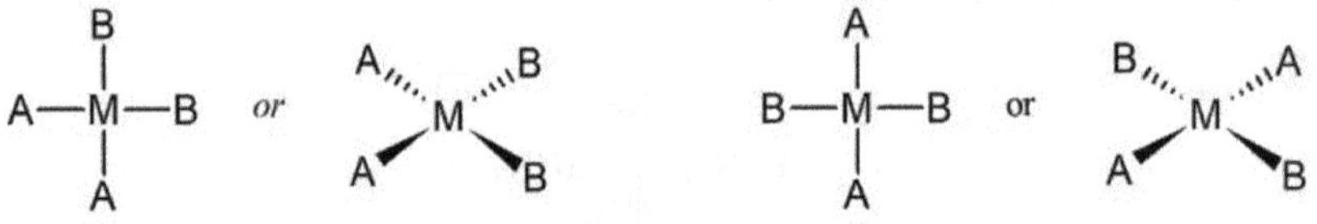

E7.5 With five-coordinate complexes, the two possible geometries are trigonal bipyramidal and square-based pyramidal, as shown below.

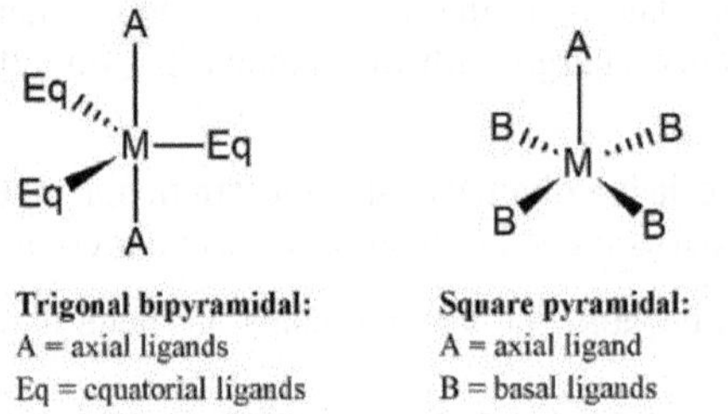

E7.6 **a)** Most of the six-coordinate complexes are octahedral, very seldom are they trigonal prismatic. Drawings of these are shown below.

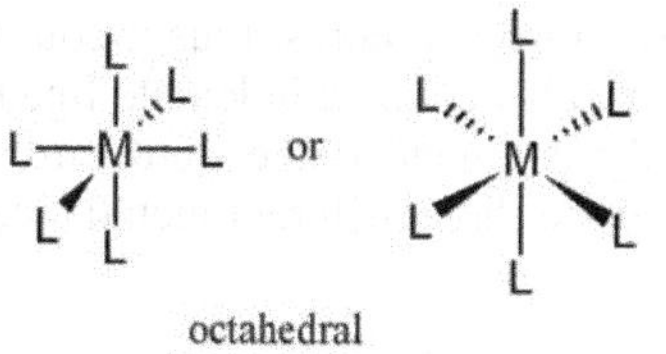

octahedral

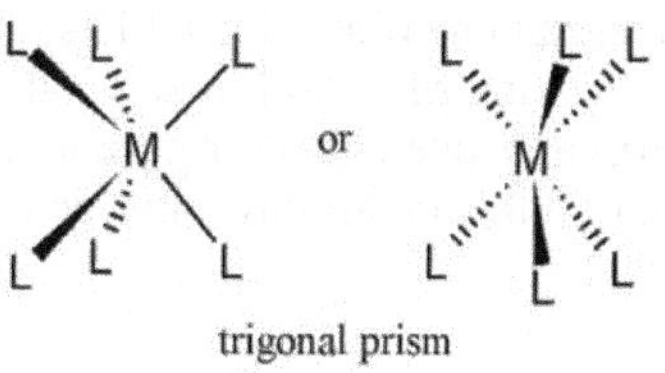

trigonal prism

b) The trigonal prism is a rare geometry. Nearly all six-coordinate complexes are octahedral.

E7.7 The terms monodentate, bidentate, and tetradentate describe how many Lewis bases you have on your ligand and whether they can physically bind to the metal (i.e., behave as electron pair donors to the metal). A monodentate ligand can bond to a metal atom only at a single atom, also called a donor atom. A bidentate ligand can bond through two atoms, and a tetradentate ligand can bond through four atoms. Examples of monodentate and bidentate ligands are shown below. Tetraazacyclotetradecane, from Table 7.1, is an example of a tetradentate ligand—all four nitrogen atoms can donate a lone pair to the metal.

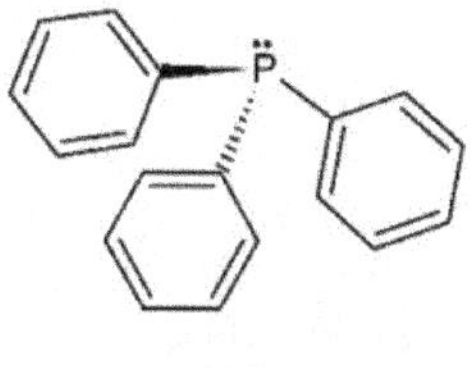

monodentate

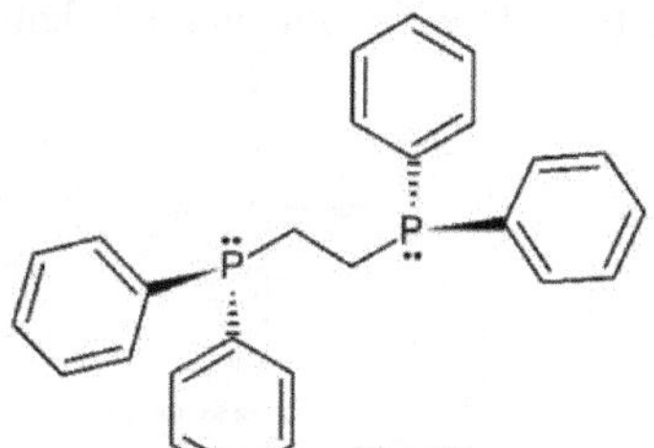

bidentate

E7.8 Linkage isomers can arise with ambidentate ligands. Ambidentate ligands are ligands that have two different atoms within the molecule that can serve as a Lewis basic donor atom. An example is the thiocyanide anion NCS^-. It can bond to a metal either through the nitrogen's lone pair or through the sulfur's lone pair, depending on the electronics of the metal. If the metal is soft, then the softer-base sulfur is preferred; if the metal is hard, then the hard-base nitrogen is preferred.

NCS–M (thiocyanato-κ*S*) SCN–M (thiocyanato-κ*N*)

E7.9 **a)** Bisdimethylphosphino ethane (dmpe) bonds to a metal through lone pairs on two phosphorus atoms; thus, dmpe is a bidentate, chelating ligand that could also be a bridging ligand due to free rotation around single bonds.

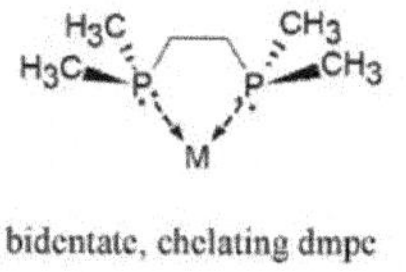

bidentate, chelating dmpe

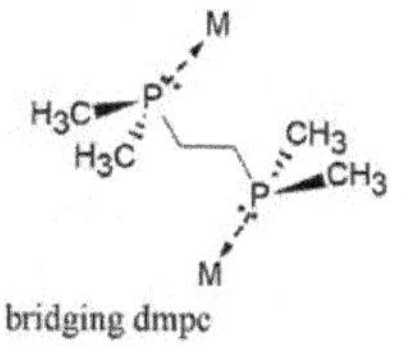

bridging dmpe

b) 2,2′-Bipyridine (bpy) is a bidentate, chelating ligand that is able to bond to a metal through both of its nitrogen atoms.

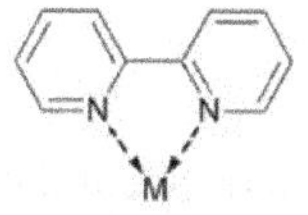

c) Pyrazine is a monodentate ligand even though it has two Lewis basic sites. Because of the location of the nitrogen atoms, the ligand can only bond to one metal; it can, however, bridge two metals, as shown below.

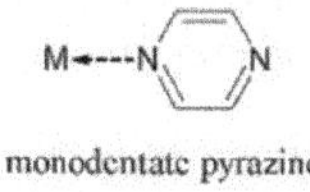

monodentate pyrazine

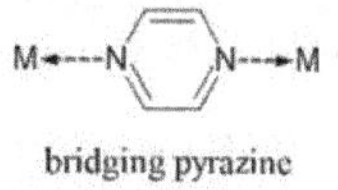

bridging pyrazine

d) Diethylenetriamine (dien) ligand has three Lewis basic nitrogen atoms, thus it could be a tridentate ligand that forms two chelating rings with one metal. It can, however, also act as a bidentate ligand (bonding using only two N atoms) forming one chelate ring and could be a bridging ligand if the third nitrogen atom is bonded to the second metal (in theory, all three nitrogen atoms could bind to three different metallic centres as well):

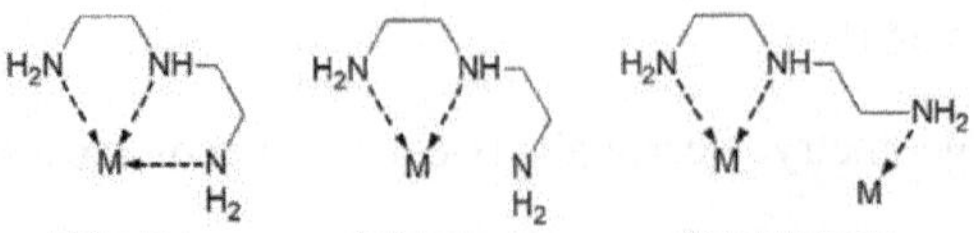

tridentate, chelating dien — bidentate, chelating dien — bidentate and bridging dien

e) Tetraazacyclododecane ligand, with its four Lewis basic nitrogen atoms, could be a tetradentate, macrocyclic ligand. Note that structure **I**, in which M and N atoms are more-or-less in one plane, would be limited to small cations (for example Li^+) because the hole of the macrocycle prevents larger cations from fitting in. For a larger cation, structure **II** in which the ligand's N atoms are above the metal atom, would be expected. This ligand could be bidentate and bridging (e.g., free N atoms in the bidentate structure are free to bind to another metallic centre).

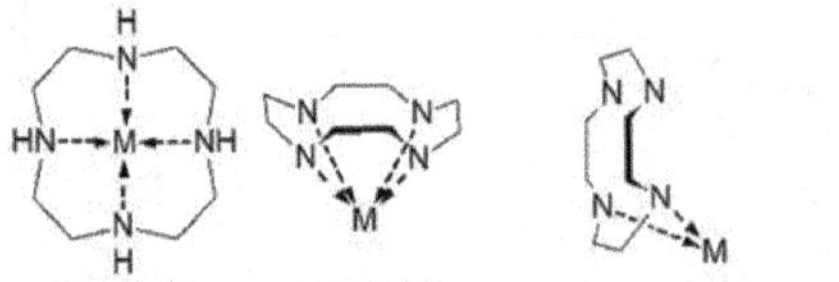

structure I — *structure* II

tetradentate, chelating — bidentate, chelating

E7.10 The generic structures of complexes containing ligands (a) to (g) are shown below. The examples are taken for various coordination numbers and geometries because the exercise does not strictly assign them. Following the structures are some important points regarding each one that you should start to recognize and think about. In each case L is a neutral monodentate ligand, and metal has a positive charge n+; the overall charge of the complex is indicated as well (note that it can be the same as the metal charge provided that all ligands are neutral).

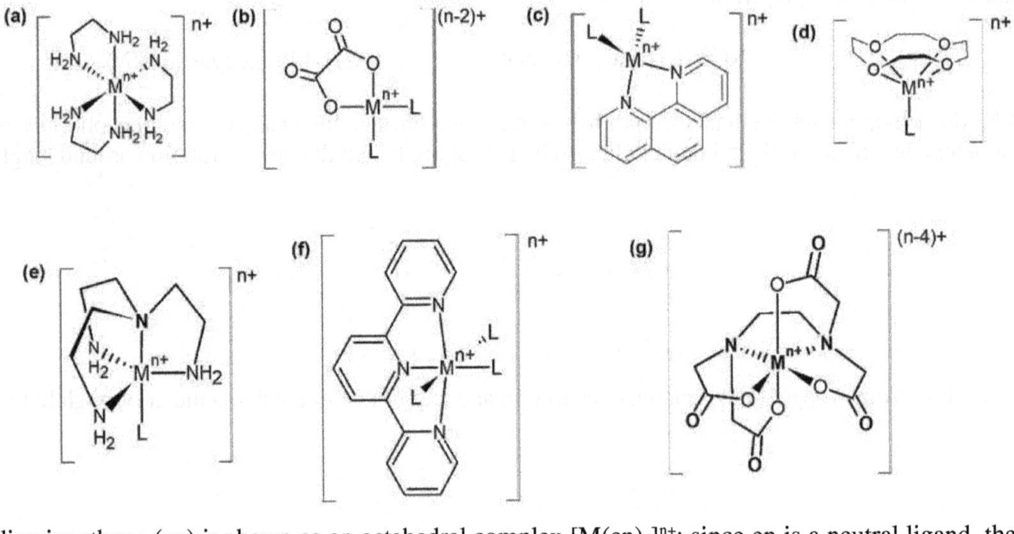

a) 1,2-diaminoethane (en) is shown as an octahedral complex $[M(en)_3]^{n+}$; since en is a neutral ligand, the complex is a cation of charge n+. Remember that octahedral complexes such as (a) are chiral.

b) Oxalate ($C_2O_4^{2-}$, ox^{2-}) is represented as a square planar complex $[ML_2(ox)]^{(n-2)+}$; since the oxalate ligand has a 2– charge, the overall charge of the complex depends on n+. Note that, although square planar, complex (b) can only exist as a *cis* isomer because rigid ox^{2-} prevents the formation of *trans* isomer.

c) Phenantroline (phen) is represented by a tetrahedral complex $[ML_2(phen)]^{n+}$. Remember that tetrahedral complexes cannot have *cis*/*trans* isomers. Again, the charge of the complex depends on the charge on the metal M.

d) 12-crown-4 is a crown ether ligand. Remember that the first number of the name (in this case 12) gives the total number of atoms forming the ring while the number at the end (in this case 4) gives the number of oxygen atoms in the ring structure. The geometry of complex (d) can be described as square pyramidal, but keep in mind that the hole in 12-crown-4 is relatively small and the majority of cations will not fit inside. Consequently, the M^{n+} cannot be in the same plane as O atoms.

e) Triaminotriethylamine (tren) is also shown in a complex with coordination number five, but this time in trigonal bipyramidal geometry. Recall that the two most common geometries for coordination number five are square pyramid and trigonal bipyramid. The two isomeric structures usually have small differences in energy meaning that 5-coordinate complexes can exchange geometries, particularly in solution (they are fluxtional). Complexes (d) and (e), however, cannot change their geometries because their tetradentate ligands are rigid and lock them in one ligand arrangement only.

f) 2,2';6',2"-terpyridine (terpy), a tridentate ligand, is shown in an octahedral complex. Although you should expect *mer*/*fac* isomers with tridentate ligands, in this case only *mer* isomer is possible due to the terpy's rigid structure.

g) $EDTA^{4-}$ is a very common hexadentate ligand that forms very stable complexes with the majority of metals. This is frequently used in analytical chemistry during complexometric titrations.

E7.11 $[RuBr(NH_3)_5]Cl$ and $[RuCl(NH_3)_5]Br$ complexes differ as to which halogen is bonded, and as to which one is the counter ion. These types of isomers are known as ionization isomers because they give different ions in the solution. When dissolved $[RuBr(NH_3)_5]Cl$ will dissociate into $[RuBr(NH_3)_5]^+$ and chloride ions, while $[RuCl(NH_3)_5]Br$ will produce $[RuCl(NH_3)_5]^+$ and bromide ions.

E7.12 For tetrahedral complexes, the only isomers found are when you have four different ligands bound to the metal. Therefore, $[CoBr_2Cl_2]^-$ and $[CoBrCl_2(OH_2)]$ have no isomers. However, $[CoBrClI(OH_2)]$ has an optical isomer shown below.

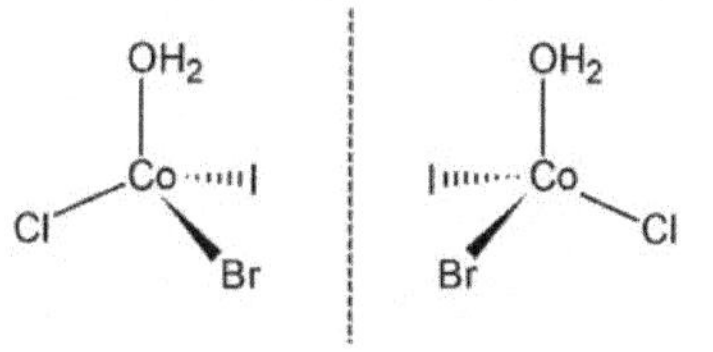

E7.13 For square planar complexes, depending on the ligands, several isomers are possible.

a) $[Pt(ox)(NH_3)_2]$ has no isomers because of the chelating oxalato ligand. The oxalate forces the ammonia molecules to be *cis* only.

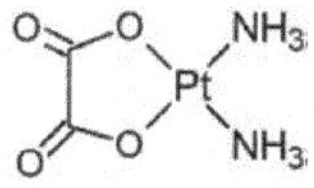

b) $[PdBrCl(PEt_3)_2]$ has two isomers, *cis* and *trans*, as shown below.

cis-$[PdBrCl(PEt_3)_2]$ *trans*-$[PdBrCl(PEt_3)_2]$

c) $[IrHCO(PR_3)_2]$ has two isomers, *cis* and *trans*, as shown below.

cis-$[IrH(CO)(PR_3)_2]$ *trans*-$[IrH(CO)(PR_3)_2]$

d) $[Pd(gly)_2]$ has two isomers, *cis* and *trans*, as shown below.

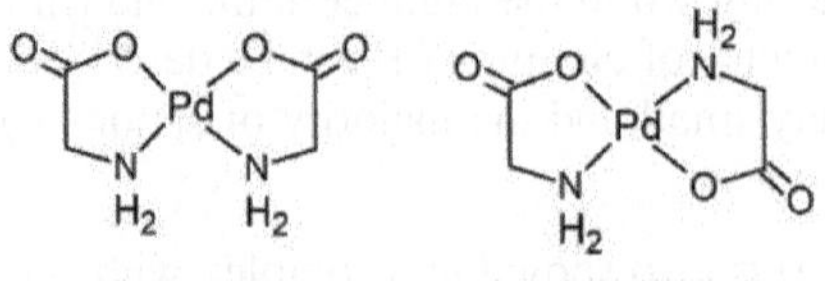

cis-$[Pd(gly)_2]$ *trans*-$[Pd(gly)_2]$

E7.14 **a) $[FeCl(OH_2)_5]^{2+}$:** This complex has no isomers.

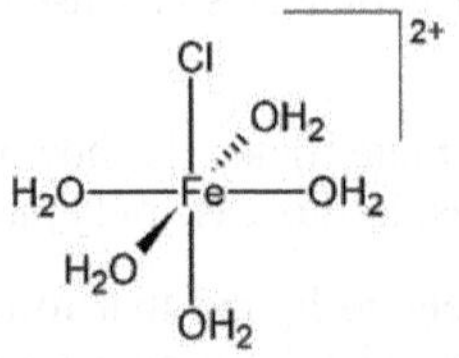

b) $[IrCl_3(PEt_3)_2]$: There are two isomers for this complex, shown below.

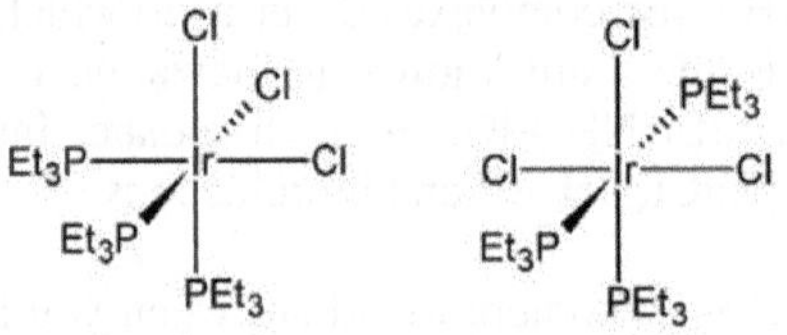

fac *mer*

c) $[Ru(bpy)_3]^{2+}$: This complex has optical isomers, shown below.

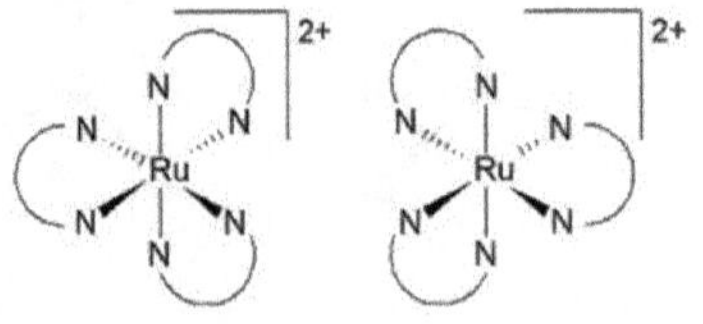

Δ isomer Λ isomer

d) $[CoCl_2(en)(NH_3)_2]^+$: This complex has *cis* and *trans*, as well as optical isomers, shown below.

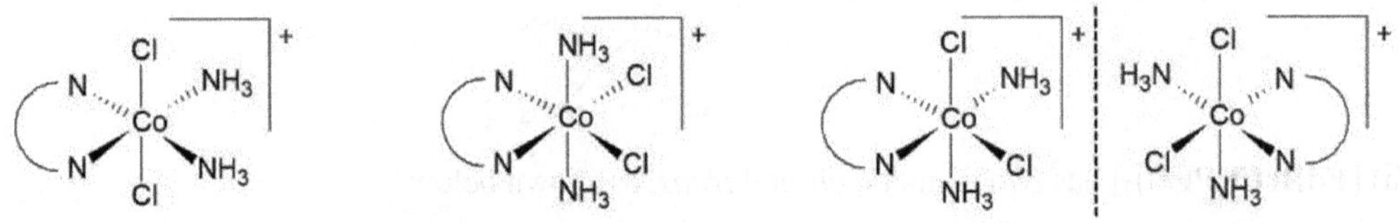

trans,cis-$[CoCl_2(NH_3)_2(en)]^+$ *cis,trans*-$[CoCl_2(NH_3)_2(en)]^+$ *cis,cis*-$[CoCl_2(NH_3)_2(en)]^+$

e) $[W(CO)_4(py)_2]$ has two isomers, *cis* and *trans*, shown below.

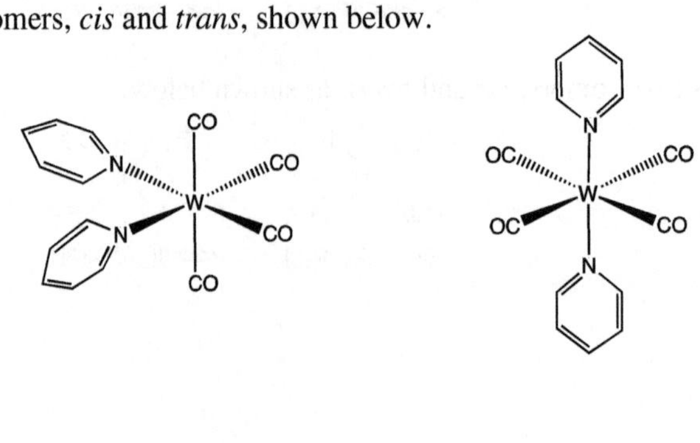

E7.15 Ignoring optical isomers, nine isomers are possible. Including optical isomers, 15 isomers are possible! Read Section 7.10(a) *Geometrical isomerism* for a better understanding of all the possible isomers.

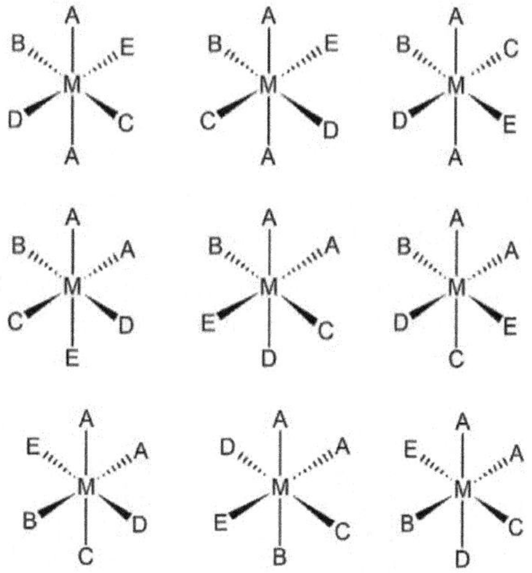

E7.16 **a) $[Cr(ox)_3]^{3-}$:** All octahedral *tris*(bidentate ligand) complexes are chiral because they can exist as either a right-hand or a left-hand propeller, as shown in the drawings of the two nonsuperimposable mirror images.

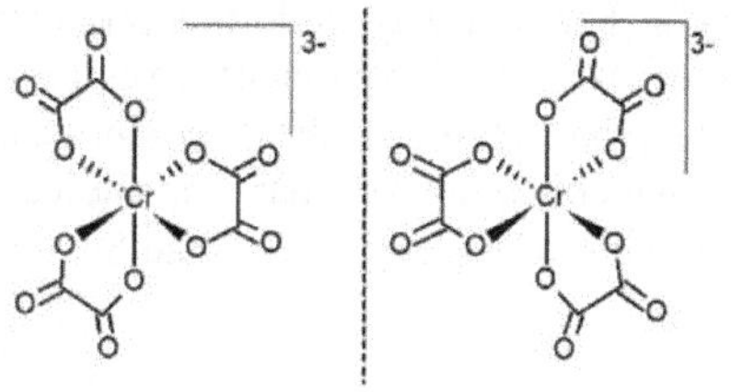

b) *cis*-$[PtCl_2(en)]$: This is a four-coordinate complex of a period 6 d^8 metal ion, so it is undoubtedly square planar. You will recall from Chapter 3 that any planar complex contains at least one plane of symmetry (the one in the plane of the molecule) and must be achiral. In this case the five-membered chelate ring formed by the 1,2-diaminoethane ligand is not planar, so, strictly speaking, the complex is not planar. It *can* exist as two enantiomers, depending on the conformation of the chelate ring, as shown below (note that chloride ligands have been removed for clarity). However, the conformational interconversion of the carbon backbone is extremely rapid, so the two enantiomers cannot be separated.

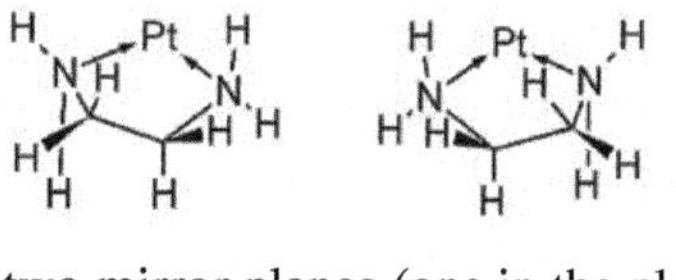

c) *cis*-$[RhCl_2(NH_3)_4]^+$: This complex has two mirror planes (one in the plane defined by 2Cl ligands and Rh metal and the second perpendicular to the first bisecting Cl-Rh-Cl and N-Rh-N bonds), so it is not chiral. The structure is shown below.

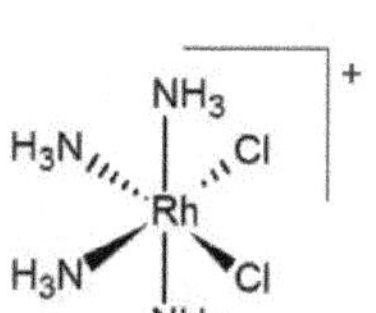

d) $[Ru(bpy)_3]^{2+}$: As stated in the answer to part (a) above, all octahedral *tris* (bidentate ligand) complexes are chiral, and this one is no exception. The two nonsuperimposable mirror images are shown below (the bipyridine ligands in abbreviated form).

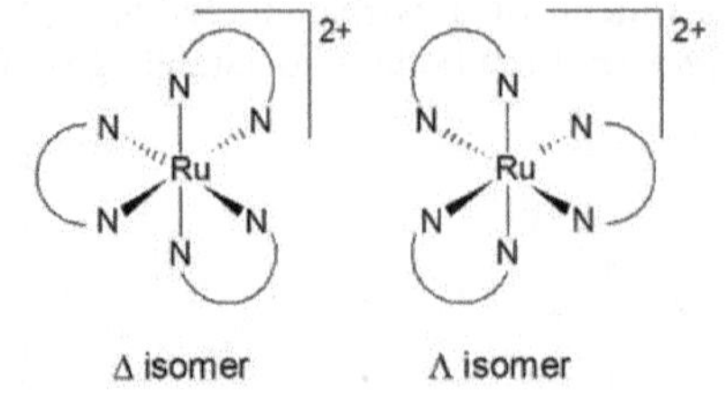

(e) *fac*-[$Co(NO_2)_3$(dien)]: There is a mirror plane through the metal, bisecting the dien ligand and (O_2N)–Co–(NO_2) angle shown below. Thus, the complex is not chiral.

(f) *mer*-[$Co(NO_2)_3$(dien)]: There are two mirror planes: one through all three coordinating N atoms of the dien ligand and the metal atom, the other through all three nitro groups and the metal atom. Thus, the complex is not chiral.

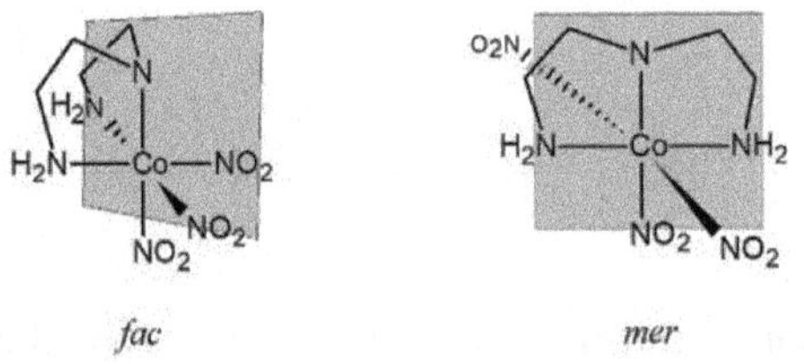

E7.17 Below is a picture of both isomers. The best way to do this problem is to draw both isomers as mirror images of each other, and look along one of the four C_3 symmetry axes found of ideal octahedral structure. If the ligand backbone is rotating clockwise then the structure is the Δ isomer, if it rotates counter-clockwise, it is the Λ isomer. When you do this, it is obvious which isomer you have, in the case of this problem, the complex drawn in the exercise is the Λ isomer. For more help with this concept read Section 7.10(b) *Chirality and optical isomerism.*

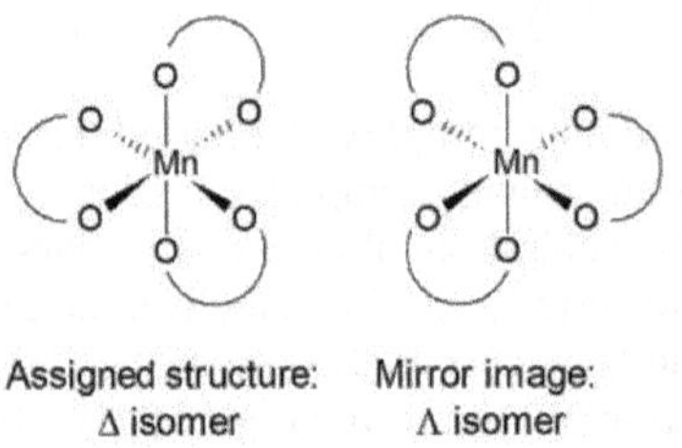

E7.18 Below is a picture of both isomers. As in Problem 7.17, the best way to do this problem is to draw both isomers as mirror images of each other. A clockwise rotation around C_3 of perfect octahedral structure would be the Δ isomer, and counter-clockwise rotation would be the Λ isomer.

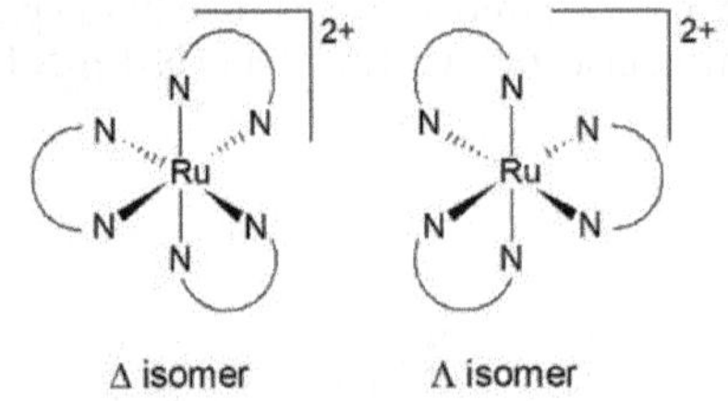

E7.19 The values of the stepwise formation constants drop from K_{f1} to K_{f4}, as expected on statistical grounds. However, the fifth stepwise formation constant is substantially lower, suggesting a change in coordination. In fact, what happens is the very stable square planar complex ion $[Cu(NH_3)_4]^{2+}$ is formed, and the addition of the fifth NH_3 ligand is very difficult. The equilibrium for this step is shifted to the left (toward reactants). This results in $K_{f5} < 1$ with a negative log value.

E7.20 The values for the 1,2-diaminoethane reaction are substantially higher, indicating more favourable complex formation, and this can be attributed to the chelate effect. Read Section 7.14 *The chelate and macrocyclic effects* for better understanding of this phenomenon.

Guidelines for Selected Tutorial Problems

T7.2 The most important part of this problem is deciding on the composition of inner-sphere and outer-sphere complexes of the two compounds mentioned (see Section *The language of coordination chemistry*). For example, the fact that three equivalents of AgCl can be quickly obtained from a solution of the pink salt indicates that three Cl^- anions are a part of the outer-sphere complex. The rest, combined together, should form the inner-sphere complex $[Co(NH_3)_5(OH_2)]^{3+}$, and the salt should be $[Co(NH_3)_5(OH_2)]Cl_3$. It can be demonstrated similarly that the purple solid is $[CoCl(NH_3)_5]Cl$. These guidelines are useful for Tutorial Problems **7.3** and **7.5** as well.

T7.4 The vital part of information for determining the structure (and hence the name) of the diaqua complex is the fact that it does not form a chelate with (otherwise chelating) 1,2-diaminoethane ligand, not so much the fact that we have started with *trans* isomer. Consult Figure 7.4 in your textbook to determine the structure of the diaqua complex. For the third isomer you have been given an *empirical* composition of $PtCl_2{\cdot}2NH_3$. The actual composition can have any multiple of 1 : 2 ratio (i.e., the actual composition might be $2PtCl_2{\cdot}4NH_3$, or $3PtCl_2{\cdot}6NH_3$, etc. as long as the simplest ratio remains 1 : 2). If you look at the products from the reaction with $AgNO_3$, you will see that one contains a cation $[Pt(NH_4)_4]^{2+}$ and the other anion $[PtCl_4]^{2-}$ as inner sphere complexes. Simply combining the two we have the third isomer $[Pt(NH_4)_4][PtCl_4]$ with composition $2PtCl_2{\cdot}4NH_3$. (This is the special case isomerism sometimes referred to as polymer isomerism because the isomers differ by the number of empirical formula units, or monomers, in their structure; for example, in this case the general formula of the polymer isomers can be written as $(PtCl_2{\cdot}2NH_3)_n$ with n = 2.) The name for the third isomer is tetraammineplatinum(II) tetrachloridoplatinate(II).

T7.9 Recall from other courses that amines with three different substituents on N atom should be chiral, but due to the fast isomerization this chirality is not observed in real situations. Once, however, the N atom coordinates to the metal, the isomerization is not possible any longer—in this case particularly because the N atoms become a part of a rigid chelating ring.

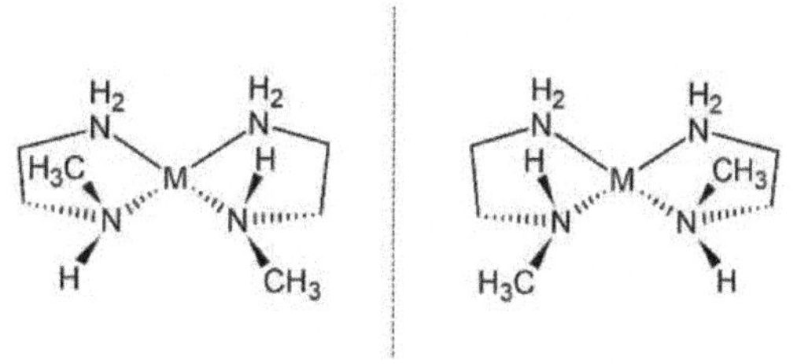

T7.11 BINAP (short for (2,2'-bis(diphenylphosphino)-1,1'-binaphthyl) has axial chirality with the chiral axis coinciding with the C–C bond connecting two naphthyl systems. The bulky diphenylhosphino groups as well as indicated H atoms on naphthyl groups prevent the rotation around this C–C bond making the structure stable with respect to racemization:

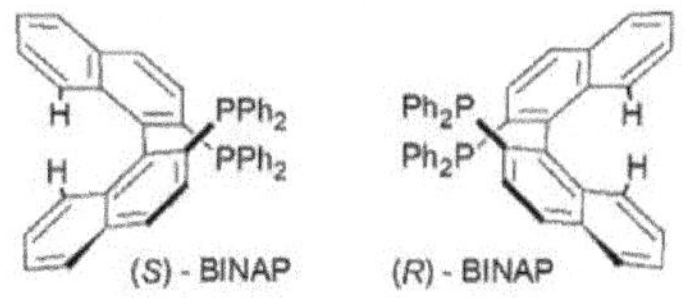

Combine this chirality of BINAP ligand with what you learned about the chirality of complexes to discuss the chirality of BINAP-containing complexes.

T7.12 The relevant sections are 7.13 and 7.14, trends in successive formation constants and chelating effect. The "copper anomaly" has been addressed in Exercises 7.19 and 7.20.

T7.13 Rotaxanes are molecular machines—a very interesting research topic since the Nobel prize in chemistry for 2017 was awarded to the three scientists pioneering the concept and syntheses of rotaxanes: Jean-Pierre Sauvage, Sir J. Fraser Stoddart and Bernard L. Feringa. Their work is a good starting point for discussion. Pseudorotaxanes are intermediates in the rotaxane synthesis. The material relevant to coordination chemistry in their synthesis is found in Box 7.4.

Chapter 8 Physical Techniques in Inorganic Chemistry

Self-Tests

S8.1 The ionic radius of Cr(IV), 69 pm, is smaller than that of Ti(IV), 75 pm. The unit cell and *d* spacings will shrink because of the smaller radius of chromium. The XRD pattern for CrO_2 will show identical reflections to those of rutile TiO_2 (see Figure 8.4) but shifted to slightly higher diffraction angles.

S8.2 **a)** The K_2Se_5 crystal with dimensions 5 μm × 10 μm × 20 μm would be too small for the single-crystal X-ray diffraction analysis that typically requires a crystal with dimensions of 50 μm × 50 μm × 50 μm or above. Smaller crystals, such as the one in this Self-Test require X-ray sources of higher intensity which are obtained using synchrotron radiation. Thus, we would need different source of X-rays (see Section 8.1(b) *Single-crystal X-ray diffraction* and Section 8.1(c) *X-ray diffraction at synchrotron sources*).

b) Recall that X-rays are scattered by the electrons and the neutrons by nuclei. In the case of any element-H bond (specifically in this Self-test, O–H bond), hydrogen's single electron is involved in bonding, and is shared with the other partner in the bond. This means that the X-ray scattering points are electrons in the bond located between the two nuclei, and what we see is short element-H bonds. On the other hand, the neutrons are going to be scattered by a proton in hydrogen's nucleus and the element-H bonds are going to be longer, and more accurate, than that determined with X-ray crystallography.

S8.3 As we can clearly see from Figure 8.13 TiO_2 particles absorb the ultraviolet radiation and prevent this type of radiation from reaching the skin offering protection from the sun's rays that can cause skin cancer.

S8.4 Using the VSEPR theory we find that Xe in XeF_2 has five electron pairs: three lone and two bonding. Thus, the molecular geometry is linear with a $D_{\infty h}$ point group symmetry. As a linear molecule, we expect XeF_2 to have $3N - 5 = 3(3) - 5 = 4$ total vibrational modes. These modes cannot be both IR and Raman active as the molecule has a centre of symmetry (the exclusion rule).

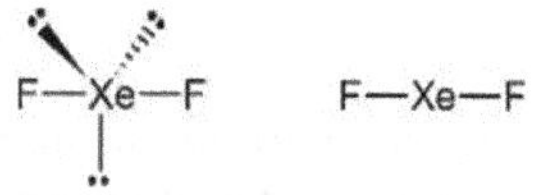

S8.5 **a)** The VSEPR theory would predict a square pyramidal geometry for the BrF_5 molecule. Recall from previous chapters that this geometry would place F atoms in two different environments: axial (F_a) and basal (F_b). Thus, we would expect two different chemical shifts in ^{19}F NMR: one for F_a and one for F_b. The intensity of these signals should have relative ratios 1 : 4 because there is only one F_a whereas there are four F_b atoms. According to the n + 1 rule for the nuclei with $I = ½$, the F_a resonance would be split into five equally spaced lines (i.e., a quintet) by four basal F atoms with line intensities 1 : 4 : 6 : 4 : 1 according to Pascal's triangle. The signal due to F_b atoms will be split into a doublet by one F_a with line intensities 1 : 1.

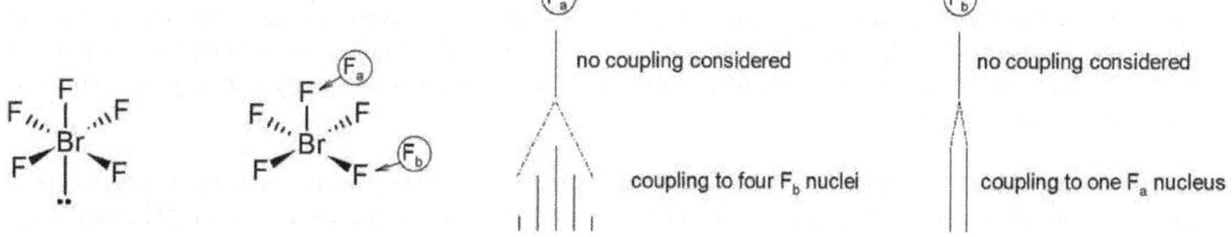

b) The hydride resonance couples to three nuclei (two different ^{31}P nuclei—one *trans* to the hydride and the other one *cis* to the hydride ligand—and the ^{103}Rh nucleus) that are 100% abundant and all of which have $I = ½$ (see Table 8.4). The first doublet is observed due to the coupling between hydride and rhodium. Since the two atoms are directly bonded, the coupling constant is going to be large. Then, through coupling with P atom *trans* to the hydride, each line of the first doublet is going to be split into a doublet creating doublet of doublets. Finally, due to the coupling to the P atom *cis* to the hydride, every line of the doublet of doublets is going to be further split into a doublet giving the observed doublet of doublets of doublets pattern. Since the three coupling constants are different, the effect is to split the signal into a doublet of doublet of doublets, thus generating eight lines in the NMR of equal intensity.

S8.6 **a)** The assigned compound, $Tm_4(SiO_4)(Si_3O_{10})$, contains two different silicate anions: the orthosilicate, SiO_4^{4-} and chain $Si_3O_{10}^{8-}$. Within $Si_3O_{10}^{8-}$ we have two different Si environments—one Si atom is in the centre of the chain with the remaining two Si atoms on each end. The Si atom in the orthosilicate anion is in its own environment. Thus, we can expect three signals in ^{29}Si-MASNMR of this compound: one for SiO_4^{4-} (with intensity 1) and two for $Si_3O_{10}^{8-}$ (with intensities 1 : 2 for central and terminal Si atoms).

b) The cyclic anion $Si_4O_{12}^{8-}$ would show a single resonance. This anion is very similar to the structure **4**, but has one more $\{SiO_3\}$ ring member.

S8.7 Consulting Table 8.4, we see that 14% of the naturally occurring tungsten is ^{183}W, which has I = ½. Owing to this spin, the EPR signal of a new material should be split into two lines. This doublet would be superimposed on a nonsplit signal that arises from the 86% of tungsten that does not have a spin. The splitting would be a characteristic of a new material containing tungsten.

S8.8 The oxidation state for iron in Sr_2FeO_4 is +4. The outermost electron configuration is $3d^4$. We would expect the isomer shift to be smaller and less positive (below 0.2 mm s^{-1}) due to a slight increase of s-electron density at the nucleus.

S8.9 Generally, as the oxidation state of an atom is increasing, the radius is decreasing and the K-shell electrons are moving closer to the nucleus resulting in an increased stability of these electrons. Thus, the energy of XAS K-edge is expected to gradually increase with an increase in oxidation state of sulfur from S(–II) in S^{2-} to S(VI) in SO_4^{2-}.

S8.10 Both chlorine (^{35}Cl 76% and ^{37}Cl 24%) and bromine (^{79}Br 51% and ^{81}Br 49%) exist as two isotopes. Consider the differences in mass numbers for the isotopes—any compound containing either Cl or Br will have molecular ions $2m_u$ apart. The lightest isotopomer of $ClBr_3$ is $^{35}Cl^{79}Br_3$ at 272*u* and the heaviest is $^{37}Cl^{81}Br_3$ at 280*u*. Three other molecular masses are possible, giving rise to a total of five peaks in the mass spectrum shown in Figure 8.43. The differences in the relative intensities of these peaks are a consequence of the differences in the percent abundance for each isotope.

S8.11 CHN analysis is used to determine percent composition (mass percentages) of C, H, and N. Because the atomic masses of 5d metals are significantly higher than the atomic masses of 3d metals, the hydrogen percentages will be less accurate as they correspond to smaller fractions of the overall molecular masses of the compounds.

S8.12 The EDAX analysis does not give very accurate quantitative results for the magnesium aluminium silicate because the three elements (Mg, Al, and Si) are next to each other in the periodic table; thus, we can expect some peak overlapping.

S8.13 7.673 mg of tin oxide corresponds to 0.0646 mmol of Sn. The 10.000 mg sample will contain 2.327 mg of oxygen (10.000 mg – 7.673 mg) or 0.145 mmol of O. We could write the initial formula of this oxide as $Sn_{0.0646}O_{0.145}$. The molecular formulas of compounds, however, are generally written using whole numbers in the subscripts (unless the compound is indeed nonstoichiometric) and not fractions or irrational numbers. To convert our subscripts to the whole numbers we divide both with the smaller of the two. We obtain 0.0646/0.0646 = 1 and 0.145/0.0646 = 2.244. The last number is still not a whole number but it becomes obvious that by multiplying both values by 2 we can obtain whole numbers 2 and 5.

S8.14 Os(III) solution at pH 3.1 contains an $Os^{III}OH$ species, whereas Os(IV) solutions at the same pH have an $Os^{IV}O$ species. From Pourbaix diagram we see that $Os^{III}OH$ complex remains protonated in the pH range 3.1 to 13. This means that at very high sweep rates electron transfer is going to be fast in comparison to proton transfer. Therefore, we should see one reversible peak in the cyclic voltamograms in this pH range.

Exercises

E8.1 The three compounds, two starting materials and the spinel product, clearly have different structures and different unit cells. The easiest way to monitor the reaction is powder X-ray diffraction. As long as we can observe the diffraction cones associated with TiO_2 and MgO together with those from Mg_2TiO_4, the reaction is not complete. Once our diffractogram shows only the pattern due to Mg_2TiO_4 the reaction is done!

E8.2 As in previous exercise, powder X-ray diffraction comes to the rescue. But a bit more care is needed here. Besides being white and insoluble, MgO and CaO share the same structure–that of NaCl–and the two diffraction patterns will be very similar. To clearly differentiate between the two, we must determine unit cell parameters of both with powders. Since Mg^{2+} is smaller than Ca^{2+}, the MgO unit cell is going to be smaller.

E8.3 If the laboratory single-crystal diffractometer can study crystals that are 50 μm × 50 μm × 50 μm, with a new synchrotron source that has a millionfold increase in source intensity, one can reduce the crystal volume by $1/10^6$, thus the size would be 0.5 μm × 0.5 μm × 0.5 μm. The opposite happens if the flux is reduced: we need a bigger crystal. For the 10^3 times weaker neutron flux we must have a crystal with 10^3 larger volume—the minimum size is then 500 μm × 500 μm × 500 μm.

E8.4 The difficulty in determining element-hydrogen bond lengths (i.e., the distance between the two nuclei) with high accuracy is because a hydrogen atom has only one electron. Hydrogen's electron is normally a part of a bond and hence displaced from the H atom. Since the single-crystal X-ray analysis relies on electron densities to determine the locations of the atoms, H atoms lack sufficient electron density around the nucleus for their location to be determined with a high accuracy. This, however, is not the case with Se-O bonds because both Se and O have a sufficient number of electrons for accurate determination of their locations in the crystal structure. Hydrogen bonding can also further reduce the accuracy because the nucleus is shifted toward H-bond acceptor (i.e., O atom in SeO_4^{2-} in this case) and away from H-bond donor (N atom in this case). See also Self-test 8.2b.

E8.5 To determine if a hydrogen bond is symmetric or not, we need a precise location of hydrogen atom. The single crystal diffraction techniques provide us with atomic positions within the unit cell. The question now remains which one to use? The neutron diffraction is much better in locating H atoms with high accuracy, but it does need a larger single crystal than the X-ray (see Exercise 8.4 and Section 8.2 *Neutron diffraction*).

In the case of a) BO(OH) we could still get a good accuracy in H atom positions using single crystal X-ray diffraction because B and O are also light elements with few electrons and of low scattering power.

Unfortunately, that is not the case with CrO(OH)—the heavy Cr centre in the structure is going to dominate the electron density obtained from the X-ray scattering. Seeing small almost "naked" H in this case would be very difficult. Thus, in this case we should hope for a large, good quality single crystal for the neutron diffraction experiment.

E8.6 **a)** Using the de Broglie relation given in Section 8.2, we have:

$$\lambda = h/mv = (6.626 \times 10^{-34}\ \text{J s})/(1.675 \times 10^{-27}\ \text{kg} \times 2600\ \text{m/s}) = 1.52 \times 10^{-10}\ \text{m} = 152\ \text{pm}.$$

This wavelength is very useful for diffraction because it is comparable to interatomic separations. This value is right in the middle of 100–200 pm range mentioned in Section 8.2 *Neutron diffraction*.

b) We can use again the de Broglie relation, but we need to convert given wavelength to SI units first (see *Note on good practice* in Section 8.1 *Diffraction Methods* which provides convenient conversion relations):

$$1\ \text{Å} = 10^{-10}\ \text{m} = 10^{-8}\ \text{cm} = 10^{-2}\ \text{pm}.$$

Thus, our 5 Å = 5×10^{-10} m. Note that we chose metres as unit to achieve dimensional homogeneity–i.e. to cancel the units during the dimensional analysis.

We can calculate the velocity now:

$$v = h/m\lambda = (6.626 \times 10^{-34}\ \text{J s})/(1.675 \times 10^{-27}\ \text{kg} \times 5 \times 10^{-10}\ \text{m}) = 791\ \text{m/s}.$$

E8.7 Unlike X-rays, which are scattered by the electrons, the neutrons are scattered by atomic nuclei in a crystalline sample. The X-ray analysis always underestimates element-hydrogen bond lengths because of very low electron density at H nucleus. H atom, having only one electron, has very low electron density that is further decreased when H bonds to the other atoms because the electron now resides between the two nuclei forming a bond. This is not an issue with neutron diffraction because neutrons will be scattered by hydrogen nucleus providing us with its very accurate location. We could, however, expect less discrepancy with measurement of C–H bonds. This is due to the lower electronegativity of carbon vs. oxygen. In O–H bonds the bonding electron pair (which contains H's electron) is shifted more toward oxygen which makes H atom appear closer in the electron density map. On the other hand, C–H bond is of low polarity and the bonding pair is more equally distributed between the two nuclei. See also Self-test 8.2b and Exercise 8.5.

E8.8 $N(SiH_3)_3$ is planar and thus N is at the centre of the molecule and does not move in the symmetric stretch. $N(CH_3)_3$ is pyramidal and the N–C symmetric stretch involves displacement of N.

E8.9 The thiocyanate anion, an ambidentate ligand, has two resonance structures:

$$N{\equiv}C{-}S^{-} \longleftrightarrow N^{-}{=}C{=}S$$

When thiocyanate coordinates through sulfur, the resonance structure on the left is favoured because a positively charged metal can stabilize negative charge on sulfur. Note that in this structure, carbon-sulfur bond is single. On the other hand, if the ligand coordinates through nitrogen atom, the resonance structure on the right is stabilized. This structure has a negative charge on N atom and a double carbon-sulfur bond. This structural analysis tells us that S-coordinated thiocyanate has a weaker C-S bond in comparison to N-coordinated SCN^-.

Looking at the compounds given, in $K_2[Co(SCN)_4]$ we can expect κ*N*-thiocyanate because Co^{2+} is a relatively hard Lewis acid and would prefer a hard Lewis base. In $K_2[Pd(SCN)_4]$ we expect the opposite: Pd^{2+} is a soft Lewis acid and would prefer a soft Lewis base–and thiocyanate should be κS.

This hard/soft analysis of bonding matches well with the analysis of SCN^- resonance structures. The stretching frequency of CS bond in Co^{2+} complex is at higher wavenumbers than that observed in Pd^{2+} complex indicating stronger C-S bond. The wavenumber observed for KSCN is in between the two because KSCN is an ionic compound with uncoordinated (free) thiocyanate. The structure of the free SCN^- is a "mix" of the two hybrid structures and hence has intermediate C-S bond strength.

Simple comparison of CS wavenumber found for $RbBi(SCN)_4$ tells us that SCN^- is coordinated through S atom because 722 cm^{-1} observed for Bi complex is very close to 730 cm^{-1} observed for Pd complex.

E8.10 The energy of a molecular vibration is determined by Equation 8.4a in your textbook:

$$E_v = \left(v + \frac{1}{2}\right)\hbar\omega = \hbar\left(v + \frac{1}{2}\right)\left(\frac{k}{\mu}\right)^{\frac{1}{2}}, \text{ and } \mu_{\text{N–D}} = \frac{m_N m_D}{m_N + m_D}, \mu_{\text{N–H}} = \frac{m_N m_H}{m_N + m_H}.$$

We can make a few reasonable assumptions that will help us to calculate the expected wavenumber for N–D stretch from a known N–H stretch. First, we can assume that the force constants, *k*, are approximately equal in two cases, that is, $k_{\text{N–D}} \approx k_{\text{N–H}}$. The second approximation is that $m_N + m_D \approx m_N + m_H \approx m_N$. Thus, keeping in mind these assumptions and all constant values, we can show that, after cancelling, the ratio of vibrational energies is:

$$\frac{E_v^{\text{N–D}}}{E_v^{\text{N–H}}} = \frac{\sqrt{\mu_{\text{N–H}}}}{\sqrt{\mu_{\text{N–D}}}} \approx \sqrt{\frac{m_H}{m_D}}.$$

And from the above ratio:

$$E_v^{\text{N–D}} = \frac{E_v^{\text{N–H}}}{\sqrt{m_D}} = \frac{E_v^{\text{N–H}}}{\sqrt{2}}.$$

Considering that the energies and wavenumber are related through $E_v = hc\tilde{\nu}$, we finally have:

$$\tilde{\nu}_{\text{N–D}} = \frac{\tilde{\nu}_{\text{N–H}}}{\sqrt{m_D}} = \frac{3400\ \text{cm}^{-1}}{\sqrt{2}} = 2404\ \text{cm}^{-1}.$$

E8.11 The bond orders for CN^- and CO are the same, but N is lighter than O, hence the reduced mass for CN^- is smaller than for CO. From Equation 8.4a, the smaller effective mass of the oscillator for CN^- causes the molecule to have the higher stretching frequency because they are inversely proportional. The bond order for NO is 2.5, and N is heavier than C, hence CO has a higher stretching frequency than NO. For the carbide anion we should expect the highest frequency because of the triple C-C bond and lightest carbon atoms.

E8.12 The bond in O_2^+ (bond order = 2.5) is stronger than in O_2 (bond order = 2). Therefore, O_2^+ has the higher force constant and the stretching frequency is expected to be above that for O_2^- (in the region of 1800 cm^{-1}). See Section 2.8 for more on bond orders in homonuclear diatomics. Considering that this is a homonuclear, diatomic cation, there is no net dipole moment (the positive charge is uniformly distributed over all cations) and thus the stretch should be Raman and not IR active.

E8.13 Consult the solution for Example 8.5 for the structure of $^{77}SeF_4$—since Se and S are in the same group of the periodic table the VSEPR theory would predict identical structures, i.e., see-saw.

The fluorine atoms in SeF_4 are in two different magnetic environments: two F atoms form one plane with the lone electron pair on selenium (equatorial F atoms), whereas the other two are perpendicular to this plane (axial F atoms). Thus, we would expect two signals in ^{19}F NMR. However, the F atoms are bonded to the NMR active ^{77}Se with $I = ½$. (Note that the exercise states that the sample contains only ^{77}Se thus the actual abundance of this Se isotope is not relevant in this case.) Thus, the axial F atoms are going to be coupled to ^{77}Se (large one-bond coupling constant) and their signal is going to be split into a doublet. They are further coupled to two equatorial F atoms and the signal is further split into a triplet producing finally a doublet of triplets (see the figure below). The same analysis for the equatorial F atoms will give another doublet of triplets. Thus, the ^{19}F NMR of $^{77}SeF_4$ would consist of two doublet of triplets with relative intensity 1 : 1.

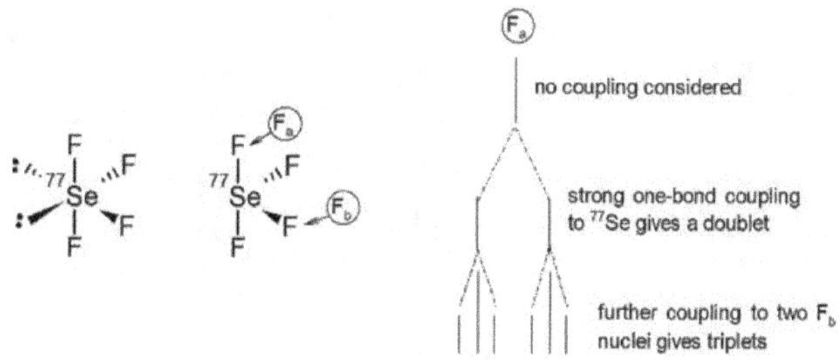

The ^{77}Se NMR would consist of triplet of triplets because ^{77}Se is coupled to two equatorial and two axial ^{19}F nuclei.

E8.14 The splitting of ^{31}P signal indicates that the F atoms are found in two different magnetic environments: one environment has two F atoms and produces a triplet, the other has one F atom and results in a doublet. One possible structure that corresponds to this analysis has two F atoms in axial positions and one in equatorial. (Using the VSEPR theory we can determine that PF_3Cl_2 has trigonal bipyramidal geometry which has, as you should recall, two different environments: axial and equatorial).

E8.15 Using VSEPR theory we can determine that XeF_5^- has a pentagonal planar molecular geometry and all five of the F atoms are magnetically equivalent. Thus, the molecule shows a single ^{19}F resonance. Approximately 25% of the Xe is present as ^{129}Xe, which has $I = ½$, and in this case the ^{19}F resonance is split into a doublet. The result is a composite: two lines each of 12.5% intensity from the ^{19}F coupled to the ^{129}Xe, and one line of 75% intensity for the remainder. The ^{129}Xe NMR would show a single resonance split into a binomial sextet by five equivalent F nuclei.

E8.16 VSEPR theory predicts a T-shape geometry for XeF_3^+ (with trigonal bipyramidal electron geometry) (Figure E8.16A):

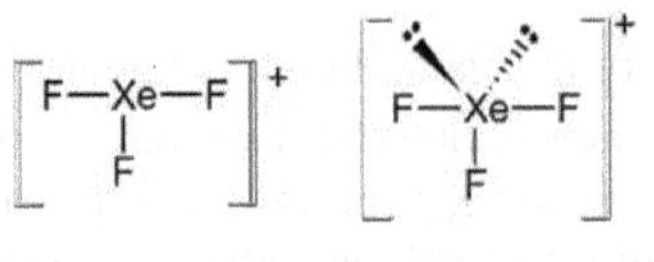

Figure E8.16A

From the point of NMR spectroscopy, this structure has two magnetic environments for fluorine atoms (F_a and F_b in Figure E8.16B), and one unique Xe environment.

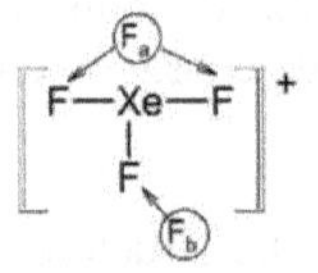

Figure E8.16B

We can first analyse the ^{129}Xe spectrum. Xenon nucleus is going to be coupled to two F_a nuclei resulting in a triplet and to one F_b resulting in a doublet. Thus, the ^{129}Xe signal is going to appear as a triplet of doublets. This corresponds well to the spectrum shown in the exercise, although that might not be obvious at the first glance. The two triplets are actually overlapping because the values of the two coupling constants, $J(^{129}\text{Xe–F}_a)$ and $J(^{129}\text{Xe–F}_b)$, are close. Figure E8.16C shows what happens when $J(^{129}\text{Xe–F}_b) > J(^{129}\text{Xe–F}_a)$ vs. $J(^{129}\text{Xe–F}_b) \approx J(^{129}\text{Xe–F}_a)$ and shows two triplets on the spectrum provided in the problem.

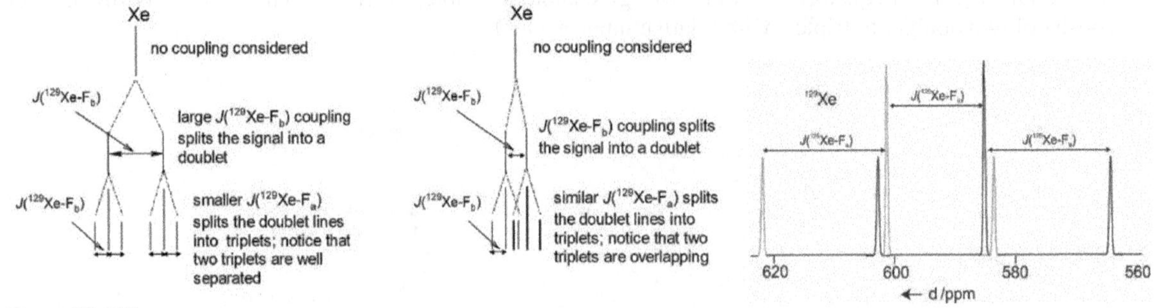

Figure E8.16C

To analyse the ^{19}F NMR we have to recognize that only 26.5% of our XeF_3^+ contains ^{129}Xe isotope that is NMR active. Considering that there are two distinct magnetic environments for F nuclei, we can expect two signals in ^{19}F NMR spectrum. The signal due to F_a is going to be split into a doublet by single F_b nucleus in the case of XeF_3^+ ions that do not contain ^{129}Xe isotope. For the $^{129}XeF_3^+$ in the sample, F_a signal is going to be first split into a doublet by one-bond coupling to ^{129}Xe and then again into a doublet by Fe_b giving a doublet of doublets as satellites flanking the stronger doublet signal. Similar reasoning applies to F_b signal that is split into a triplet by two F_a nuclei, this triplet is flanked by a doublet of triplets due to additional ^{129}Xe-F_b one-bond coupling. Note that in this case there is no ambiguity in coupling magnitudes: one bond couplings are always stronger than two bond couplings: i.e. $J(^{129}\text{Xe-F}_b) >> J(\text{F}_a\text{-F}_b)$. Analysis of the spectrum given in the problem is shown in Figure E8.16D.

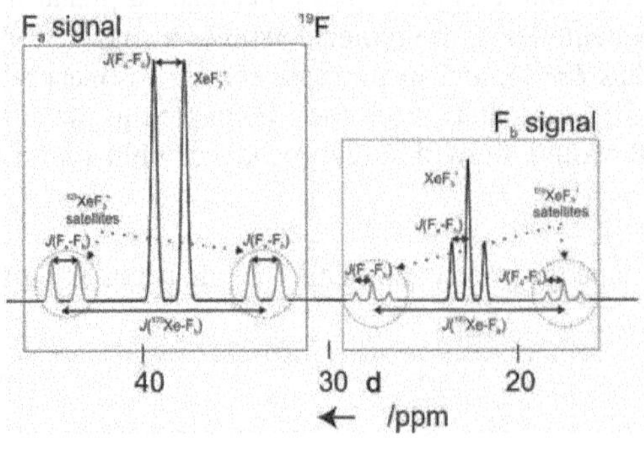

Figure E8.16D

E8.17 The structure of triiron dodecacarbonyl, $[Fe_3(CO)_{12}]$ is shown below:

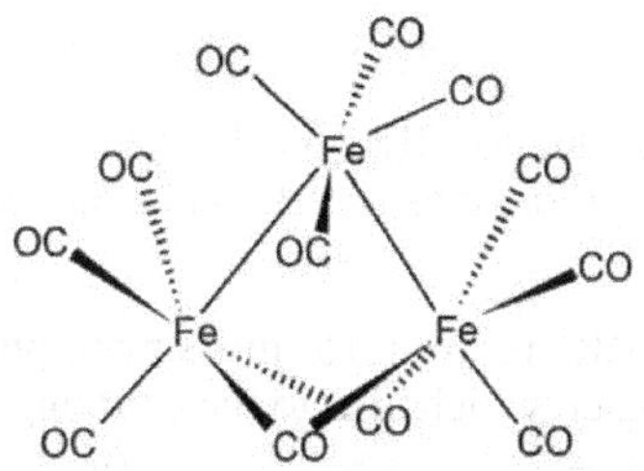

Although this complex has two magnetically distinct carbonyl environments—terminal and bridging carbonyls—the structure is fluxional at room temperature in the solution and the CO ligands are exchanging bridging and terminal position sufficiently fast so that an average signal is seen and only one peak is observed. In the solid state, however, the molecule is locked in the crystal structure and this exchange is either not possible or very slow. Consequently, we can see both terminal and bridging CO ligands in the solid state IR spectrum.

E8.18 The salt $CsPCl_6$ is an ionic compound built of Cs^+ cations and octahedral PCl_6^- anions. If the ^{31}P MASNMR of PCl_5 shows two resonances, one of which is similar to the one found in the ^{31}P MASNMR of the salt, that means solid PCl_5 contains PCl_6^- anions. The second resonance must belong to some other P-containing species in PCl_5, which in this case is the cation PCl_4^+. PCl_5 in a solid state is a ionic compound $[PCl_4][PCl_6]$; the formation of ions can be understood as a Lewis acid-base reaction:

$$PCl_5 + PCl_5 \rightarrow [PCl_4][PCl_6].$$

E8.19 The compound has four phosphorus atoms in its structure, and two peaks—a doublet and a quartet—along with their intensities, three and one respectively, tell us that there are two magnetic environments; one has three P atoms (a doublet with intensity 3) and the other has one P atom (a quartet with intensity 1). A possible structure of P_4S_3 is shown below:

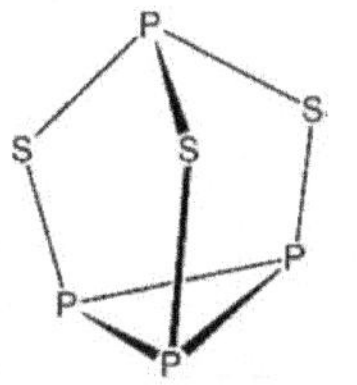

E8.20 a) For the tetrahedral anion BH_4^- we have to consider two boron isotopes: ^{10}B with abundance of 20% and $I = 3$ and ^{11}B with abundance of 80% and $I = 3/2$. In our sample of KBH_4 we will have $^{10}BH_4^-$ and $^{11}BH_4^-$. Although in both cases hydrogens are in the same environment, we have to consider the two separately because two boron isotopes have different I values.

The four protons in $^{10}BH_4^-$ are in the same magnetic environment and are coupled to a nucleus with $I = 3$. According to equation $2I + 1$, the ^{10}B nucleus will split the proton signal into $2 \times 3 + 1 = 7$ lines. This would be a non-binomial septet with all the lines of equal intensity (review Section 8.6(c) *Spin-spin coupling* if necessary):

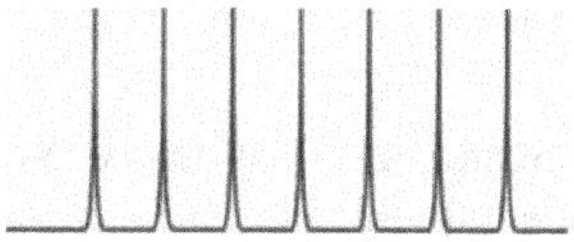

Similar analysis would lead us to the conclusion that the proton resonance in $^{11}BH_4^-$ is split into $2 \times 3/2 + 1 = 4$ lines. Again, the quartet lines would have equal intensity:

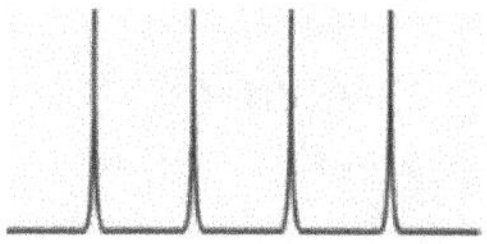

The actual spectrum of BH_4^- is a combination of spectra shown above because our sample contains both B isotopes:

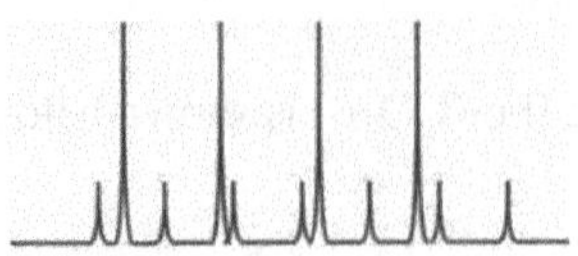

Note: i) that the intensity of lines due to $^{10}BH_4^-$ is about ¼ of those due to $^{11}BH_4^-$ because of isotope abundances and ii) the coupling to ^{10}B is not the same as to ^{11}B, hence the separation between mutiplet lines is different in two isotopomers.

b) We can consider first ^{195}Pt NMR spectrum for this square planar complex. This is a straight forward case: one platinum nuclei would give a singlet, which, due to the coupling to the hydride ligand, is split into a doublet.

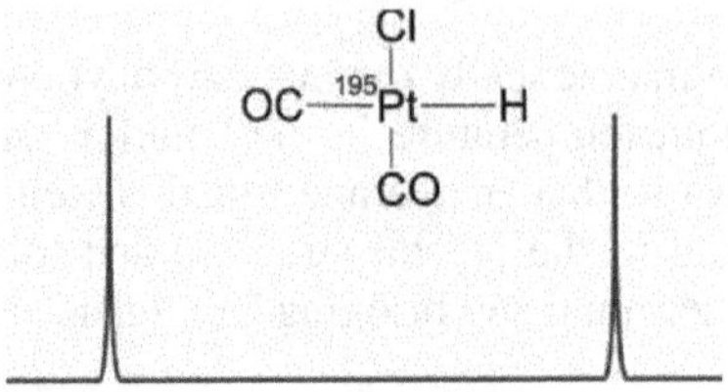

For ^1H NMR we need to consider two isotopomeric molecules of our complex: one contains NMR inactive isotopes of Pt, the other contains NMR active ^{195}Pt isotope. The first case would produce a singlet in ^1H NMR. The second would produce a doublet due to the coupling between the hydride and ^{195}Pt nucleus with $I = ½$. The actual NMR is a combination of the two—one strong singlet flanked by two lines of the doublet:

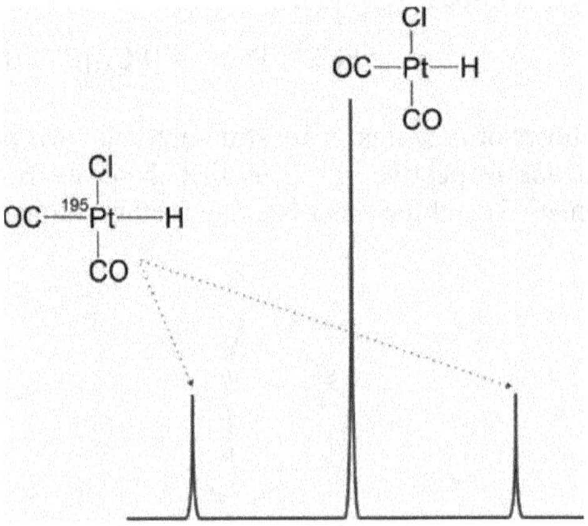

E8.21 Rearranging Equation 8.7 we have $g = \Delta E/\mu_B B_0$. To calculate the g values, we obtain the B_0 values from the EPR spectrum shown in Figure 8.56 (the three lines drawn to the x-axis correspond to 348, 387, and 432 mT) and set the $\Delta E = h\nu$ where ν is equal to the given microwave frequency. The three g values are thus 1.94, 1.74, and 1.56, respectively. See below for an example of the calculation for the first line.

$$g = [(6.626 \times 10^{-34}\ \text{J s})(9.43 \times 10^{9}\ \text{s}^{-1})]/(9.27401 \times 10^{-24}\ \text{J/T})(0.348\ \text{T}) = 1.94.$$

E8.22 Consulting Table 8.2 in the text, we can see that EPR is a somewhat faster technique than NMR, thus EPR would be more sensitive.

E8.23 The structure of tris(2-pyridyl)methane ligand is shown below. Note that the ligand has three Lewis basic nitrogen atoms and hence can be potentially a tridentate ligand.

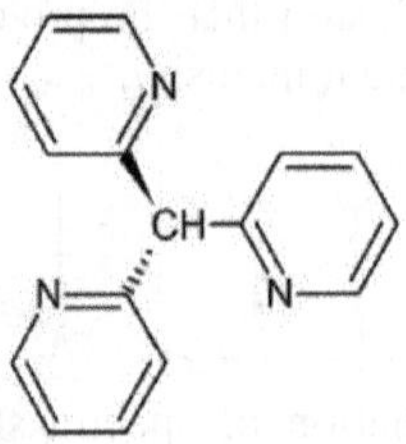

The complex structure is shown below. The counterion, nitrate, is not important because it is not coordinated to the metal and is EPR silent (it has no unpaired electrons). The Cu^{2+} centre has electronic configuration d^9 and its complexes are prone to Jahn-Teller distortions (see Chapter 7). In the structure, two axial Cu–N bonds are marked as "long" to indicate the distortion from the ideal octahedral structure.

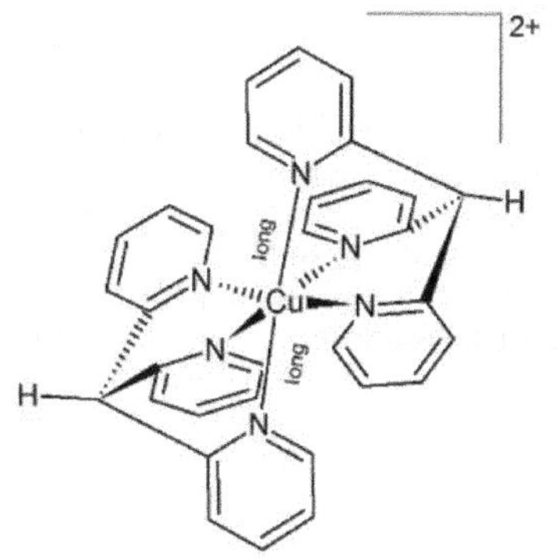

At room temperature, there is a fast exchange between "long" and normal bonds, so fast that EPR instrument observes six bonds of equal, average lengths and consequently an isotropic octahedral geometry. As the temperature is lowered, there is less energy for fluxional movement of the atoms in the molecule and the exchange is slowed down to a point that it becomes slower than EPR measurement. The spectrum becomes axial, as one would expect from the structure.

E8.24 In aqueous solution at room temperature, molecular tumbling is fast compared to the timescale of the EPR transition (approximately $g m_b B_0/h$), and this removes the effect of the *g*-value anisotropy. We expect a derivative-type spectrum, possibly exhibiting hyperfine structure that is centred at the average *g* value. In frozen solution, *g*-value anisotropy can be observed because the spectrum records the values of *g* projected along each of the three axes, and the averaging from molecular tumbling does not apply.

E8.25 The oxidation state for iron in $Na_3Fe^{(V)}O_4$ results in a $3d^3$ configuration. As the 3d electrons are removed the isomer shift becomes less positive as the 3d electrons partly screen the nucleus from the inner s electrons. Compared to Fe(III) with isomer shift between +0.2 and +0.5 mm s^{-1}, we would expect a smaller positive shift for Fe(V) below +0.2 mm s^{-1}. The shift for Na_3FeO_4 will be less positive than for Sc_2FeO_4 due to the higher oxidation state of Fe (+5 in Na_3FeO_4 vs. +4 in Sc_2FeO_4).

E8.26 The three iron centres have different oxidation states so we should expect unique isomer shifts for each. Each signal should show quadrupole splitting due to the nuclear quadrupole moment of ^{57}Fe nucleus. If the material has Fe(IV) only we would expect to see one doublet in the Mössbauer spectrum. However, if both Fe(III) and Fe(V) are present, two doublets should be observed. The position of Fe(V) peak should be close to 0 mm s^{-1} with Fe(III) peak at more positive isomer shifts.

At low temperatures, $CaFeO_3$ becomes magnetic, and further hyperfine coupling is observed due to the Zeeman effect. Since two six-line patterns are observed, most likely composition of the material is $CaFe(III)_{0.5}Fe(V)_{0.5}O_3$. If it were $CaFe(IV)O_3$ only one six-line pattern would be observed.

E8.27 The XANES spectrum could be very useful in deciding on Fe oxidation states found in $CaFeO_3$. The near edge structure can differentiate between different oxidation states and geometries. In this case, the geometry is octahedral at Fe (regardless of the oxidation state) making the interpretation of the spectrum easier.

E8.28 The EXAFS region of X-ray absorption spectrum is very sensitive to electron density at, and the chemical environment of, the absorbing atom. The aqueous solution of $CoSO_4 \times 7H_2O$ contains (among other species) hexaaquacobalt(II) ion, $[Co(H_2O)_6]^{2+}$. This octahedral species is going to be converted into tetrahedral anion tetrachloridocobaltate(II), $[CoCl_4]^{2-}$. The major change observed in EXAFS is due to the change in electron density (from cation to anion) as well as overall environment (H_2O ligands vs. Cl^- ligands).

E8.29 Silver has two isotopes, ^{107}Ag (51.82%) and ^{109}Ag (48.18%). You should look up this information yourself, which can be found in many reference books (e.g., *CRC Handbook of Chemistry and Physics*). Thus the average mass is near 108, but no peak exists at this location because there is no isotope with this mass number. Compounds that contain silver will have two mass peaks flanking this average mass.

E8.30 Considering systematic loss of individual ligands on the organometallic complex, the major peaks would be at 258 (due to the molecular ion, $Mo(C_6H_6)(CO)_3^+$), 230 ($Mo(C_6H_6)(CO)_2^+$), 200 ($Mo(C_6H_6)(CO)^+$), 186 ($Mo(C_6H_6)^+$), and 174 ($Mo(CO)_3^+$).

E8.31 The molar mass of the dehydrated zeolite, $CaAl_2Si_6O_{16}$, is 518.5 g mol^{-1}. The molar mass of water is 18.0 g mol^{-1}. We can solve for n using the mass percentage of water.

$$\text{Mass \% of H}_2\text{O} = \frac{\text{mass of H}_2\text{O in the sample}}{\text{mass of the sample}} \times 100\% = \frac{18n}{518.5 + 18n} \times 100\% = 20\%$$

$18n = 0.2(518.5 + 18n)$

$14.4n = 103.7$

$n = 7.2.$

Isolating n as an integer we can estimate that $n = 7$.

E8.32 First we will calculate the mass percentage of hydrogen in $LiBH_4$:

$$\text{Mass \% of H} = \omega(\text{H}) = \frac{4 \times Ar(\text{H})}{Mr(\text{LiBH}_4)} \times 100\% = \frac{4 \times 1.0079}{21.784} \times 100\% = 18.5\%.$$

This percentage matches the observed weight loss of 18.5%, and we can conclude that all hydrogen has been released during heating process. Note as well that other two elements, Li and B, are non-volatile solids and remain behind.

E8.33 The complex undergoes a reversible one-electron reduction with a reduction potential of +0.21 V. This can be calculated by taking the mean of E_{pa}(+0.240 V) and E_{pc}(+0.18 V), see Figure 8.57. Above +0.720 V the complex is oxidized to a species that undergoes a further chemical reaction, and thus is not re-reduced. This step is not reversible.

E8.34 The regular crystalline periodicity is required for Bragg diffraction and constructive interference of the X-rays producing observable peaks in the XRD pattern. Glass, however, is a network covalent material with no long-range periodicity or order—it is an amorphous rather than crystalline solid—as such it will not diffract X-rays. The observation of X-ray diffraction pattern after the glass sample was heated indicates that at least a part of the sample crystallized. The fact that the exothermic event was observed suggests either change of phase (from amorphous to crystalline) or chemical reaction that yielded a crystalline product.

E8.35 The ratio of cobalt to acetylacetone in the complex is 3:1. Converting the given mass percentages of Co and C to moles of Co and C, we find that for every mole of Co in the product we have 15 moles of C: moles of Co = 0.28 mol; moles of C = 4.2 mol. Considering that every acac$^-$ ligand has five carbon atoms, it is obvious that for each mole of Co we have three moles of acac$^-$. This ratio is consistent with the formula $[Co(acac)_3]$. (Consult Section 7.1 for more detail on the acetylacetonate ligand and cobalt coordination complexes.)

E8.36 **a)** If Os(IV) complex decomposed rapidly, then the voltammogram would not be reversible, and would be unsymmetrical.

b) If Os(III) complex is oxidized rapidly to Os(V), we would see only one peak in voltammogram with twice the height of single peaks shown in Figure 8.53(a).

E8.37 Microscopy techniques, particularly SEM and TEM, are very useful for determining the nanoparticle size down to 1 μm. The other methods used are based on light scattering. In this case, the particle size must be much smaller than the wavelength of the light used. The light sources are monochromatic (usually lasers) and the wavelength of the light emitted places the limits of the measurement.

E8.38 A good way to determine the composition of this sample (all heavy elements) would be X-ray emission spectroscopy.

To determine the composition of the compound we convert first the mass percentages into moles:

$$n(\text{Cu}) = \frac{26.74\,\text{g}}{63.55\,\text{g}\,\text{mol}^{-1}} = 0.43\,\text{mol},$$

$$n(\text{Zn}) = \frac{15.83\,\text{g}}{65.38\,\text{g}\,\text{mol}^{-1}} = 0.24\,\text{mol},$$

$$n(\text{Sn}) = \frac{10.90\,\text{g}}{118.71\,\text{g}\,\text{mol}^{-1}} = 0.09\,\text{mol},$$

$$n(\text{S}) = \frac{46.54\,\text{g}}{32.065\,\text{g}\,\text{mol}^{-1}} = 1.45\,\text{mol}.$$

From these values we can set the first formula of the compound as $Cu_{0.43}Zn_{0.24}Sn_{0.09}S_{1.45}$. To convert the indexes into whole numbers, we divide each number by the smallest value:

$$Cu_{(0.43/0.09)}Zn_{(0.24/0.09)}Sn_{(0.09/0.09)}S_{(1.45/0.09)}$$

We have:

$$Cu_{4.78}Zn_{2.67}Sn_1S_{16}.$$

We can keep the formula as is or we can multiply all by 2 and get approximate whole numbers:

$$Cu_9Zn_5Sn_2S_{32}.$$

The sample is likely going to increase in mass. Heavy metals can be hygroscopic and the sulfide anion, S^{2-}, can be oxidized to SO_4^{2-}.

Guidelines for Selected Tutorial Problems

T8.2 Review Section 8.2 of the main text.

T8.3 You must think about two important factors when considering hydrogen storage materials. First, the candidates for hydrogen storage materials are based on both compounds containing light elements (most notably B and N) and transition metals. These two groups can require different methods of analysis. Second, ultimately the good hydrogen storage material should "pack" a lot of hydrogen (a good measure of this is the weight percentage of H stored in the material—you do not want to carry around a lot of weight that contains very little useful material), and the analysis of hydrogen content itself presents some challenges.

T8.5 Hydrogen bonding usually results in the broadening of O-H stretching frequencies. A very good text for the IR in general is Nakamoto's book provided in the "Further Reading" section of this chapter.

T8.9 Some suggestions (with details on *how* in the chapter): a) and b) atomic absorption or emission spectroscopy, c) IR (BrF_5 is a liquid so X-ray diffraction at room temperature is not possible), d) X-ray diffraction on a single crystal, e) NH percentage analysis could be helpful.

T8.10-12 Any analytical chemistry textbook will describe the analysis of the data collected using AAS. From your "Further Reading" list Skoog's text is a good reference point.

Chapter 9 Periodic Trends

Self-Tests

S9.1 **a)** Refer to Figure 9.9 in your textbook. Cd and Pb are classified as chalcophiles and give soft cations. Thus, they will be found as sulfides. Rb and Sr are lithophiles and are hard; they can be found in aluminosilicate minerals. Cr and Pd are siderophiles and give cations of intermediate hardness. As such, they can be found as both oxides and suldfides. Palladium can also be found in elemental form.

b) Thallium is expected to be a chalcophile, and it should occur as a sulfide ore. This can be concluded based on Figure 9.9 or by thallium's position in the periodic table of elements: it is close to lead and bismuth—typical chalcophiles with sulfide ores.

S9.2 **a)** In this case you have to consider two potential reactions: (i) possible oxidation by H^+ and (ii) possible oxidation by O_2. The first case has been covered in the text, but we can look at the two half-reactions, assuming standard conditions:

$$H^+(aq) + e^- \rightarrow \tfrac{1}{2}H_2(g), \qquad E^\circ = 0.00\ V$$

$$V^{3+}(aq) + e^- \rightarrow V^{2+}(aq), \qquad E^\circ = -0.255\ V.$$

Thus, the overall reaction, $V^{2+}(aq) + H^+(aq) \rightarrow V^{3+}(aq) + \tfrac{1}{2}H_2(g)$, would have a positive potential difference (E = +0.255 V) and the reaction is spontaneous. (Recall that reactions with a positive potential difference have negative Δ_rG and are thus spontaneous.) Further oxidation of $V^{3+}(aq)$ by $H^+(aq)$ is not feasible because the potential difference would be negative. For example, consider half-reaction for oxidation of $V^{3+}(aq)$ to $V^{4+}(aq)$ by $H^+(aq)$:

$$H^+(aq) + e^- \rightarrow \tfrac{1}{2}H_2(g), \qquad E^\circ = 0.00\ V$$

$$VO^{2+} + e^- + 2H^+ \rightarrow V^{3+}(aq) + H_2O(l), \; E^\circ = +0.337\ V.$$

Combining these two reactions we would get a negative potential difference (E = –0.337 V) and a nonspontaneous reaction.

The reduction potential for $O_2(g)$ in acidic medium is +1.23 V (for half-reaction $\tfrac{1}{2}O_2(g) + 2e^- + 2H^+ \rightarrow H_2O(l)$), meaning that O_2 is a better oxidizing agent than H^+. Consequently, $O_2(g)$ will be able to oxidize $V^{2+}(aq)$ to $V^{3+}(aq)$ as well, but we have to check if oxidation of $V^{3+}(aq)$ to VO^{2+} is now possible: E = +1.23 V – (+0.337 V) = +0.893 V. The potential difference is positive, and $VO^{2+}(aq)$ should be more stable than $V^{3+}(aq)$ and $V^{2+}(aq)$ under these conditions. Similar analysis shows that $VO^{2+}(aq)$ could be further oxidized to $VO_2^+(aq)$ with $O_2(g)$ with potential difference of +0.23 V. Overall, this indicates that $V^{2+}(aq)$ should be oxidized by $O_2(g)$ in the acidic solution all the way to the thermodynamically favoured species $VO_2^+(aq)$. Note that this does not necessarily mean that the process will occur—the kinetics might be slow, and particularly important factor to consider when dealing with $O_2(g)$ oxidation are overpotentials. Consequently, $V^{3+}(aq)$ and/or VO^{2+} might be reasonably long- lived species in acidic medium when exposed to oxygen from air.

b) Mn can have oxidation states from +2 to +7. Looking at the Latimer diagram we can conclude that the most stable oxidation state is +2. Higher oxidation states have positive reduction potential from +7 all the way to +2 indicating that all are good oxidizing agents. The reduction potential for Mn^{2+}/Mn couple is negative telling us that Mn can be easily reduced to +2.

S9.3 **a)** Oxygen forms a double bond that is three times more stable than its single bond. Owing to this tendency to form strong double bonds, it is very unlikely that longer-chain polyoxygen anions would exist. Sulfur is much less likely to form π bonds and therefore more likely to generate catenated polysulfide anions.

b) To answer this self-test, we can extrapolate from the discussion on O and S single and double bonds. A general trend can be stated that the enthalpy of π component of the double bond significantly decreases going down a group in the periodic table. Following this trend, we can expect that Si=Si bond is relatively weak in comparison with C=C and can be easily broken. Another trend worth mentioning here is that large atoms do not form p bonds easily—the separation between p atomic orbitals on two atoms prevents formation of efficient orbital overlap and consequently π bond is weak.

S9.4 **a)** The energy required to convert solid elements into the gaseous atomic form drops going down the group. This is expected because the bond strengths decrease in the same order and it becomes easier to break the molecules in solids. It is evident from the values that as we move down the group from S to Se to Te, the steric crowding of the fluorine atoms is minimized owing to the increasing radius of the central atom. As a result, the enthalpy values become larger and more negative (more exothermic) and the higher steric number compounds (such as TeF_6) are more likely to form.

b) We can consider two reactions to get the best estimate:

$$Te(s) + 2F_2(g) \rightarrow TeF_4$$

and

$$Te(s) + 3F_2 \rightarrow TeF_6.$$

$\Delta_f H$ for two tellurium fluorides and $\Delta H_{sub}(Te)$ are given in the table for part a) and bond dissociation energy for F_2 is provided in the text of Section 9.10(b) *Covalent halides*: 155 kJ mol^{-1}.

For TeF_4:

$$\Delta_f H(TeF_4) = \Delta H_{sub}(Te) + 2 \times B(F\text{–}F) - 4 \times B(Te\text{–}F)$$

$$-1036 \text{ kJ mol}^{-1} = +199 \text{ kJ mol}^{-1} + 2 \times 155 \text{ kJ mol}^{-1} - 4 \times B(Te\text{–}F)$$

$$B(Te\text{–}F) = (199 \text{ kJ mol}^{-1} + 310 \text{ kJ mol}^{-1} + 1036 \text{ kJ mol}^{-1})/4$$

$$B(Te\text{–}F) = 386 \text{ kJ mol}^{-1}.$$

Similar procedure can be applied to TeF_6:

$$\Delta_f H(TeF_6) = \Delta H_{sub}(Te) + 3 \times B(F\text{–}F) - 6 \times B(Te\text{–}F)$$

$$-1319 \text{ kJ mol}^{-1} = +199 \text{ kJ mol}^{-1} + 3 \times 155 \text{ kJ mol}^{-1} - 6 \times B(Te\text{–}F)$$

$$B(Te\text{–}F) = (199 \text{ kJ mol}^{-1} + 465 \text{ kJ mol}^{-1} + 1319 \text{ kJ mol}^{-1})/6$$

$$B(Te\text{–}F) = 330 \text{ kJ mol}^{-1}.$$

Finally, we can find the average of two values:

$$B(Te\text{–}F) = (330 \text{ kJ mol}^{-1} + 386 \text{ kJ mol}^{-1})/2 = 358 \text{ kJ mol}^{-1}.$$

S9.5 We would have to know the products formed upon thermal decomposition for P_4O_{10} and thermodynamic data for the product. We could use data for V_2O_5 from the Exercise 9.5.

S9.6 **a)** Xe is the central atom. With 8 valence electrons from Xe and 24 electrons (6 from each O) we have 32 total electrons and 16 electron pairs. We would predict a tetrahedral geometry to minimize electron pair repulsions. Note that to minimize formal charge, the xenon will form double bonds with each oxygen. Because the atomic number of Xe is 54, the $Z + 22$ is Os ($Z = 76$). The compound with the same structure is OsO_4.
b) The atomic number of Si is $Z = 14$ and the $Z + 8$ element is Ti ($Z = 22$). Indeed, just like Si, Ti forms a very stable oxide of the formula TiO_2.

Exercises

E9.1 (a) Ba, +2; (b) As, +5; (c) P, +5; (d) Cl, +7; (e) Ti, +4; (f) Cr, +6

E9.2 The group is the alkaline earth metals or Group 2 elements. Consult Sections 12.8 and 12.9 for detailed reactions.

E9.3 These elements are Group 15. This group is very diverse. N and P are nonmetals; As and Sb are metalloids, and Bi is metallic. The +5 and +3 oxidation states are common for the group electron configuration of ns^2p^3. Phosphine and arsine are well-known toxic gases.

E9.4 If the maximum oxidation state at the bottom is +6, that means the group in question must have six valence electrons. That narrows the answer down to Groups 6 and 16. Group 16 is headed by oxygen which is a non-metallic element and has –2 as the most stable oxidation state. That leaves us with Group 6 in the d-block. In this group all elements are metallic and chromium has +3 as its most stable oxidation state. Thus, the group sought is the chromium group of Group 6.

E9.5 The key steps in the Born–Haber cycle for $NaCl_2$ are:

Na(s)	→	Na(g)	sublimation	$\Delta_{sub}H^o(Na)$
$Cl_2(g)$	→	2Cl(g)	dissociation	$\Delta_{dis}H^o(Cl_2)$
Na(g)	→	$Na^+(g) + 1e^-$	first ionization	$I_1(Na)$
$Na^+(g)$	→	$Na^{2+}(g) + 1e^-$	second ionization	$I_2(Na)$
$2Cl(g) + 2e^-$	→	$2Cl^-$ (g)	electron gain	$\Delta_{eg}H^o(Cl)$
$Na^{2+}(g) + 2Cl^-(g)$	→	$NaCl_2(s)$	lattice enthalpy	$-\Delta_L H^o(NaCl_2)$

You can calculate the lattice enthalpy by moving around the cycle and noting that the enthalpy of formation $\Delta_f H^o(NaCl_2) = \Delta_{sub}H^o(Na) + \Delta_{dis}H^o(Cl_2) + I_1(Na) + I_2(Na) - \Delta_{eg}H^o(Cl) - \Delta_L H^o(NaCl_2)$. The second ionization energy of sodium is 4562 kJ mol^{-1} and is responsible for the fact that the compound does not exist as it would result in a large, positive enthalpy of formation.

E9.6 The relative stability of an oxidation state that is lower for 2 than the group number is an example of the inert pair effect. This is a recurring theme in the heavier p-block elements. Beyond Group 15, where we see the inert pair effect favouring +3 oxidation state for Bi and Sb, we also find stable +5 oxidation state for the halogens in Group 17. Examples include BrO_3^- and IO_3^-. Because of this intermediate oxidation state, these compounds can function as both oxidizing and reducing agents. Group 16 elements (except oxygen) also form several stable compounds with +4 oxidation states including SF_4 and SeF_4. The inert pair effect would manifest itself the most for Po, the heaviest member of Group 16.

E9.7 Metallic character and ionic radii decrease across a period and down a group. Ionization energy increases across a period and decreases down a group. Large atoms typically have low ionization energies and are more metallic in character.

E9.8 **a)** Be has the higher first ionization energy. Although B is on the right-hand side of Be, boron's characteristic electronic configuration results in a lower first ionization energy than expected (see Section 1.7(b) *Ionization energy*).

b) C has the higher first ionization energy. Recall that the first ionization energy generally decreases going down a group in the periodic table.

c) Mn has the higher first ionization energy. Recall that the first ionization energy generally increases if we move from left to right along a period in the periodic table.

E9.9 **a)** Na is the more electronegative one. Recall that electronegativity decreases down a group in the periodic table (see Section 1.7(d) *Electronegativity*).

b) O is the more electronegative one—after fluorine, oxygen is the most electronegative element; also, with respect to Si, it is located closer to the top right corner where the highest electronegativity is found (i.e., that for fluorine).

E9.10 Refer to Section 9.9(a) *Hydrides of the elements*:
(a) LiH is saline; **(b)** SiH_4 is molecular; **(c)** B_2H_6 is molecular; **(d)** UH_3 is saline; **(e)** PdH_x is metallic hydride.

E9.11 Refer to Sections 9.9(b) *Oxides of the elements* and 4.4 *Anhydrous oxides*:
(a) Na_2O is basic; **(b)** P_2O_5 is acidic; **(c)** ZnO is amphoteric; **(d)** SiO_2 is acidic; **(e)** Al_2O_3 is amphoteric; **(f)** MnO is basic.

E9.12 **(a)** Na_2O saline, **(b)** P_4O_{10} molecular, **(c)** ZnO saline/intermediate, **(d)** SiO_2 intermediate, **(e)** Al_2O_3 saline, **(f)** MnO saline.

E9.13 The covalent character increases with increasing oxidation state of the metal; thus the covalent character is increasing in order $CrF_2 < CrF_3 < CrF_6$. Refer to Section 9.9(c) *Halides of the elements.*

E9.14 **(a)** Mg, $MgCO_3$ magnesite; **(b)** Al, Al_2O_3 bauxite; **(c)** Pb, PbS galena; **(d)** Fe, magnetite Fe_3O_4.

E9.15 The atomic number of P is 15. The Z + 8 element has an atomic number of 23 and is V (vanadium). Both form compounds with varying oxidation states up to a maximum value of +5. Both form stable oxides including ones in +5 oxidation state (V_2O_5 and P_2O_5). Like phosphorus, vanadium forms oxoanions including ortho-, pyro-, and meta-anions. Consult Section 15.15 for analogous phosphorus oxoanions.

E9.16 Tin's atomic number is 50 and element with $Z + 22$ is Hf. Both elements form oxides MO_2, fluorides MF_4 and chlorides MCl_4. Oxidation state +2 is relatively unstable for Sn, but very unstable for Hf.

E9.17 What we are looking for are Se–F bond dissociation energies in SeF_4 and SeF_6. To determine these values, we have to use the Born–Haber cycle. Looking first at SeF_6, we have:

$$Se(s) + 3F_2(g) \rightarrow SeF_6(g),\ \Delta_f H = -1030\ kJ/mol$$

We can break this reaction into three elementary steps:

1. $Se(s) \rightarrow Se(g);\ \Delta_a H(Se) = +227\ kJ/mol$
2. $3F_2(g) \rightarrow 6F(g);\ \Delta_a H(F) = +159\ kJ/mol \times 3$
3. $Se(g) + 6F(g) \rightarrow SeF_6(g);\ \Delta_r H = X$

Note that $\Delta_a H$ for reaction 2 has to be multiplied by three because we need to atomize (break bonds in) three moles of $F_2(g)$. Also note that in reaction 3 we form six Se–F bonds and no other physical or chemical transformation takes place. If we sum reactions 1–3 we get our first reaction given above. Thus, we can write:

$\Delta_f H = \Delta_a H(Se) + 3\times\Delta_a H(F) + \Delta_r H$. Now substitute given values:

$$-1030\ kJ/mol = 227\ kJ/mol + 3 \times 159\ kJ/mol + X.$$

From here:

$$X = -1030\ kJ/mol - 227\ kJ/mol - 3 \times 159\ kJ/mol = -1575\ kJ/mol.$$

Since X gives energy released per six Se–F bonds and we need only one Se–F, we have to divide X by 6, and get -262.5 kJ/mol. Further, we have to change the sign to finally obtain $B(Se - F)$ in SeF_6 as 262.5 kJ/mol, because B(Se–F) is bond dissociation energy (energy required to break the bond) not bond formation energy (energy released when a bond is formed).

In a similar way we can determine the B(Se–F) for SeF_4.

$$Se(s) + 2F_2(g) \rightarrow SeF_4(g),\ \Delta_f H = -850\ kJ/mol$$

Elementary steps are:

1. $Se(s) \rightarrow Se(g);\ \Delta_a H(Se) = +227\ kJ/mol$
2. $2F_2(g) \rightarrow 4F(g);\ \Delta_a H(F) = +159\ kJ/mol \times 2$
3. $Se(g) + 4F(g) \rightarrow SeF_4(g);\ \Delta_r H = X$

Note that this time $\Delta_a H(F)$ is multiplied by 2 and that in reaction 3 four Se–F bonds are formed. We have:

$$-850 \text{ kJ/mol} = 227 \text{ kJ/mol} + 2 \times 159 \text{ kJ/mol} + X.$$

And

$$X = -850 \text{ kJ/mol} - 227 \text{ kJ/mol} - 2 \times 159 \text{ kJ/mol} = -1395 \text{ kJ/mol}.$$

And the bond dissociation energy *B*(Se–F) in SeF_4 is (–1395 kJ/mol × –1)/4 = 348.7 kJ/mol.

Comparing the Se–F bond strength in SeF_4 and S–F bond strength in SF_4 we see that the Se–F bond is slightly stronger (for about 8 kJ/mol) than the S–F bond. However, the Se–F bonds in SeF_6 are significantly weaker in comparison to the S–F bonds in SF_6. This reflects the general periodic trend according to which the E–X bond decreases in strength on going down the group.

E9.18 The relevant bond enthalpies are:

$B(\text{N}\equiv\text{N}) = 946 \text{ kJ mol}^{-1}$,

$B(\text{F–F}) = 155 \text{ kJ mol}^{-1}$,

$B(\text{Cl–Cl}) = 242 \text{ kJ mol}^{-1}$,

$B(\text{N–F}) = 270 \text{ kJ mol}^{-1}$ and

$B(\text{N–Cl}) = 200 \text{ kJ mol}^{-1}$.

First, we consider the formation of NF_3 and its $\Delta_f H$:

$$N_2(g) + 3F_2(g) \rightarrow 2NF_3$$

$$\Delta_f H(NF_3) = B(\text{N}\equiv\text{N}) + 3 \times B(\text{F–F}) - 6 \times B(\text{N–F})$$

$$\Delta_f H(NF_3) = 946 \text{ kJ mol}^{-1} + 3 \times 155 \text{ kJ mol}^{-1} - 6 \times 270 \text{ kJ mol}^{-1}$$

$$\Delta_f H(NF_3) = -209 \text{ kJ mol}^{-1}.$$

Now we consider the formation of NCl_3 and its $\Delta_f H$:

$$N_2(g) + 3Cl_2(g) \rightarrow 2NCl_3$$

$$\Delta_f H(NCl_3) = B(\text{N}\equiv\text{N}) + 3 \times B(\text{Cl–Cl}) - 6 \times B(\text{N–Cl})$$

$$\Delta_f H(NCl_3) = 946 \text{ kJ mol}^{-1} + 3 \times 242 \text{ kJ mol}^{-1} - 6 \times 200 \text{ kJ mol}^{-1}$$

$$\Delta_f H(NCl_3) = 472 \text{ kJ mol}^{-1}.$$

We can see that the synthesis of NF_3 from elements is a thermodynamically favourable process and this fluoride is stable, but that the synthesis of NCl_3 from elements is not a thermodynamically favourable process and NCl_3 is unstable. The reasons for the instability (and reactivity) of NCl_3 are strong Cl–Cl bond (in comparison to F–F bond) and weak N–Cl bond (in comparison to N–F).

Guidelines for Selected Tutorial Problems

T9.1 Follow the logic outlined by the section of this chapter. For each property extrapolate from known values to get the prediction for the new elements.

T9.5 A good starting reference for this exploration is Eric R. Scerri. (1998). "The Evolution of the Periodic System." *Scientific American,* 279, 78–83 and references provided at the end of this article. The website "The Internet Database of Periodic Tables" (http://www.meta-synthesis.com/webbook/35 pt/pt database.php) is also useful. From the modern designs (post-Mendeleev), interesting ones to consider are Hinrichs' spiral periodic table, Benfey's oval table, Janet's left-step periodic table, and Dufour's tree.

Chapter 10 Hydrogen

Self-Tests

S10.1 Table 1.7 lists electronegativity values for the s and p block elements. We have to compare the electronegativity of hydrogen with electronegativity for Group 15 elements keeping in mind that hydridic compounds have hydrogen bound to a less electronegative element. Thus, BiH_3 is going to be the most hydridic because of bismuth's low electronegativity (2.02 vs. 2.20 for H, both Pauling's electronegativity values). This is to be expected as well based on the periodic trends: electronegativity is decreasing going down the group and is generally low for metallic elements. Bismuth is at the bottom of Group 15 (thus has the lowest electronegativity in the group) and is also metallic element.

S10.2 A good guide as to whether a compound EH_4 will act as a proton donor or hydride donor is provided by the electronegativity difference between H and E (see Self-Test 10.1). If E is more electronegative than H, cleavage of an E–H bond releases a proton, H^+. If E is less electronegative than H, cleavage results in transfer of a hydride, H^-. Given the following compounds, CH_4, SiH_4, and GeH_4, carbon is the only E that is more electronegative than H, thus it would be the strongest Brønsted acid. In other words, CH_4 would be more likely to release protons than SiH_4 and GeH_4. Ge has the least electronegative E, thus GeH_4 would be the best hydride donor of the three.

S10.3 **a)** $Ca(s) + H_2(g) \rightarrow CaH_2(s)$. This is the reaction of a reactive s-block metal with hydrogen, which is the way that saline metal hydrides are prepared.

b) $NH_3(g) + BF_3(g) \rightarrow H_3N\text{–}BF_3(g)$. This is the reaction of a Lewis base (NH_3) and a Lewis acid (BF_3). The product is a Lewis acid–base complex. (Review Sections 5.6 and 5.7(b)).

c) $LiOH(s) + H_2(g) \rightarrow$ NR. Although dihydrogen can behave as an oxidant (e.g., with Li to form LiH) or as a reductant (e.g., with O_2 to form H_2O), it does not behave as a Brønsted or Lewis acid or base. It does not react with strong bases, like LiOH, or with strong acids.

S10.4 The given reaction,

$$[PdH(PNP)_2]^+ + [Pt(PNP)_2]^{2+} \rightleftharpoons [Pd(PNP)_2]^{2+} + [PtH(PNP)_2]^+ \qquad \Delta_r G$$

can be broken into two separate processes:

$$[PdH(PNP)_2]^+ \rightleftharpoons [Pd(PNP)_2]^{2+} + H^- \qquad \Delta_H G(Pd)$$

and

$$[Pt(PNP)_2]^{2+} + H^- \rightleftharpoons [PtH(PNP)_2]^+ \qquad -\Delta_H G(Pt).$$

According to Hess' law:

$$\Delta_r G = \Delta_H G(Pd) + (-\Delta_H G(Pt))$$

where $\Delta_r G$ is the Gibbs energy of the total reaction. $\Delta_r G$ can be calculated from the equilibrium constant that is given and the temperature of 298 K:

$$\Delta_r G = -2.3RT\log K = -2.3 \times 8.314 \text{ K J}^{-1} \text{ mol}^{-1} \times 298 \text{ K} \times \log(450) = -15.1 \text{ kJ mol}^{-1}.$$

From here:

$$\Delta_H G(Pd) = \Delta_r G - (-\Delta_H G(Pt))$$

$$= -15.1 \text{ kJ mol}^{-1} + 232 \text{ kJ mol}^{-1}$$

$$\Delta_H G(Pd) = 227 \text{ kJ mol}^{-1}.$$

(Note: The value for $\Delta_H G$(Pt) has been taken from Example 10.4.)

S10.4 A possible procedure is as follows:

$$Et_3SnH + 2Na \rightarrow 2Na^+ \, Et_3Sn^- + H_2$$

$$2Na^+ \, Et_3Sn^- + CH_3Br \rightarrow MeEt_3Sn + NaBr.$$

Exercises

E10.1 **a) Hydrogen in Group 1:** Hydrogen has one valence electron like the group 1 metals and is stable as H^+, especially in aqueous media. The other group 1 metals have one valence electron and are stable as M^+ cations in solution and in the solid state as simple ionic salts. In most periodic charts, hydrogen is generally put with this group, given the above information.

b) Hydrogen in Group 17: Hydrogen can fill its 1s orbital and make a hydride H^-. Hydrides are isoelectronic with He, a noble gas configuration, thus are relatively stable. Group 1 and Group 2 metals, as well as transition metals, stabilize hydrides. The halogens form stable X^- anions, obtaining a noble gas configuration, both in solution and in the solid state as simple ionic salts. Some periodic charts put hydrogen both in Group 1 and in Group 17 for the reasons stated above. The halogens are diatomic gases just like hydrogen, so physically hydrogen would fit well in Group 17.

c) Hydrogen in Group 14: All elements in Group 14, however, have four valence electrons and thus are half way to obtaining the octet of a corresponding noble gas (i.e., neon for carbon, Ar for Si, etc.). Hydrogen has only one electron and thus is half way toward obtaining a stable electronic configuration of its corresponding noble gas, He. Looking at the physical properties, the addition of hydrogen to Group 14 would add a gaseous non-metal to the group producing a gradual change of the properties to non-metallic (but solid) carbon, then somewhat more metallic solid silicon all the way to purely metallic lead. However, none of these reasons are compelling enough to warrant placing hydrogen in Group 14.

E10.2 **a) H_2S:** When hydrogen is less electronegative than the other element in a binary compound (H is 2.20, S is 2.58; see Table 1.7), its oxidation number is +1. Therefore, the oxidation number of sulfur in hydrogen sulfide (sulfane) is –2.

b) KH: In this case, hydrogen (2.20) is more electronegative than potassium (0.82), so its oxidation number is –1. Therefore, the oxidation number of potassium in potassium hydride is +1.

c) $[ReH_9]^{2-}$: The electronegativity of rhenium is not given in Table 1.7. However, since it is a metal (a member of the d-block), it is reasonable to conclude that it is less electronegative than hydrogen. Therefore, if hydrogen is –1 and considering the overall 2– charge of this species, then the rhenium atom in the $[ReH_9]^{2-}$ ion has an oxidation number of +7.

d) H_2SO_4: The structure of sulfuric acid is shown below. Since the hydrogen atoms are bound to very electronegative oxygen atoms, their oxidation number is +1. Furthermore, since oxygen is always assigned an oxidation number of –2 (except for O_2 and peroxides, as well as relatively rare superoxides and ozonides), sulfur has an oxidation number of +6.

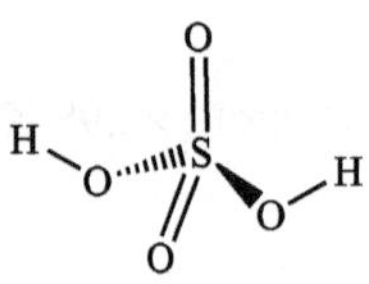

H_2SO_4

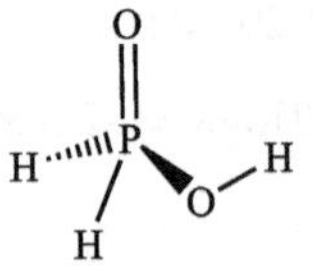

$H_2PO(OH)$

e) $H_2PO(OH)$: The structure of hypophosphorous acid (also called phosphinic acid) is shown above. There are two types of hydrogen atoms. The one that is bonded to an oxygen atom has an oxidation number of +1. The two that are bonded to the phosphorus atom present a problem because phosphorus (2.19) and hydrogen (2.20) have nearly equal electronegativities. Considering that P atom is less electronegative than H atom (although not by very much) the two hydrogen atoms have an oxidation number of +1, and oxygen, as above, is assigned an oxidation number of –2, then the phosphorus atom in $H_2PO(OH)$ has an oxidation number of +1.

E10.3 As discussed in Section 10.4 *Production of dihydrogen* the three industrial methods of preparing H_2 are: (i) steam (or hydrocarbon) reforming, (ii) coal gasification, and (iii) the water-gas shift reaction. The balanced equations are:

$$\text{(i)} \quad CH_4(g) + H_2O \rightarrow CO(g) + 3H_2(g)\ (1000°C)$$

$$\text{(ii)} \quad C(s) + H_2O \rightarrow CO(g) + H_2(g)\ (1000°C)$$

$$\text{(iii)} \quad CO(g) + H_2O \rightarrow CO_2(g) + H_2(g).$$

Note that in the countries where electricity is cheap, the electrolysis of water may be an important process as well.

These large-scale reactions are not very convenient for the preparation of small quantities of hydrogen in the laboratory. Instead, (iv) treatment of an acid with an active metal (such as zinc) or (v) treatment of a metal hydride with water would be suitable. The balanced equations are:

$$\text{(iv)} \quad Zn(s) + 2HCl(aq) \rightarrow Zn^{2+}(aq) + 2Cl^-(aq) + H_2(g)$$

$$\text{(v)} \quad NaH(s) + H_2O \rightarrow Na^+(aq) + OH^-(aq) + H_2(g).$$

E10.4 **a) Position in the periodic table:** Your result should look like Figure 10.2.

b) Trends in $\Delta_f G°$: Good guiding principles to keep in mind here are the differences in electronegativity and atomic and ionic sizes. You can consult Figure 10.3 that shows E–H bond energies as well. Your result should reflect the data shown in Table 10.1.

c) Different molecular hydrides: Molecular hydrides are found in Groups 13 through 17. Those in Group 13 are electron-deficient, those in Group 14 are electron-precise, and those in Groups 15 through 17 are electron-rich.

E10.5 If water did not have hydrogen bonds, it most likely would be a gas at room temperature like its heavier homologues H_2S, H_2Se, and H_2Te; see Figure 10.6. Extrapolating the boiling point values, we can estimate a boiling point of about –50°C or below for H_2O without H-bonding. Also, most pure compounds are denser as a solid than as a liquid. Because of hydrogen bonding, water is actually less dense as a solid and has the structure shown in Figure 10.7. Therefore, ice floats. If there were no hydrogen bonds, it would be expected that ice would be denser than water. Consult Section 10.6(a)(iii) *The Influence of Hydrogen bonding* and Figure 10.7 if you need to review this material.

E10.6 Since S is less electronegative than O, the partial positive charge on the H in O–H is higher than that on the H atom in S–H; therefore, S–H···O has a weaker hydrogen bond than O–H···S because the electrostatic interaction between the more positive H in H_2O and a lone electron pair on S is stronger.

E10.7 **a) BaH_2:** This compound is named barium hydride. It is a saline hydride.

b) SiH_4: This compound is named silane. It is an electron-precise molecular hydride.

c) NH_3: This familiar compound is known by its common name, ammonia, rather than by the systematic names azane or nitrane. Ammonia is an electron-rich molecular hydride.

d) AsH_3: This compound is generally known by its common name, arsine, rather than by its systematic name, arsane. It is also an electron-rich molecular hydride.

e) $PdH_{0.9}$: This compound is named palladium hydride. It is a metallic hydride.

f) HI: This compound is known by its common name, hydrogen iodide, rather than by its systematic name, iodane. It is an electron-rich molecular hydride.

E10.8 **a) Hydridic character:** Barium hydride from Exercise 10.7 is a good example because it reacts with proton sources such as H_2O to form H_2:

$$BaH_2(s) + 2H_2O(l) \rightarrow 2H_2(g) + Ba(OH)_2(s)$$

$$\text{Net reaction: } 2H^- + 2H^+ \rightarrow 2H_2.$$

b) Brønsted acidity: Hydrogen iodide is a good example because it transfers its proton to a variety of bases, and in water is behaving as a strong Brønsted acid:

$$HI + H_2O \rightarrow I^- + H_3O^+.$$

c) Variable composition: The compound $PdH_{0.9}$ is a good example. Generally, we can expect hydrides of heavier d-block and f-block elements to display variable composition.

d) Lewis basicity: Ammonia is a good example because it forms acid–base complexes with a variety of Lewis acids, including BMe_3:

$$NH_3(g) + BMe_3(g) \rightarrow H_3NBMe_3(g).$$

Note that AsH_3 could also behave as a Lewis base (As and N are both in Group 15 and both AsH_3 and NH_3 are electron-rich hydrides), however NH_3 is a better Lewis base.

E10.9 Of the compounds listed in Exercise 10.7, BaH_2 and $PdH_{0.9}$ are solids; none is a liquid; and SiH_4, NH_3, AsH_3, and HI are gases (see Figure 10.6). Only $PdH_{0.9}$ is likely to be a good electrical conductor.

E10.10 Recall from Section 1.7 that the size of an ion is not really a fixed value; rather it depends largely on the size of the hole the ion in question fills; in a larger hole, the ion will appear bigger than in the smaller one. We have also to recall periodic trends in cation variation. All hydrides of the Group 1 metals have the rock-salt structure (see Section 10.6(b) *Saline hydrides*). This means that in Group 1 hydrides H^- has a coordination number 6 and is surrounded by cations in an octahedral geometry. In other words, H^- occupies an octahedral hole within the face-centred cubic lattice of M^+ cations. The sizes of cations increase down the group (i.e., $r(Li^+) < r(Na^+) < r(K^+) < r(Cs^+)$); as a consequence both the sizes of octahedral holes and the apparent radius of H^- will increase in the same direction.

If we look at the Group 2 hydrides, we'll note that there is no clear trend as in the case of Group 1. The reason for this is that the structures here are different; whereas CaH_2 and BaH_2 adopt the fluorite structure, MgH_2 has a rutile structure. In the case of CaH_2 and BaH_2, the tetrahedral hole in which H^- resides is increasing in size from CaH_2 to BaH_2 and so is the H^- radius. In the case of MgH_2, the H^- is bonded to three Mg in a trigonal planar arrangement.

We can compare briefly the H^- radii in KH and CaH_2. Since $r(K^+)$ is larger than $r(Ca^{2+})$, and H^- lowers its coordination from octahedral (in KH) to tetrahedral (in CaH_2), the H^- radius drops significantly.

E10.11 Reactions (a) and (c) involve the production of both H and D atoms at the surface of a metal. The recombination of these atoms will give a statistical distribution of H_2 (25%), HD (50%), and D_2 (25%). However, reaction (b) involves a source of protons that is 100% $^2H^+$ (i.e., D^+) and a source of hydride ions that is 100% $^1H^-$:

$$D_2O + NaH(s) \rightarrow HD(g) + NaOD(s)$$

$$\text{Net reaction: } D^+ + H^- \rightarrow HD.$$

Thus, reaction (b) will produce 100% HD and no H_2 or D_2.

E10.12 Of the compounds H_2O, NH_3, $(CH_3)_3SiH$, and $(CH_3)_3SnH$, the tin compound is the most likely to undergo radical reactions with alkyl halides. This is because the Sn–H bond in $(CH_3)_3SnH$ is weaker than either O–H, N–H, or Si–H bonds. The formation of radicals involves the homolytic cleavage of the bond between the central element and hydrogen, and a weak nonpolar bond undergoes homolysis most readily (see Section 10.6(a)(ii) *Reactions of molecular hydrides*).

E10.13 Since Al has the lowest electronegativity of the three elements, B (2.04), Al (1.61), and Ga (1.81, see Table 1.7 and Self-test 10.1 for reference), the Al–H bonds of AlH_4^- are more hydridic than the B–H bonds of BH_4^- or the Ga–H bonds of GaH_4^-. Therefore, since AlH_4^- is more "hydride-like," it is the strongest reducing agent. The reaction of GaH_4^- with aqueous HCl is as follows:

$$GaH_4^- + 4HCl(aq) \rightarrow GaCl_4^-(aq) + 4H_2(g).$$

E10.14 One important difference between period 2 and period 3 hydrogen compounds is their relative stabilities. The period 2 hydrides, except for BeH_2 and B_2H_6, are all exergonic (see Table 10.1). Their period 3 homologues either are much less exergonic or are endergonic (cf. HF and HCl, for which $\Delta_fG° = -273.2$ and -95.3 kJ mol^{-1}, and NH_3 and PH_3, for which $\Delta_fG° = -16.5$ and $+13.4$ kJ mol^{-1}). Another important chemical difference is that period 2 compounds tend to be weaker Brønsted acids and stronger Brønsted bases than their period 3 homologues. Diborane, B_2H_6, is a gas whereas AlH_3 is a solid. Methane, CH_4, is inert to oxygen and water, whereas silane, SiH_4, reacts vigorously with both. The bond angles in period 2 hydrogen compounds reflect a greater degree of sp^3-hybridization than the homologous period 3 compounds (compare the H–O–H and H–N–H bond angles of water and ammonia, which are 104.5° and 106.6°, respectively, with the H–S–H and H–P–H bond angles of hydrogen sulfide and phosphane, which are 92° and 93.8°, respectively). Several hydrides of period 2 elements exhibit strong hydrogen bonding, namely, HF, H_2O, and NH_3, whereas their period 3 homologues do not (see Figure 10.6). Because of hydrogen bonding, the boiling points of HF, H_2O, and NH_3 are all higher than their respective period 3 homologues.

E10.15 As the data in Table 10.1 illustrate, the thermodynamic stability of hydrides decreases down the group. This fact is reflected in the Δ_fH values for SbH_3 and BiH_3 provided in the problem. Therefore, it is expected that it would be very difficult to prepare a sample of BiH_3, even more difficult than SbH_3. The direct combination of elements (i.e., Bi + H_2) is not a practical approach because the reaction is thermodynamically strongly unfavourable. A good synthetic procedure could be a reaction of Bi_2Mg_3 with water:

$$Bi_2Mg_3(s) + 3H_2O(l) \rightarrow 2BiH_3(g) + 3Mg(OH)_2(aq).$$

However, the current method for the synthesis of bismuthine, BiH_3, is by the redistribution of methylbismuthine, BiH_2Me.

$$3BiH_2Me \rightarrow 2BiH_3 + BiMe_3.$$

The starting material BiH_2Me can be prepared by the reduction of $BiCl_2Me$ with $LiAlH_4$. The precursor, BiH_2Me, is unstable as well. The difficulty in synthesizing BiH_3 is because it involves a multi-step process, all of which deal with extremely reactive and unstable compounds.

BiH_3 is unstable above –45°C, yielding hydrogen gas and elemental bismuth. The original preparation was reported in 1961, but it was not until 2002 that this procedure was repeated and yielded enough BiH_3 to do spectroscopic characterization!

E10.16 The reaction product is called a clathrate hydrate. Clathrate hydrate consists of cages of water molecules found in ice, all hydrogen bonded together, each surrounding a single krypton atom. Strong dipole–dipole forces (hydrogen bonds) hold the cages together, whereas weaker van der Waals forces hold the krypton atoms in the centres of their respective cages.

E10.17 **Potential energy surfaces for hydrogen bonds, see Figure 10.9:** There are two important differences between the potential energy surfaces for the hydrogen bond between H_2O and Cl^- ion and for the hydrogen bond in bifluoride ion, HF_2^-. The first difference is that the surface for the H_2O, Cl^- system has a double minimum (as do most hydrogen bonds, see the potential energy surface for $[ClHCl]^-$ on Figure 10.9) because it is an unsymmetrical H bond, whereas the surface for the bifluoride ion has a single minimum (characteristic of only the symmetrical hydrogen bonds). The second difference is that the surface for the H_2O, Cl^- system is not symmetric because the proton is bonded to two different atoms (oxygen and chlorine), whereas the surface for bifluoride ion is symmetric. The two surfaces are shown below.

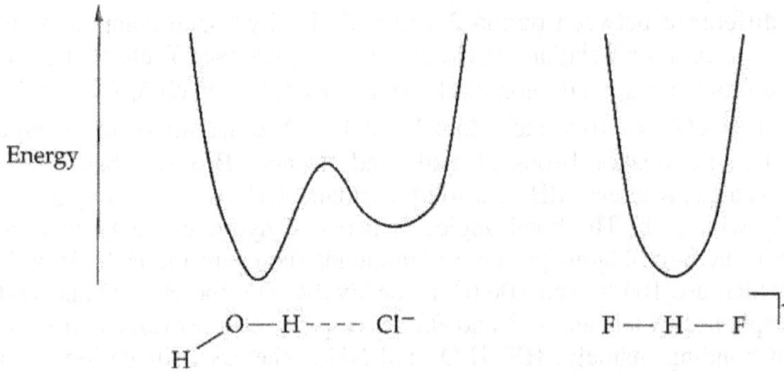

E10.18 Dihydrogen can behave both as a reducing and oxidizing agent because hydrogen can have two oxidation states in its compounds: +1 and –1. When the hydrogen in the product has oxidation state –1, the dihydrogen gas behaves as an oxidizing agent. The classic example of H_2 as an oxidizing agent is its reaction with an active s-block metal such as sodium. In the reaction below, elemental sodium gets oxidized, whereas elemental hydrogen gets reduced. Thus, H_2 is an oxidizing agent in this case.

$$2Na(s) + H_2(g) \rightarrow 2NaH(s)$$

When hydrogen has an oxidation state +1 in the final product, dihydrogen behaves as a reducing agent in the reaction. The most common example of H_2 as a reducing agent is its reaction with dioxygen (an oxidizing agent) to produce water:

$$2H_2(g) + O_2(g) \rightarrow 2H_2O \text{ (l or g).}$$

E10.19 The overall reaction is shown below:

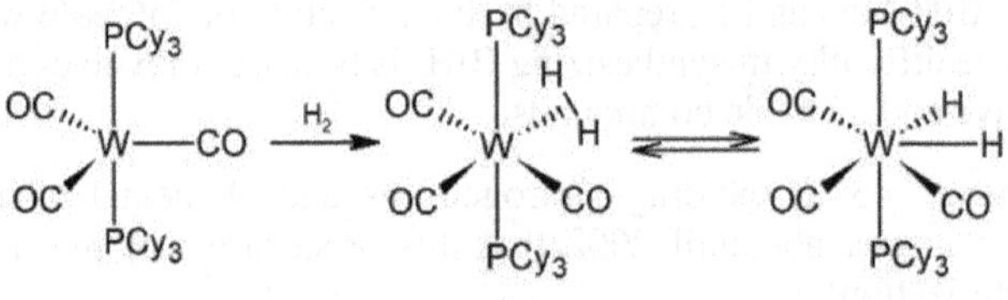

The starting material *trans*-[$W(CO)_3(PCy_3)_2$] initially adds H_2 molecule to produce a dihydrogen complex *mer*, *trans*-[$W(CO)_3(H_2)(PCy_3)_2$] **(1)**. In this complex tungsten is still formally in oxidation state 0. Complex **(1)** is in equilibrium with complex **(2)** which is a seven coordinate dihydride W(II) complex. Note the important difference between dihydrogen and dihydride complexes: dihydrogen complexes contain an H_2 molecule bonded to the metal as a ligand with the electron pair coming from the σ_g molecular orbital of H_2 molecule and with H–H bond still present. On the other hand, dihydride complexes contain two hydride, H^-, ligands. Hence, the oxidation state of W in complex **(1)** is 0 whereas in complex **(2)** it is +2. The formation of **(2)** is also an excellent example of *oxidative addition* reactions (a very common type of reaction in organometallic chemistry): the oxidation state of W has increased (i.e., it was oxidized) and the coordination number increased by two (from five in the starting material to seven in **(2)**); that is, two ligands have been added—overall an *oxidative addition.*

E10.20 Going statement by statement:

"Hydrogen, the lightest element, forms thermodynamically stable compounds with all of the nonmetals and most metals."

Incorrect: "all of the nonmetals"—recall that some p-block hydrides are not thermodynamically stable (e.g., HI).

In the second statement, the isotope of mass number 3 is radioactive, not of mass number 2.

"The structures of the hydrides of the Group 1 and 2 elements are typical of ionic compounds because the H^- ion is compact and has a well-defined radius."

Incorrect: almost everything—not all hydrides of Group 2 are ionic (BeH_2 is covalent and even MgH_2 has a high degree of covalent character); H^- is not compact (there is no such thing as a compact atom or ion because the electronic clouds in them do not have "edges") and therefore it does not have a well-defined radius.

"The structures of the hydrogen compounds of the nonmetals are adequately described by VSEPR theory."

Incorrect: not all structures can be predicted by VSEPR theory, only those of electron-precise and electron-rich hydrides; this theory cannot explain B_2H_6, an electron-poor hydride.

"The compound $NaBH_4$ is a versatile reagent because it has greater hydridic character than the simple Group 1 hydrides such as NaH."

Incorrect: while $NaBH_4$ is a common reagent, it does not have greater hydridic character (hydridicity) than a saline hydride such as NaH.

"The boron hydrides are called electron-deficient compounds because they are easily reduced by hydrogen."

Incorrect: boron hydrides are electron-deficient because they lack the electrons required to fill the bonding and non-bonding molecular orbitals.

E10.21 Recall from Chapter 8 that the energy of a molecular vibration is determined by (Equation 8.4a in your textbook):

$$E_v = \left(v+\frac{1}{2}\right)\hbar\omega = \hbar\left(v+\frac{1}{2}\right)\left(\frac{k}{\mu}\right)^{\frac{1}{2}}, \text{ with } \mu_{^1\text{H}-^{35}\text{Cl}} = \frac{m_{^1\text{H}}m_{^{35}\text{Cl}}}{m_{^1\text{H}}+m_{^{35}\text{Cl}}}, \mu_{^3\text{H}-^{35}\text{Cl}} = \frac{m_{^3\text{H}}m_{^{35}\text{Cl}}}{m_{^3\text{H}}+m_{^{35}\text{Cl}}}.$$

We can make a few reasonable assumptions that will help us to calculate the expected wavenumber for ^{3}H-^{35}Cl stretch from a known ^{1}H-^{35}Cl stretch. First, we can assume that the force constants, k, are approximately equal in the two cases. The second approximation is that $m(^1\text{H}) + m(^{35}\text{Cl}) \approx m(^3\text{H}) + m(^{35}\text{Cl}) \approx m(^{35}\text{Cl})$. Thus, keeping in mind these assumptions and all constant values, we can show that, after cancelling, the ratio of vibrational energies is:

$$\frac{E_v^{^3\text{H}-^{35}\text{Cl}}}{E_v^{^1\text{H}-^{35}\text{Cl}}} = \frac{\sqrt{\mu_{^1\text{H}-^{35}\text{Cl}}}}{\sqrt{\mu_{^3\text{H}-^{35}\text{Cl}}}} \approx \sqrt{\frac{m_{^1\text{H}}}{m_{^3\text{H}}}}.$$

And from the above ratio:

$$E_v^{^3\text{H}-^{35}\text{Cl}} = \frac{E_v^{^3\text{H}-^{35}\text{Cl}}}{\sqrt{m_{^3\text{H}}}} = \frac{E_v^{^3\text{H}-^{35}\text{Cl}}}{\sqrt{3}}.$$

Considering that the energies and wavenumber are related through $E_v = hc\tilde{\nu}$, we finally have:

$$\tilde{\nu}^{^3\text{H}-^{35}\text{Cl}} = \frac{\tilde{\nu}^{^3\text{H}-^{35}\text{Cl}}}{\sqrt{3}} = \frac{2991\ \text{cm}^{-1}}{\sqrt{3}} = 2074\ \text{cm}^{-1}.$$

E10.22 The two NMR spectra are outlined below:

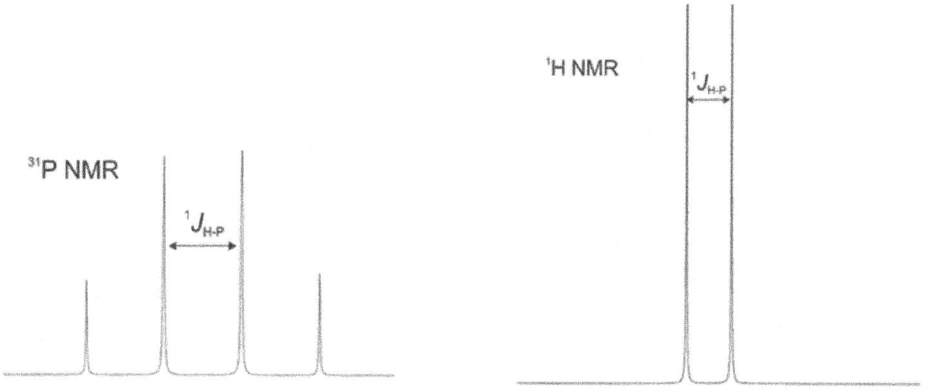

Recall that both ^{1}H and ^{31}P have a nuclear spin of ½. The PH_3 molecule has one P atom. The resonance for this phosphorus nucleus is split into a quartet with relative intensities of about 1 : 3 : 3 :1 by three equivalent H atoms (following the splitting formula n + 1 where n is the number of coupling spin ½ nuclei). The three equivalent H atoms are appearing as a doublet with relative intensities 1 : 1 due to the coupling to one P nucleus. Note that the coupling constants should be rather large (because it is a one bond coupling) and identical in two spectra.

Chapter 11 The Group 1 Elements

Self-Tests

S11.1 The ionic radius for F^- is 133 pm (Resource Section 1). Considering that the radius ratio values are less than 1, we have to consider the r_-/r_+ ratio:

$$r_-/r_+ = 133 \text{ pm}/196 \text{ pm} = 0.679.$$

This ratio suggests an NaCl-type of structure and 6 : 6 coordination.

S11.2 Polarizable anions have electronic clouds that can be easily deformed. If the hydride radius suggests a structure with 8 : 8 coordination but actual structures have a 6 : 6 coordination then hydride anion must be very polarizable—fitting a large anion into a relatively small octahedral hole (small in comparison to the cubic hole) must be possible only if the anion can be "squished," i.e. deformed.

S11.3 The ionic radii of Fr^+ and I^- are 196 pm and 220 pm respectively. We can use $2(r_+ + r_-)/3^{1/2}$ to calculate the length of the unit cell expected from these radii:

$$d = 2(196 + 220)/3^{1/2} \text{ pm} = 481 \text{ pm}.$$

This value is reasonably close to the observed 490 pm, thus the data are consistent with the prediction.

S11.4 The Kapustinskii equation, given in Section 4.14, is:

$$\Delta H_L^0 = \frac{N_{ion}|z_+z_-|}{d}\left(1-\frac{d^*}{d}\right)\kappa .$$

We can approximate the interionic distances, d, with the sum of ionic radii for the two compounds in question. Both LiF and CsF have the rock-salt type structure with (6,6) coordination; thus, we have to look for radii in this specific coordination environment.

For LiF we have:

$$\Delta H^\circ_L(\text{LiF}) = \frac{2|(+1)\times(-1)|}{(76+133)\,\text{pm}}\left(1-\frac{34.5\,\text{pm}}{(76+133)\,\text{pm}}\right)\times 1.21\times 10^5 \text{ kJ pm mol}^{-1} = 966.8\,\text{kJ mol}^{-1}.$$

For CsF we have

$$\Delta H^\circ_L(\text{CsF}) = \frac{2|(+1)\times(-1)|}{(167+133)\,\text{pm}}\left(1-\frac{34.5\,\text{pm}}{(167+133)\,\text{pm}}\right)\times 1.21\times 10^5 \text{ kJ pm mol}^{-1} = 713.9\,\text{kJ mol}^{-1}.$$

As expected, the lattice enthalpy for LiF is higher than for CsF. LiF is insoluble, because the hydration enthalpy for Li^+ (956 kJ mol^{-1}) is insufficient to compensate for the lattice enthalpy of LiF. On the other hand, hydration enthalpy for Cs^+ (710 kJ mol^{-1}) is about the lattice enthalpy for CsF, and it is expected that CsF should be more soluble than LiF. Note that both LiF and CsF contain the same anion, F^-, and that the hydration of this anion should also be considered. However, the hydration enthalpy for F^- is insignificant (particularly when compared to the error produced by the Kapustinskii equation) and constant for both compounds, thus it can be ignored for a qualitative analysis like this one.

S11.3 Ozonides decompose to oxygen and superoxides ($MO_3(s) \rightarrow MO_2(s) + \frac{1}{2}O_2(g)$). All simple ozonides are known except for Li. We can expect a similar trend for ozonides like the one observed for peroxides and superoxides. For example, going down the group the lattice energies for oxides, peroxides, and superoxides decrease; the difference in lattice energies also decreases down the group. Because of this decrease in the difference, the heavy superoxides have less tendency to decompose. By extension, we can expect that the difference between the ozonide and superoxide lattice energies decreases going down the group; consequently, the stability of ozonides would increase in the same direction. Thus, we would expect that, on warming, NaO_3 would decompose first and CsO_3 last. This indeed is the case: NaO_3 decomposes at 37°C while CsO_3 does so at about 53°C.

S11.6 The answer follows similar reasoning as in Example 11.6 and description of thermal stabilities of Group 1 carbonates in the text. Larger K^+ is going to stabilize SO_3^{2-} anion better than small Li^+. Therefore, K_2SO_3 is going to decompose at higher temperatures.

S11.7 KNO_3 has two different decomposition temperatures, 350°C and 950°C. These correspond to the following changes in the composition:

$$KNO_3(s) \rightarrow KNO_2(s) + {}^1/_2O_2(g)$$

$$2KNO_2(s) \rightarrow K_2O(s) + 2NO_2(g) + {}^1/_2O_2(g).$$

Whereas $LiNO_3$ decomposes in one step at the temperature of 600°C:

$$LiNO_3(s) \rightarrow {}^1/_2Li_2O(s) + NO_2(g) + {}^1/_4O_2(g).$$

Both end-products are the binary oxides of lithium and potassium. It is important to remember here that large cations can stabilize large anions whereas small cations can stabilize small anions. Since K^+ is larger than Li^+, the nitrate is more stable in the lattice with K^+ than with Li^+, hence the final decomposition temperature for KNO_3 is higher. Larger size of K^+ is also responsible for the stabilization of the nitrite intermediate (without this added stability, we probably would not be able to observe this intermediate). On the side of the products, Li_2O is more stable, has higher lattice energy, than K_2O. (Recall that the lattice energies dictate the oxidation products: Li with O_2 produces Li_2O, whereas K produces KO_2 with larger superoxide O^{2-} anion). Therefore, $LiNO_3$ goes straight to Li_2O in one step at lower temperature whereas the decomposition of KNO_3 requires higher temperature and includes an intermediate.

S11.8 Li_3N has an unusual crystal structure (Figure 11.11), that consists of two types of layers. The first layer has the composition Li_2N^-, containing six-coordinate Li^+ centres. The second layer consists only of lithium cations. Therefore, the Li^+ ions are in two different chemical and magnetic environments. Therefore, one would expect two peaks in the ^7Li NMR spectrum. However, the lithium ion is highly mobile (as discussed in Section 11.12) and, because there are vacant sites in the structure, Li^+ can hop from one site to another. At higher temperatures, all the lithium ions will be fluxional, giving only one averaged resonance in the ^7Li NMR. At lower temperatures, the structure would freeze, and two resonances due to the two different environments in the structure would be observed.

Exercises

E11.1 **a)** All Group 1 metals have one valence electron in the ns^1 subshell. They also have relatively low first ionization energies; therefore, loss of one electron to form a closed-shell electronic configuration is favourable.

b) The Group 1 metals are large, electropositive metals and have little tendency to act as Lewis acids. They also lack empty orbitals to which the ligand lone electron pair could be donated. However, they do complex well with hard Lewis bases, such as oxides, hydroxides, and many other oxygen-containing ligands.

E11.2 Caesium metal is extracted from silicate minerals, such as pollucite $Cs_2Al_2Si_4O_{12}\cdot nH_2O$. The caesium-bearing silicate minerals are digested with sulfuric acid and Cs and Al are precipitated as alum $Cs_2SO_4\cdot Al_2(SO_4)_3\cdot 24H_2O$. The simple sulfate, obtained after roasting with carbon, is then converted to chloride using anion exchange. The isolated CsCl is then reduced with Ca or Ba.

E11.3 From the given value for $r(Au^-)$ and ionic radii for Rb^+ (148 pm) and Cs^+ (167 pm):

RbAu: $r(Rb^+)/r(Au^-)$ = (148 pm)/(196 pm) = 0.755; within 0.732 – 1 range with stoichiometry 1 : 1, hence CsCl type structure;

CsAu: $r(Cs^+)/r(Au^-)$ = (167 pm)/(196 pm) = 0.852; within 0.732 – 1 range with stoichiometry 1 : 1, hence CsCl type structure.

E11.4 The results are shown in the table below. The values have been calculated using Equation 11.1 and the data provided in the tables in the textbook.

	$-\Delta_f H^o/(\text{kJ mol}^{-1})$		$-\Delta_f H^o/(\text{kJ mol}^{-1})$
LiF	625	LiCl	470
NaF	535	NaCl	411
KF	564	KCl	466
RbF	548	RbCl	458
CsF	537	CsCl	456

Your plot should look like a part of Figure 11.6, which plots the enthalpy of formation of the Group 1 metal halides. Fluoride is a hard Lewis base and will form strong complexes with hard Lewis acids. The lithium cation is the hardest Lewis acid of the Group 1 metals, so it makes sense that it has the largest enthalpy of formation with fluoride compared to the rest of the Group 1 metals. However, the trends reverse for the chloride ion; since it is a softer Lewis base than the fluoride ion, it will form stronger bonds with the heavier Group 1 metals.

E11.5 It would be a good idea to review Section 9.11. Look at the reasons why diagonal relationships between elements occur within the periodic table. Then look for the relevant values of atomic properties for Li and Mg (Chapter 12 or resource sections) and compare them—you will find significant similarities. The similarities in chemistry are related to the similar ionic radii.

E11.6 Figure 11.3 in the text clearly shows that the values of reduction potential depend on thermodynamic values for several processes. The hydration enthalpy for small Li^+ is large enough to (almost completely) compensate for lithium's high ionization energy. This $\Delta_{hyd}H$ makes Li as reducing as Cs. This exercise illustrates one important point: periodic trends are observed clearly for basic atomic properties (those covered in Chapters 2 and 9), but care needs to be exercised when comparing and/or looking for trends in more complex processes, i.e., those that can be broken into more fundamental steps using Born–Haber thermodynamic cycle.

The electrode potential for Li/Li^+ in dimethylsulfoxide would not change much. This is a very polar solvent (like water) hence its molecules would still interact very strongly with a cation of high charge density such as Li^+. The reduction potential would significantly increase in a weakly coordinating ionic liquid whose components are not interacting strongly with Li^+: little interaction implies low enthalpy of solvation, possibly insufficient to compensate for high ionization potential.

E11.7 **a)** $Edta^{4-}$ is expected to give a more stable complex with Cs^+ compared to acetate. In general the Group 1 cations form stable complexes with polydentate ligands like $edta^{4-}$. This ligand has 6 potential donor atoms all of which are hard (four oxygen and two nitrogen donor atoms). On the other hand, acetate has two hard oxygen donor atoms and can act only as a bidentate ligand resulting in a much lower formation constant. The acetate complex is further destabilized due to acetate's small bite angle—the four-member chelating ring formed with relatively large Cs^+ would be rather unstable.

b) K^+ is more likely to form a complex with the cryptate ligand C2.2.2 than Li^+. The difference has to do with the match between the interior size of the cryptate and the ionic radius of the alkali metal ion. According to Figure 11.14, bigger K^+ forms a stronger complex, by about two orders of magnitude, with C2.2.2 than does smaller Li^+.

E11.8 **(i)** A = LiOH; B = Li_2O; C = none (Li_2O does not decompose); D = $LiNH_2$.
(ii) A = RbOH; B = RbO_2; C = Rb_2O; D = $RbNH_2$.

E11.9 Section 11.8 *Oxides and Related Compounds* briefly describes the synthesis of ozonides. The best method of synthesis uses metal superoxide and dry ozone/oxygen mixture (hydroxides are less desirable starting materials because of potential formation of H_2O which rapidly reacts with the product):

$$CsO_2 + O_3 \rightarrow CsO_3 + O_2.$$

When choosing the solvent for the recrystallization we have to think about the properties of CsO_3: it is an ionic compound that contains a very basic anion. This limits the choice to polar, basic solvents—ideal example of such a solvent is liquid ammonia.

E11.10 In LiF and CsI the cation and anion have similar radii. Solubility is lower if cation and anion have similar radii. On the other hand, there is a large difference between radii of cation and anion in CsF and LiI, and these two salts are soluble in water (see also Section 4.15).

E11.11 It is very likely that FrI would have low solubility in water; similar radii for Fr^+ and I^- (196 pm and 220 pm respectively) would reduce the solubility of this salt (parallel with solubility of Cs salts).
Particularly useful for separation of Fr^+ from Na^+ should be $Na_3[Co(NO_2)_6]$. This yellow sodium salt is soluble in water, but potassium, rubidium and caesium salts with the same anion are not. Thus, it can be expected that $Fr_3[Co(NO_2)_6]$ is not soluble either.

E11.12 Hydrides decompose to elements. The MH lattice energy decreases down the group as the radius of cation increases; thus, LiH has the highest lattice energy and requires the highest decomposition temperature. $LiHCO_3$ is only known in solution. Small size of Li^+ is insufficient to stabilize HCO_3^- anion, which easily breaks into H_2O, CO_2 and CO_3^{2-}.

E11.13 Your NaCl structure should look like the one shown in Figure 11.4 and CsCl like the one shown in 11.5. Na^+ is six-coordinate, whereas Cs^+ is eight-coordinate. The compounds have different structures because Na^+ is smaller than Cs^+ resulting in a different r_+/r_- for the same anion. Caesium is so large that it can fill a cubic hole formed in simple cubic packing of Cl^- ions.

E11.14 This exercise is similar to the Example 11.3. RbBr has a NaCl type of structure (face centred lattice, with 6 : 6 coordination). Increase in pressure would likely result in the structure (phase) change to primitive lattice with 8 : 8 coordination.

E11.15 **(a)** The driving force behind this reaction is the formation of lithium bromide (very large lattice energy and insoluble in organic solvents in which the reaction is performed). The same reasoning applies to reaction **(b)**. For reaction **(c)**, the driving force is the loss of ethane gas and higher acidity of C_6H_6 compared to C_2H_6.

(a) $CH_3Br + Li \rightarrow Li(CH_3) + \mathbf{LiBr}$

(b) $MgCl_2 + LiC_2H_5 \rightarrow Mg(C_2H_5)Br + \mathbf{LiCl}$

(c) $C_2H_5Li + C_6H_6 \rightarrow LiC_6H_5 + \mathbf{C_2H_6}$

E11.16 Review Section 11.17 *Organometallic compounds*. Whether a molecule is monomeric or polymeric is based on the streric size of the alkyl group. Less bulky alkyl groups lead to polymerization; methyl groups are tetrameric or hexameric (depending on solvent used), whereas a bulkier ${}^tBu$ group yields tetramers as the largest possible species. The larger aggregates can be broken down into dimmers and monomers using strong Lewis bases such as TMEDA.

Guidelines for Selected Tutorial Problems

T11.7 Sodide ion is not produced by simply dissolving Na in $NH_3(l)$. See section 11.14 *Solutions in Liquid Ammonia* for details. Dissolved Na produces Na^+ in ammonia. This cation will not be involved in hydrogen bonding because hydrogen in NH_3 is delta-positive. Na^+ is going to interact with lone electron pair on NH_3 molecule.

Chapter 12 The Group 2 Elements

Self-Tests

S12.1 Water cannot be used to extinguish a magnesium fire. Hot magnesium reacts with water liberating explosive H_2 gas. Burning magnesium metal can be covered with sand and left to cool down before handling.

S12.2 The oxygen atom in the structure of diethyl ether can behave as Lewis base because it has two lone electron pairs. Thus, diethyl ether can form adducts with Lewis acidic $BeCl_2$. Due to its small size, Be^{2+} likes tetrahedral geometry. Considering that it is already bonded to two Cl^- ions, we can bond two Et_2O molecules to one $BeCl_2$, and the product is expected to be $[BeCl_2(OEt_2)_2]$.

S12.3 The electronegativities of elements are provided in Table 1.7, and consult the Ketelaar triangle (Fig. 2.28). The Pauling electronegativities for Be, Ba, Cl, and F are 1.57, 0.89, 3.16, and 3.98 respectively. The average electronegativity for $BeCl_2$ is (1.57 + 3.16)/2 = 2.36, and the electronegativity difference is 3.16 – 1.57 = 1.59. These values on the Ketalaar triangle correspond to covalent bonding. For BaF_2 the average electronegativity is (3.98 + 0.89)/2 = 2.43 and the difference is (3.98 – 0.89) = 3.09, corresponding to the ionic bond on the Ketalaar triangle.

Considering the bonding type in $BeCl_2$ and Lewis acid character of Be^{2+}, we can predict that $BeCl_2$ will have a polymeric, molecular structure. On the other hand, BeF_2 with its ionic bonding is very likely to adopt CaF_2-type ionic structure.

S12.4 Since Ca is between Mg and Ba in Group 2 we should expect, based on the periodic trends, that the difference in lattice enthalpies for CaO and CaO_2 lies between that of Mg and Ba oxides and peroxides and that the CaO_2 lattice is less stable than the CaO lattice. Using the Kapustinskii equation (Equation 4.4), the ionic radii from Table 12.1, and the appropriate constants—which are $\kappa = 1.21 \times 10^5$ kJ pm mol^{-1}, $N_{ion} = 2$, $z^+_{Ca} = +2$, $z^-_O = -2$ — the sum of the thermochemical radii, d_0, 140 pm + 100 pm = 240 pm, and $d^* = 34.5$ pm, find the lattice enthalpy of CaO. For calcium peroxide use $N_{ion} = 2$, $z^+_{Ca} = +2$, $z^-_{O_2} = -2$, and d_0 = 180 pm + 100 pm = 280 pm. The lattice enthalpies of calcium oxide and calcium peroxide are 3465 kJ mol^{-1} and 3040 kJ mol^{-1}, respectively. The trend that the thermal stability of peroxides decreases down Group 2 and that peroxides are less thermally stable than the oxides is confirmed.

S12.5 The ionic radius of Be^{2+} is 45 pm (consult Resource Section 1 in your textbook) and the ionic radius of Se^{2-} is 198 pm. The radius ratio is 45 pm/198 pm = 0.23, which according to Table 4.6 should be close to ZnS-like structure.

S12.6 Heating $BeSO_4 \cdot 4H_2O$, the product of reaction (i), would likely result in the hydrolysis and loss of SO_3 (or SO_2 and oxygen), similarly to Be nitrate, and result in formation of $Be_2(OH)_2(SO_4)$ or related basic compound. On the other hand, heating $SrBe(OH)_4$, the product of reaction (ii), would produce SrO and BeO with elimination of two equivalents of H_2O:

$$SrBe(OH)_4(s) \xrightarrow{500°C} SrO(s) + BeO(s) + 2H_2O(g).$$

S12.7 Using the Kapustinskii equation (Equation 4.4) and the values of $\kappa = 1.21 \times 10^5$ kJ pm mol^{-1}, $N_{ion} = 3$, $z^+_{Mg} = +2$, $z^-_F = -1$, the sum of the thermochemical radii, d_0, 133 pm + 72 pm = 205 pm and $d^* = 34.5$ pm, the calculated lattice enthalpy of MgF_2 is 2945 kJ mol^{-1}. Similarly, for $MgBr_2$, using 196 pm as the ionic radius for Br^-, we can calculate the lattice enthalpy to be 2360 kJ mol^{-1}, whereas for MgI_2 (with 220 pm as I^- radius) we have 2193 kJ mol^{-1} as calculated lattice enthalpy.

Comparing the three calculated values, we see that, as expected, the lattice enthalpies decrease with an increase in the ionic radius of the halide anion; we can expect the increase in solubility in the same direction (i.e., from MgF_2 to MgI_2).

Exercises

E12.1 Be^{2+} has large polarizing power and a high charge density due to its small radius. Also, beryllium has the highest electronegativity among Group 2 elements. Descending Group 2 the atoms of elements increase in size, they are more electropositive (electronegativity drops), and predominantly ionic character of bonds results.

E12.2 The similarity between Be and Al is arising from their similar atomic and ionic radii (note that the two display diagonal similarity). They also have similar Pauling electronegativities. The similarity between zinc and beryllium is less pronounced than between aluminium and beryllium and it lies also in similar radii and the same charge (Be^{2+} and Zn^{2+}).

E12.3 All chlorides of Group 2 elements are ionic solids except $BeCl_2$ which is a polymeric covalent solid. Ionic compounds (in general) always have higher melting points than covalent compounds.

E12.4 The mean electronegativity for Be and Cs is (1.57 + 0.79)/2 = 1.18, whereas the electronegativity difference is 1.57 – 0.79 = 0.78. These two values correspond to the point in the metallic bond area.

E12.5 $Ba + 2H_2O \rightarrow Ba(OH)_2 + H_2$; **A** = $Ba(OH)_2$

$Ba(OH)_2 + CO_2 \rightarrow BaCO_3$; **B** = $BaCO_3$

$2BaCO_3 + 5C \rightarrow 2BaC_2 + 3CO_2$; **C** = BaC_2

$BaC_2 + 2H_2O \rightarrow Ba(OH)_2$ (= compound **A**) + C_2H_2

$Ba(OH)_2 + 2HCl \rightarrow BaCl_2 + 2H_2O$; **D** = $BaCl_2$

E12.6 Unlike the other alkaline earth metal fluorides, MF_2 (M = Mg, Ca, Sr, Ba) which crystallizes with CaF_2-like structure, BeF_2 adopts SiO_2-like arrangement. As with SiO_2, which forms vitreous matter when melted because of the strength of the Si-O bonds (460 kJ mol^{-1}) that are not fully broken in the vitreous phase, BeF_2, with strong B-F bonds (~600 kJ mol^{-1}) also forms a glass (a disordered arrangement of molecular fragments) when cooled.

E12.7 If you had a problem with this exercise, have a look again at Example 12.2 and Self-Test 12.2. Be^{2+} is a strong Lewis acid while NH_3 is a good Lewis base since nitrogen atom has a lone electron pair. Thus, we can expect formation of Lewis acid–base adduct between Be^{2+} and NH_3. This is very similar to the reaction of BeF_2 with H_2O described in Section 12.11. The likely product will be $[Be(NH_3)_4]F_2$ with a tetrahedral geometry at Be^{2+} centre—four NH_3 ligands with fluorides as counterions in the secondary coordination sphere of $[Be(NH_3)_4]^{2+}$.

E12.8 The slight acidity of Mg^{2+} solution can be explained through hydrolysis. $Mg(OH)_2$ is weaker base than $Ba(OH)_2$, and Mg^{2+} cation is going to be slightly hydrolysed in water according to equation:

$$[Mg(H_2O)_6]^{2+}(aq) + H_2O(l) \rightarrow [Mg(OH)(H_2O)_5]^{+} + H_3O^{+}$$

The formation of hydronium cation lowers the pH value.

E12.9 The weight percent of hydrogen in BeH_2 and MgH_2 are:

$$\omega(H)_{BeH_2} = \frac{2\times M(H)}{M(BeH_2)}\times 100\% = \frac{2\times 1}{9.01+2\times 1}\times 100\% = 18.2\%$$

$$\omega(H)_{MgH_2} = \frac{2\times M(H)}{M(MgH_2)}\times 100\% = \frac{2\times 1}{24.3+2\times 1}\times 100\% = 7.6\%.$$

Although BeH_2 has a higher percentage of H in its composition, it is not investigated as a possible hydrogen storage material because BeH_2 cannot be prepared directly from the elements (unlike MgH_2) but rather from alkyl-beryllium compounds—the fact that complicates its potential use.

E12.10 The hydroxides of Group 1 metals have higher solubilities in water; in fact, they are all strong bases completely dissociating in water. In contrast, $Be(OH)_2$ is amphoteric; $Mg(OH)_2$ is weakly soluble in water; and $Ca(OH)_2$, while being more soluble than $Mg(OH)_2$, still cannot compare to any Group 1 hydroxide. The increase in OH^- concentration, due to higher solubility, increases the corrosiveness of Group 1 hydroxides. Only $Sr(OH)_2$ and $Ba(OH)_2$ match Group 1 hydroxides in their solubility and thus basicity.

E12.11 $Ba(OH)_2$ is a very strong base that gives two equivalents of OH^- per mol of the base. NaOH, although equally strong base provides only one equivalent of hydroxide ion per mol. Also, $Ba(OH)_2$ is less hygroscopic than NaOH making it easier to obtain more precise weight of the base in moist (humid) environment.

E12.12 Recall from Chapter 4 that the closer cation and anion are in size, the lower the solubility. Since Mg^{2+} has a higher ionic radius than Ba^{2+}, $MgSeO_4$ is expected to have higher solubility in water than $BaSeO_4$ (Table 4.10 lists thermochemical radius for SeO_4^{2-} to be 243 pm, while Mg^{2+} and Ba^{2+} radii are 72 pm and 135 respectively).

E12.13 Decomposition of $CaCO_3$ produces one mol of CaO and one mole of CO_2. In grams, 56.08 g of CaO is produced for every 44 g of CO_2 released. Multiplying all by "billion tonnes" (note that 1 tonne = 1000 kg) we see that 56.08 billion tonnes of CaO comes with 44 billion tonnes of CO_2. Simple ratio can give us the final answer: 56.08/44 = x/35 or x = 44.6 billion tonnes.

E12.14 Ra^{2+} could be selectively precipitated by addition of a large anion. Large anions, such as IO_4^- (249 pm) or TeO_4^{2-} (254 pm), could form insoluble salts with Ra^{2+} (170 pm).

E12.15 Anhydrous Mg and Ca sulfates are preferred as drying agents over the other alkaline earth sulfates. This is because of the higher affinity of Mg and Ca sulfates for water compared to the other alkaline earth sulfates, which are nearly insoluble in water. It may be interesting to note here that $BeSO_4$, although expected to be more hydrophilic, it is not considered as a desiccant because of its higher cost and toxicity. Anhydrous calcium chloride is also a useful drying agent, it is very hygroscopic due to the high hydration energy of Ca^{2+}.

E12.16 The ionic radius of Be^{2+} is 45 pm (Resource Section 1) and the ionic radius of Te^{2-} is 207 pm. The radius ratio is 45 pm/207 pm = 0.22. According to Table 4.6, this should be close to ZnS-like structure. The ionic radius of Ba^{2+} is 135 pm and the radius ratio is 135 pm/207 pm = 0.65, and BaTe should have NaCl-like structure.

E12.17 The Pauling electronegativity values of Be, Mg, Ba, and Br are 1.57, 1.31, 0.89, and 2.96 respectively.

The average electronegativity of $BeBr_2$ is therefore 2.27 and the difference is 1.39. The values on the Ketelaar triangle indicate that $BeBr_2$ should be covalent.

The average electronegativity of $MgBr_2$ is 2.13 and the difference is 1.65. The values on the Ketelaar triangle indicate that $MgBr_2$ should be ionic.

The average electronegativity of $BaBr_2$ is therefore 1.93 and the difference is 2.07. The values on the Ketelaar triangle indicate that $BaBr_2$ should be ionic.

E12.18 In solution, C_2H_5MgBr will be tetrahedral with two molecules of solvent coordinated to Mg^{2+}. The bulky organic group in 2,4,6-$(CH_3)_3C_6H_2MgBr$ leads to a coordination number of two:

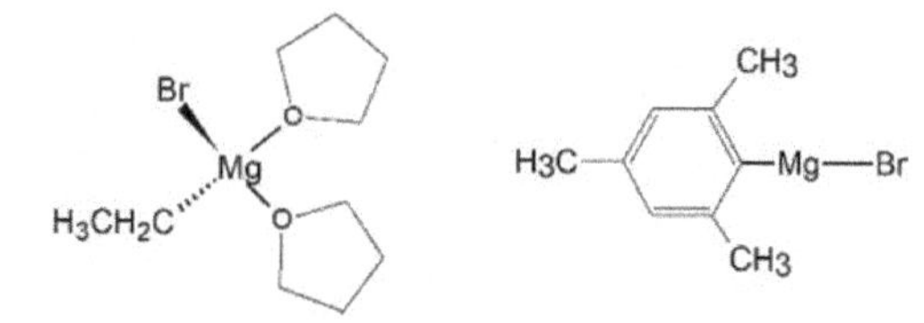

E12.19 **(a)** The products depend on stoichiometry chosen:

$MgCl_2 + LiC_2H_5 \rightarrow LiCl + (C_2H_5)MgCl$

or

$MgCl_2 + 2LiC_2H_5 \rightarrow 2LiCl + Mg(C_2H_5)_2$

(b) $Mg + (C_2H_5)_2Hg \rightarrow Mg(C_2H_5)_2 + Hg$

(c) $Mg + C_2H_5HgCl \rightarrow C_2H_5MgCl + Hg$

Guidelines for Selected Tutorial Problems

T12.4 BeF_2 is in many ways similar to SiO_2. A good starting point for the discussion of BeF_2 glasses is structural and chemical similarity between BeF_2 and SiO_2. A good reading on the topic is Greenwood and Earnshaw (1997) *Chemistry of the Elements*, 2nd ed. Elsevier (Chapter 5).

T12.7 When discussing these structures, it is useful to consider simple properties, such as ionic radii of Be^{2+}, Al^{3+}, Si^{4+} and P^{5+} for coordination number 4. If these are not similar to each other, then there is limited possibility for exchange (i.e., formation of beryllophosphates). There is an abundance of information on alumosilicates, including your textbook. For specific minerals, a good starting point is www.webmineral.com website, a trusted source on mineral data, which can be searched by element. A well-characterized beryllophosphate mineral is pahasapaite, with zeolite-like framework composed of BeO_4 and PO_4 tetrahedra. This structure is also a good starting point for comparison between beryllophosphates and alumosilicates.

Chapter 13 The Group 13 Elements

Self-Tests

S13.1 Following periodic trends, we could expect nihonium to form Nh_2O. Considering the decrease in formation of trioxides going down the group (i.e., –923 kJ mol^{-1} for In_2O_3 and only –94 kJ mol^{-1} for Tl_2O_3) we should expect Nh_2O_3 to be even less stable (more positive heat of formation). Also, Nh +1 oxidation state should be more stable than +3 (just as observed with Tl) due to the inert pair effect.

S13.2 The structure should be based on face-centred lattice of Y^{3+} with all octahedral holes filled with B_{12}^{3-} anions. The cuboctahedron is a polygon with six square faces and eight triangular faces:

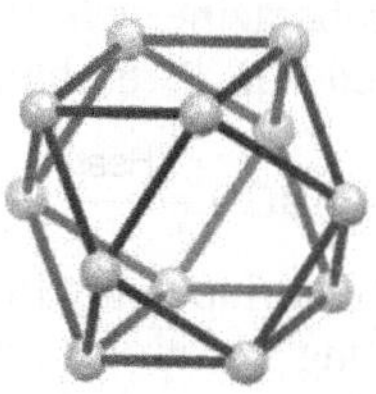

S13.3 VSEPR theory predicts that a species with four bonding pairs of electrons should be tetrahedral.

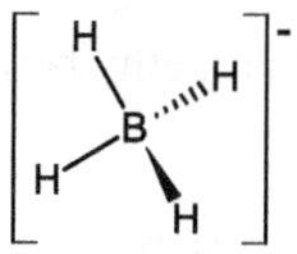

All protons are in equal environment and without coupling to the NMR active ^{11}B nucleus would produce only a singlet. However, coupling to one ^{11}B nucleus splits this singlet in $2nI + 1$ lines or $2 \times 1 \times 3/2 + 1 = 4$ lines. Due to this coupling a quartet is produced with four lines of equal intensity (1 : 1 : 1 : 1). (See also Exercise 8.20(a) and its solution)

S13.4 Look first at the reaction between B_2H_6 and propene ($CH_3CH{=}CH_2$) in ether. B_2H_6 is a dimer which can easily be cleaved. Since the solvent is ether, the first reaction would involve the cleavage of B_2H_6 with R_2O (ether molecules) to produce symmetric monomers $H_3B{-}OR_2$:

$$B_2H_6 + 2R_2O \rightarrow 2H_3B{-}OR_2.$$

In the presence of propene, and with heating, the monomer is going to add across C=C bond (hydroboration reaction):

$$H_3B{-}OR_2 + H_2C{=}CHCH_3 \rightarrow H_2BCH_2{-}CH_2CH_3 + OR_2.$$

In the second reaction, between B_2H_6 and NH_4Cl, THF solvent is also ether. The first step is the same as above with OR_2 replaced with THF:

$$B_2H_6 + 2THF \rightarrow 2H_3B{-}THF.$$

NH_4Cl contains a protic (acidic) cation (NH_4^+), whereas $H_3B{-}THF$ is a hydridic hydride. Thus, the reaction between NH_4^+ and $H_3B{-}THF$ is going to produce hydrogen gas and:

$$H_3B{-}THF + NH_4Cl \rightarrow H_2 + H_2B{-}NH_2 + HCl + THF.$$

S13.5 **a) BCl_3 and ethanol:** As mentioned in the example, boron trichloride is vigorously hydrolysed by water. Therefore, a good assumption is that it will also react with protic solvents such as alcohols, forming HCl and B–O bonds:

$$BCl_3(g) + 3EtOH(l) \rightarrow B(OEt)_3(l) + 3HCl(g).$$

b) BCl_3 and pyridine in hydrocarbon solution: Neither pyridine nor hydrocarbons can cause the protolysis of the B–Cl bonds of boron trichloride, so the only reaction that will occur is a complex formation reaction, such as the one shown below. Note that pyridine, C_5H_5N, with a lone electron pair on N atom is a good Lewis base, whereas hydrocarbons with no lone electron pairs are very weak Lewis bases. Thus, pyridine, and not hydrocarbon, is going to react with BCl_3 as Lewis acid:

$$BCl_3(g) + C_5H_5N(l) \rightarrow Cl_3B\text{–}NC_5H_5(s).$$

c) BBr_3 and $F_3BN(CH_3)_3$: Since boron tribromide is a stronger Lewis acid than boron trifluoride, it will displace BF_3 from its complex with $N(CH_3)_3$:

$$BBr_3(l) + F_3BN(CH_3)_3(s) \rightarrow BF_3(g) + Br_3BN(CH_3)_3(s).$$

S13.6 *N,N′,N″*-trimethyl-*B,B′,B″*-trimethylborazine is hexamethylborazine. The reaction of ammonium chloride with boron trichloride yields *B,B′,B″*-trichloroborazine, whereas the reaction of a primary ammonium chloride with boron trichloride yields *N*-alkyl substituted *B,B′,B″*-trichloroborazine, as shown below:

$$3[CH_3NH_3]Cl + 3BCl_3 \xrightarrow{\text{Heat}} 9HCl + Cl_3BNCH_3.$$

Therefore, if you use methylammonium chloride you will produce $Cl_3BN(CH_3)_3$ (i.e., R = CH_3). This product can be converted to the desired one by treating it with an organometallic methyl compound of a metal that is more electropositive than boron. Either methyllithium or methylmagnesium bromide could be used, as shown below:

$$Cl_3BNCH_3 + 3CH_3MgBr \rightarrow (CH_3)_3B_3N_3(CH_3)_3 + 3Mg(Br, Cl)_2.$$

The structure of *N,N′,N″*-trimethyl-*B,B′,B″*-trimethylborazine:

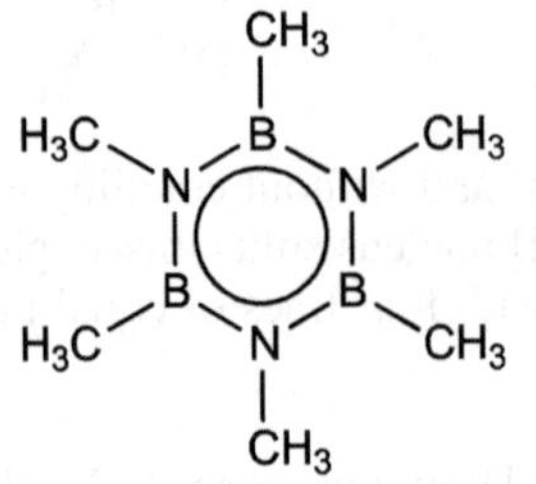

S13.7 **a)** The formula B_4H_{10} belongs to a class of borohydrides having the formula B_nH_{n+6}, which is characteristic of a *arachno* species. Assume one B–H bond per B atom, there are 4 BH units, which contribute 4 × 2 = 8 electrons, and the six additional H atoms, which contribute a further 6 electrons, giving 14 electrons, or seven electron pairs per cluster structure, which is n + 3 with n = 4; this is characteristic of *arachno* clusters. The resulting seven pairs are distributed—two are used for the additional terminal B–H bonds, four are used for the four BHB bridges, and one is used for the central B–B bond.
The structure of B_4H_{10}:

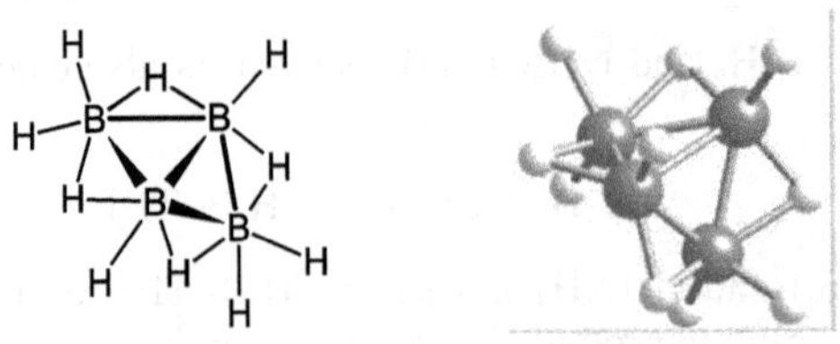

b) $[B_5H_8]^-$ has 5 B–H units contributing 5 × 2 = 10 electrons to the skeleton. There are three additional H atoms; each contributes one electron, for a total of three electrons. Finally, there is one negative charge on the cluster providing an additional electron. In total there are 10 + 3 + 1 = 14 skeletal electrons or seven skeletal electron pairs. This gives $n + 2$ (for n = 5) and a *nido* structure type. The structure is based on an octahedron with one missing vertex:

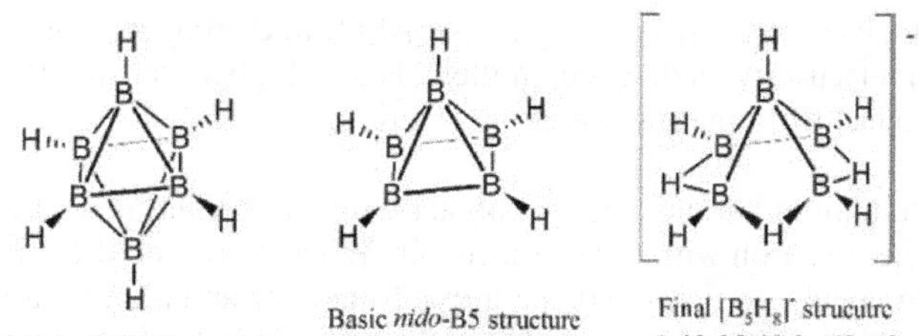

Parent *closo*-B6 structure | Basic *nido*-B5 structure (after removal of one vertex from *closo*-B6) | Final $[B_5H_8]^-$ strucutre (with 3 bridging H and cluster charge added to *nido*-B5 structure)

S13.8 General equation for hydrolysis of borohydrides is:

$$B_nH_m + 3nH_2O \rightarrow nB(OH)_3 + (3n + m)/2\ H_2$$

In our case, five moles of $B(OH)_3$ have been produced, meaning that n in above equation is 5. From this value and the fact that 12 moles of H_2 have been liberated, we can find m:

$$(3n + m)/2 = 12, \text{ hence } 3 \times 5 + m = 12 \times 2, \text{ and } m = 9$$

thus, the formula is B_5H_9.

This borohydride has 14 skeletal electrons (10 from five BH units, and four from additional four H atoms) or seven electron pairs. This means the cluster is *nido* and is based on *closo*-B6 octahedral cage:

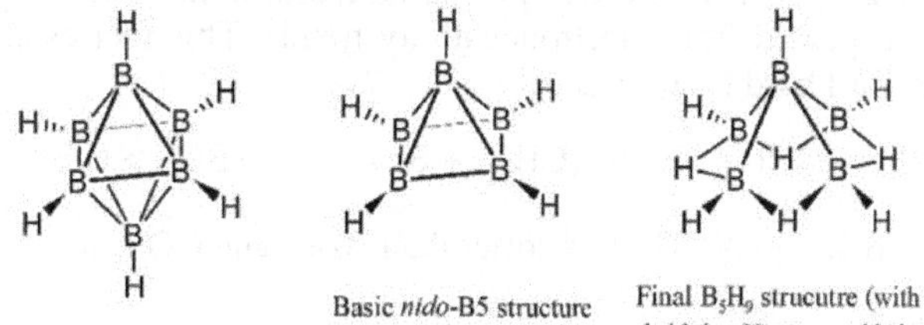

Parent *closo*-B6 structure | Basic *nido*-B5 structure (after removal of one vertex from *closo*-B6) | Final B_5H_9 strucutre (with 4 bridging H atoms added to *nido*-B5 structure)

S13.9 $C_2B_4H_6$ has $(6 \times 2) + 2 = 14$ skeletal electrons or seven skeletal electron pairs. This corresponds to $n + 1$, or a *closo* structure (octahedral).

S13.10 **a) $(CH_3)_2SAlCl_3$ and $GaBr_3$:** $GaBr_3$ is the softer Lewis acid, so the reaction is $(Me)_2SAlCl_3 + GaBr_3 \rightarrow Me_2SGaBr_3 + AlCl_3$ (Soft Lewis acid migrates from a hard Lewis acid to a soft one).

b) $TlCl_3$ and formaldehyde (HCHO) in acidic aqueous solution: Thallium trihalides are very unstable and are easily reduced, as shown below:

$$2\ TlCl_3 + H_2CO + H_2O \rightarrow 2TlCl + CO_2 + 4H^+ + 4Cl^-.$$

Exercises

E13.1 Boron is recovered from the mineral borax, $Na_2B_4O_5(OH)_4{\cdot}8H_2O$. Borax is hydrolised in the presence of a strong acid to boric acid, $B(OH)_3$:

$$Na_2B_4O_5(OH)_4(aq) + 2H_3O^+(aq) + H_2O(l) \rightarrow 2Na^+(aq) + 4B(OH)_3(s).$$

Boric acid precipitates from cold solution. This acid is then converted to B_2O_3 under heat:

$$2B(OH)_3(s) \xrightarrow{\text{heat}} B_2O_3(s) + 3H_2O(g).$$

Finally, the oxide is reacted with magnesium to liberate elemental boron:

$$B_2O_3 + 3Mg \rightarrow 2B + 3MgO.$$

E13.2 Molten cryolite dissolves and solvates very stable Al_2O_3 liberating Al^{3+} in form of $[AlF_n]^{(3-n)+}$ species. It is also a good conductor of electricity and, even molten, has a higher density than aluminium allowing for an easy collection of the product from the surface of the solution.

E13.3 Generally, when aluminium surface gets in contact with air, a thin layer of Al_2O_3 is formed that actually protects the metal form further reaction with oxygen from air. When a droplet of Hg is added to a clean aluminium surface, however, aluminium amalgam is formed, an alloy of mercury and aluminium. The Al atoms in the amalgam react with O_2 producing Al_2O_3 which is insoluble in mercury. This leads to further dissolution of Al in Hg and the reaction proceeds until there is no Al to react.

E13.4 **a) BF_3:** Both elements are non-metallic so a polar covalent bond can be expected with a strong π component of the B–F bond.

b) $AlCl_3$: In this case ionic bond between metallic Al and non-metallic Cl is predicted (you can confirm this using the Ketalaar triangle and electronegativity values for Al and Cl). In the solid state, $AlCl_3$ has a layered structure. At melting point, $AlCl_3$ converts to dimeric form, Al_2Cl_6, with *2c, 2e* bridging bonds (i.e., electron-precise structure).

c) B_2H_6: Again, two non-metallic elements and covalent bond is expected. Further, structure 1 shows that B_2H_6 is an electron-deficient dimer with 3*c*, 2*e* bridging bonds.

E13.5 For a given halogen, the order of acidity for Group 13 halides toward hard Lewis bases (such as dimethylether or trimethylamine) is $BX_3 > AlX_3 > GaX_3$, whereas the order toward soft Lewis bases (such as dimethylsulfide or trimethylphosphine) is $BX_3 < AlX_3 < GaX_3$. For boron halides, the order of acidity is $BF_3 < BCl_3 < BBr_3$, exactly opposite to the order expected from electronegativity trends. This fact establishes the order of Lewis acidity as $BCl_3 > BF_3 > AlCl_3$ toward hard Lewis bases:

(a) $BF_3N(CH_3)_3 + BCl_3 \rightarrow BCl_3N(CH_3)_3 + BF_3$ $(BCl_3 > BF_3)$

(b) $BH_3CO + BBr_3 \rightarrow$ *NR* (BH_3 is softer than BBr_3, and CO is a soft base).

E13.6 You should start by determining the number of moles for each reactant. The equivalent number of moles (2.5 mmol) is used. You should also recall from the discussion in Section 13.14 that although thallium(III) halides are unstable, they can be stabilized in the presence excess of halide by the complex formation. Thus, NaBr and $TlBr_3$ react to make $Na[TlBr_4]$, which is the formula of A.

$$TlBr_3 + NaBr \rightarrow Na[TlBr_4].$$

From this formula, Na^+ as a cation and $[TlBr_4]^-$ as the anion are easily recognized.

E13.7

$4BF_3 + 3LiAlH_4 \rightarrow 2B_2H_6 + 3LiBF_4$; **A** = B_2H_6

$B_2H_6 + 6H_2O \rightarrow 2B(OH)_3 + 6H_2$; **B** = $B(OH)_3$

$2B(OH)_3$ + heat $\rightarrow B_2O_3 + 3H_2O$; **C** = B_2O_3

$B_2O_3 + 3CaF_2 \rightarrow 2BF_3 + 3CaO$

E13.8 B_2H_6 does not survive in air; instead it reacts spontaneously with the oxygen in air and forms the solid oxide, as shown below:

$$B_2H_6 + 3O_2 \rightarrow B_2O_3 + 3H_2O$$

or boric acid:

$$B_2H_6 + 3O_2 \rightarrow 2B(OH)_3.$$

If the air is moist, B_2H_6 can react with $H_2O(g)$ as well.

E13.9 a) The structure of B_5H_{11}:

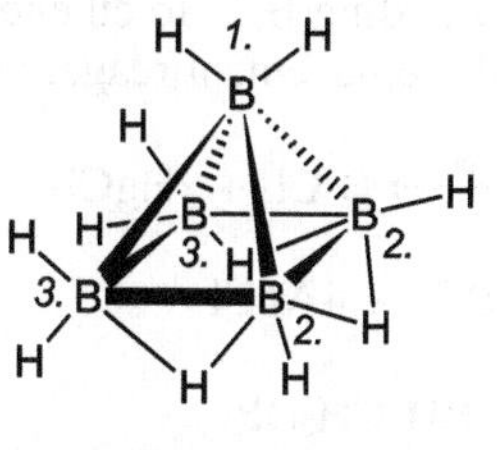
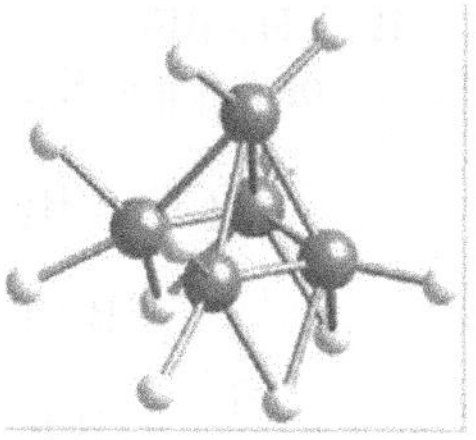

This cluster would have three different environments in the proton-decoupled ^{11}B NMR: the apical B atom (marked as *1.*), two basal B atoms bonded to two bridging and one terminal hydrides each (B atoms *2.*), and two B atoms bonded to one bridging and two terminal H each (B atoms *3.*).

b) The structure of B_4H_{10}:

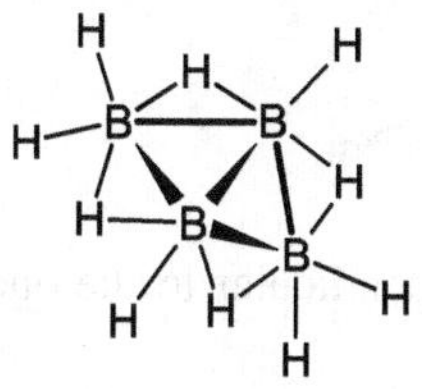
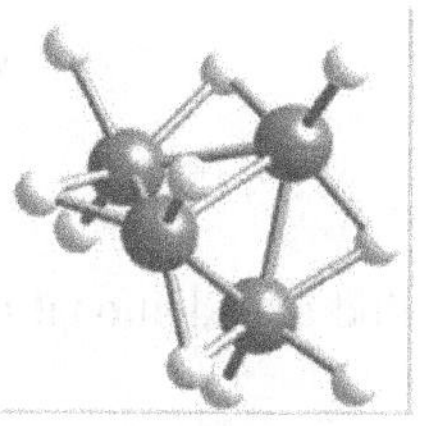

Two different boron environments are present in the proton-decoupled ^{11}B NMR of this compound: there are two equivalent B atoms in the middle of the structure (bonded to only one terminal H atom) and two B atoms on each side (bonded to two terminal H atoms).

E13.10 **a)** $BH_3 + 3(CH_3)_2C{=}CH_2 \rightarrow B[CH_2\text{-}CH(CH_3)_2]_3$

b) $BH_3 + 3CH{\equiv}CH \rightarrow B(CH{=}CH_2)_3$

E13.11 Give combustion reaction as:

$$B_2H_6(g) + 3O_2(g) \rightarrow B_2O_3(s) + 3H_2O(g).$$

In the first step, we will calculate the $\Delta_r H$ using the relationship:

$$\Delta_r H = \Sigma\Delta_f H_{products} - \Sigma\Delta_f H_{reactants}.$$

The enthalpies of formation of B_2H_6, H_2O, and B_2O_3 are given and the enthalpy of formation for O_2 is zero as per the definition. Thus:

$$\Delta_r H = [\ 3(-242\text{ kJ mol}^{-1}) + (-1264\text{ kJ mol}^{-1})] - [31\text{ kJ mol}^{-1}]$$

$$= -2021\text{ kJ mol}^{-1}.$$

To calculate the heat released, when 1 kg (1000 g) of B_2H_6 is used, we use the following procedure (using 27.62 g/mol as molecular weight of B_2H_6):

$$\frac{1000\text{g}}{27.62\text{g mol}^{-1}} \times \left(-2021\frac{\text{kJ}}{\text{mol}}\right) = -73{,}172\text{kJ}.$$

Diborane is extremely toxic. It is an extremely reactive gas and hence should be handled in a special apparatus. Moreover, a serious drawback to using diborane is that the boron-containing product of combustion is a solid, B_2O_3. If an internal combustion engine is used, the solid will eventually coat the internal surfaces, increasing friction, and will clog the exhaust valves.

E13.12 Following the general bond strength trends, we would expect that element–H bond enthalpy decreases going down the group. Tl–H bond, thus, should be the weakest of all. Although three Tl–H are formed in the synthesis of TlH_3, the liberated heath is insufficient to compensate for the bond dissociation energy of H_2 molecules, sublimation of Tl(s) *and* third ionization potential of Tl. This means that free Tl(s) and H_2 are more stable than TlH_3.

E13.13 The chelating agent, $F_2B–C_2H_4–BF_2$, can be prepared by adding B_2F_4 to ethene. Starting with BCl_3, the first step is to prepare B_2Cl_4 and then convert it to B_2F_4 with a double replacement reagent such as AgF or HgF_2, as follows:

$$2BCl_3 + 2Hg \xrightarrow{\text{electron impact}} B_2Cl_4 + 2HgCl_2$$

$$B_2Cl_4 + 4AgF \longrightarrow B_2F_4 + 4AgCl$$

$$B_2F_4 + C_2H_4 \longrightarrow F_2BH_2CH_2BF_2$$

E13.14 Follow the decision tree in Figure 3.9 in your textbook to determine the point groups of these molecules.

a) In the gas phase B_2Cl_4 has a staggered conformation. Following the decision tree, we confirm that the molecule is non-linear, it has no C_n with $n > 2$, but it has C_2:

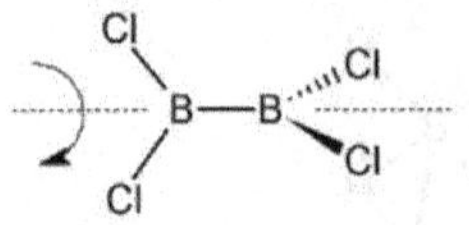

A little bit harder to find are additional two C_2 axes, perpendicular to the one above but closing a 45° angle with B–Cl bonds:

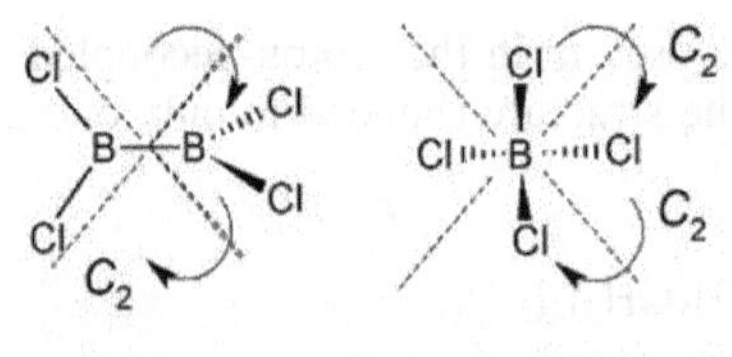

There are no horizontal mirror planes but there are dihedral mirror planes; two can easily be observed looking at the structure above right (view along B–B bond): they are perpendicular to the plane of the paper and each contains one BCl_2 fragment and cuts Cl–B–Cl angle of the other fragment in half.

Thus, this molecule belongs to D_{2d} point group.

b) Following similar procedure, we can determine that the point group of the planar structure is D_{2h}. This structure has a mirror plane that is perpendicular to a C_2 axis and lies in the plane of the molecule:

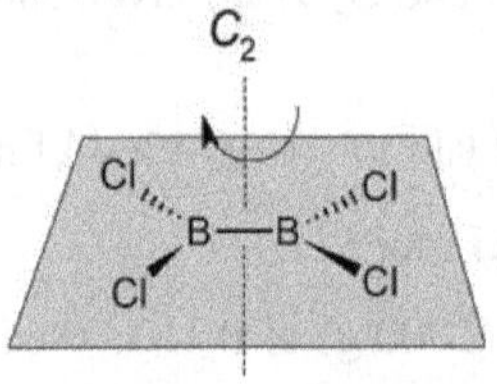

E13.15 **a)** IUPAC name for boron hydrides is boranes. B_2H_6 is simply called diborane. In the name of neutral boranes, the number of B atoms is indicated by a Greek prefix while the number of H atoms in the cluster is given at the end in brackets. The name should be preceded by *closo*, *nido* or *arachno* as appropriate.
Thus, $B_{10}H_{14}$ is called *nido*-decaborane(14).

b) The BH anions are named by providing the number of hydrogen atoms and number of boron atoms in Greek. The name has *-ate* ending to indicate negative charge with the actual charge given at the end in brackets: *closo*-dodecahydrododecaborate(2-). Note that hydrogen becomes "hydro" in the name and that the Greek numbers are added before "hydro" for hydrogen and before "borate" for boron. The prefix *closo* can also be placed before "borate" part: dodecahydro-*closo*-dodecaborate(2-).

c) *Arachno*-tetradecahydrododecaborate(2-) or tetradecahydro-*arachno*-dodecaborate(2-).

E13.16 The procedure is outlined in Section 13.12 *Metallaboranes and Carboranes*. The first step of the reaction is removal of H_2 molecule from the starting decaborane(14) with a thioether such as SEt_2:

$$B_{10}H_{14} + 2SEt_2 \longrightarrow B_{10}H_{12}(SEt_2)_2 + H_2.$$

Soft and bulky thioether Lewis base can be easily replaced by smaller alkyne Lewis acid, C_2H_2, with loss of another H_2 equivalent:

$$B_{10}H_{12}(SEt_2)_2 + HC{=}CH \longrightarrow B_{10}C_2H_{12} + H_2 + 2SEt_2.$$

The mildly acidic C–H bonds can be deprotonated and lithiated with buthyllithium:

$$B_{10}C_2H_{12} + 2LiC_4H_9 \longrightarrow Li_2[B_{10}C_2H_{10}] + 2C_4H_{10}.$$

This creates two nucleophilic carbon centres that can easily react with a good nucleophile (as all lithium organometallic compounds do). Considering our substituents on carbon atoms, a good choice would be $ClSi(CH_3)_3$, chlorotrimethylsilane:

$$Li_2[B_{10}C_2H_{10}] + 2ClSi(CH_3)_3 \longrightarrow B_{10}C_2H_{10}(Si(CH_3)_3)_2 + 2LiCl.$$

E13.17 **a)** $NaBH_4$ can be converted to BCl_3 using a solution of HCl in ether (with hydrogen gas and solid NaCl as by-products):

$$NaBH_4 + 4HCl(\text{ether sol.}) \longrightarrow BCl_3 + NaCl(s) + 4H_2(g).$$

After NaCl is removed by filtration, the remaining filtrate (which is solution of BCl_3 in ether) is reacted with ethyl Grignard reagent:

$$BCl_3 + 3C_2H_5MgBr \longrightarrow B(C_2H_5)_3 + 3MgBrCl.$$

b) Starting from $NaBH_4$ and $[HN(C_2H_5)_3]Cl$**,** we should expect NaCl, with its high lattice enthalpy, to be a likely product. If this is the case we shall be left with BH_4^- and $[HN(C_2H_5)_3]^+$. The interaction of the hydridic BH_4^- ion with the protic $[HN(C_2H_3)_5]^+$ ion will evolve dihydrogen gas to produce triethylamine and BH_3. In the absence of other Lewis bases, the BH_3 molecule would coordinate to THF; however, the stronger Lewis base triethylamine is produced in the initial reactions, so the overall reaction will be :

$$[HN(C_2H_5)_3]Cl + NaBH_4 \longrightarrow H_2 + H_3BN(C_2H_5)_3 + NaCl.$$

E13.18 C_2-axis viewpoint of B_{12} is

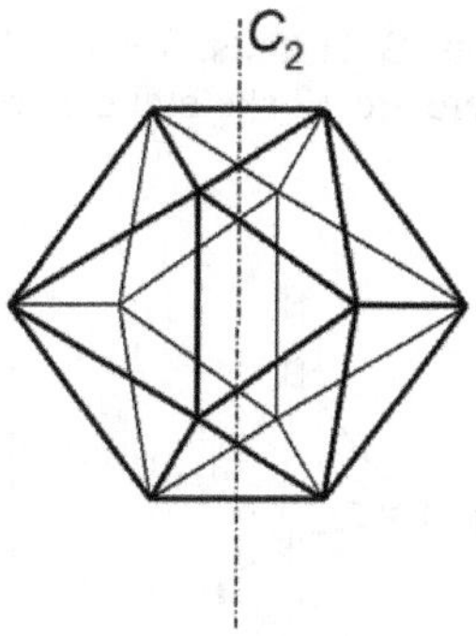

E13.19 In general, boron hydrides with more hydrogen atoms are thermally unstable. The compounds B_6H_{10} and B_6H_{12} belong to two different classes of boranes, B_nH_{n+4} and B_nH_{n+6}, respectively, with the first class having fewer H atoms per boron. In this case, n is six for both compounds; hence we expect B_6H_{10} to be stable relative to B_6H_{12}.

E13.20 The formula B_5H_9 belongs to a class of borohydrides having the formula B_nH_{n+4}, which is characteristic of a *nido* species. There are five BH units, which contribute $5 \times 2 = 10$ electrons (assuming one B–H bond per B atom), and the four additional H atoms, which contribute a further four electrons, giving 14 skeleton electrons in total, or seven skeleton electron pairs, which is $n + 2$ with $n = 5$. This is characteristic of *nido* clusters.

E13.21 **a)** The formula $B_{10}H_{14}$ belongs to a class of borohydrides with the general formula B_nH_{n+4}, characteristic of a *nido* species.

b) There are 10 BH units, which contribute 10 × 2 = 20 electrons (assuming one B–H bond per B atom), and the four additional H atoms, which contribute four additional electrons, giving a total of 24 electrons, or 12 electron pairs, which is $n + 2$ for $n = 10$. This is characteristic of *nido* clusters.

c) The total number of valence electrons for $B_{10}H_{14}$ is (10 × 3) + (14 × 1) = 44. Since there are ten 2*c*-2*e* B–H bonds, which account for 20 of the total number of valence electrons, the number of cluster electrons is the remainder of 44 – 20 = 24.

E13.22 **a)** For B_5H_{11}: 5 × 2 = 10 electrons from 5 B–H units, 6 × 1 = 6 electrons from additional H atoms; in total there are 16 skeletal electrons, or 8 electron pairs. This corresponds to $n + 3$ (for $n = 5$) or *arachno*-type structure.

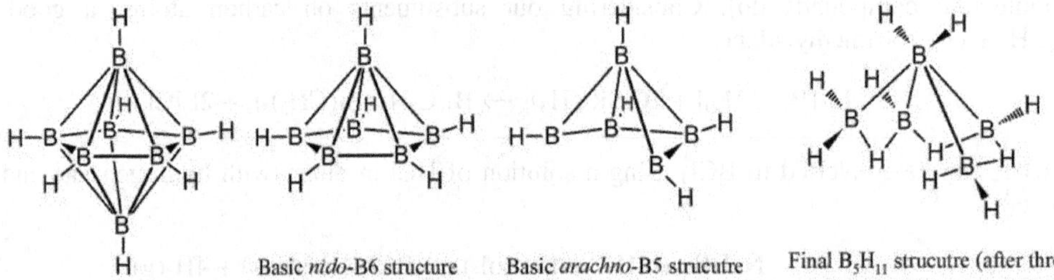

Parent *closo*-B7 structure — Basic *nido*-B6 structure (after removal of one vertex from *closo*-B7) — Basic *arachno*-B5 strucutre (after removal of another vertex from *nido*-B6 structure) — Final B_5H_{11} strucutre (after three bridging and three terminal H atoms have been added)

b) For B_5H_9: 5 × 2 = 10 electrons from five B–H units, 4 × 1 = 4 electrons from additional H atoms, no charge on the cluster; in total there are 14 skeletal electrons or 7 pairs. This corresponds to $n + 2$ (for $n = 5$) or *nido*-type structure:

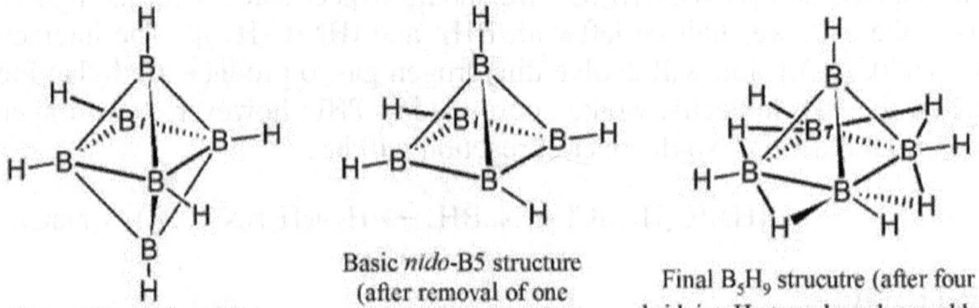

Parent *closo*-B6 structure — Basic *nido*-B5 structure (after removal of one vertex from *closo*-B6) — Final B_5H_9 strucutre (after four bridging H atoms have been added)

c) For $B_4H_7^-$: 4 × 2 = 8 electrons from four B–H units, 3 × 1 = 3 electrons from additional H atoms and one extra electron from a minus charge; in total there are 12 skeletal electrons, or six electron pairs. This corresponds to n + 2 (for $n = 4$) or *nido*-type structure:

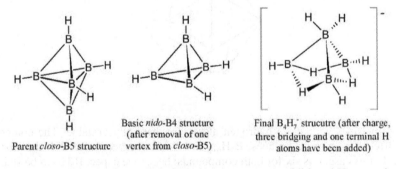

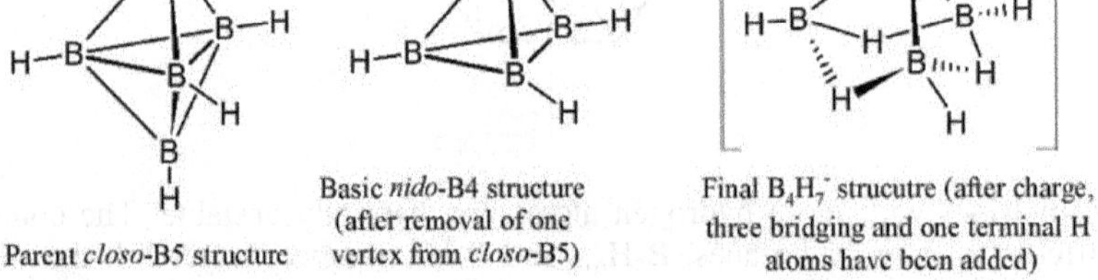

Parent *closo*-B5 structure — Basic *nido*-B4 structure (after removal of one vertex from *closo*-B5) — Final $B_4H_7^-$ strucutre (after charge, three bridging and one terminal H atoms have been added)

d) For $B_6H_6^{2-}$: 6 × 2 = 12 electrons from six B–H units, there are no additional H atoms but there is a 2– charge; in total there are 14 skeletal electrons, or seven electron pairs. This corresponds to $n + 1$ (for $n = 6$) or *closo*-type structure:

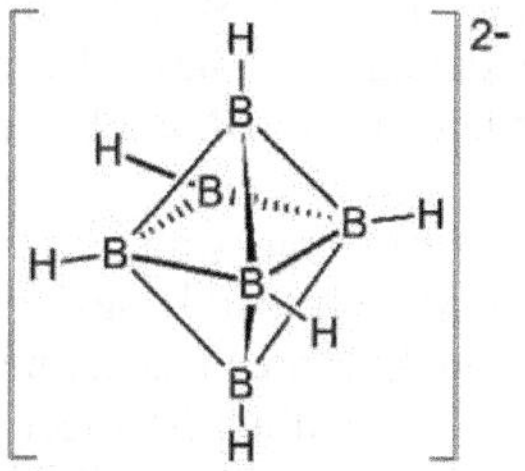

E13.23 General equation for hydrolysis of borohydrides is:

$$B_nH_m + 3nH_2O \rightarrow nB(OH)_3 + (3n + m)/2\ H_2.$$

In our case, six moles of $B(OH)_3$ have been produced, meaning that n in above equation is 6. From this value and the fact that 15 moles of H_2 have been liberated, we can find m:

$$(3n + m)/2 = 15, \text{ hence } 3 \times 6 + m = 15 \times 2, \text{ and } m = 12;$$

thus, the formula is B_6H_{12}.

This borohydride has 18 skeletal electrons (12 from six BH units, and six from additional six H atoms) or nine electron pairs. This means the cluster is *arachno* and is based on *closo*-B8 structure cage.

E13.24 The target molecule, $[Fe(\textit{nido}\text{-}B_9C_2H_{11})_2]^{2-}$, can be prepared in five steps as follows:

(1) $B_{10}H_{14} + 2SEt_2 \rightarrow B_{10}H_{12}(SEt_2)_2 + H_2$

(2) $B_{10}H_{12}(SEt_2)_2 + C_2H_2 \rightarrow B_{10}C_2H_{12} + 2SEt_2 + H_2$

(3) $2B_{10}C_2H_{12} + 2EtO^- + 4EtOH \rightarrow 2B_9C_2H_{12}^- + 2B(OEt)_3 + 2H_2$

(4) $Na[B_9C_2H_{12}] + NaH \rightarrow Na_2[B_9C_2H_{11}] + H_2$

(5) $2Na_2[B_9C_2H_{11}] + FeCl_2 \rightarrow 2NaCl + Na_2[Fe(B_9C_2H_{11})_2]$

The structure of the anion is shown below with H atoms omitted for clarity:

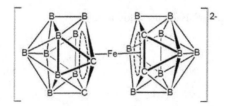

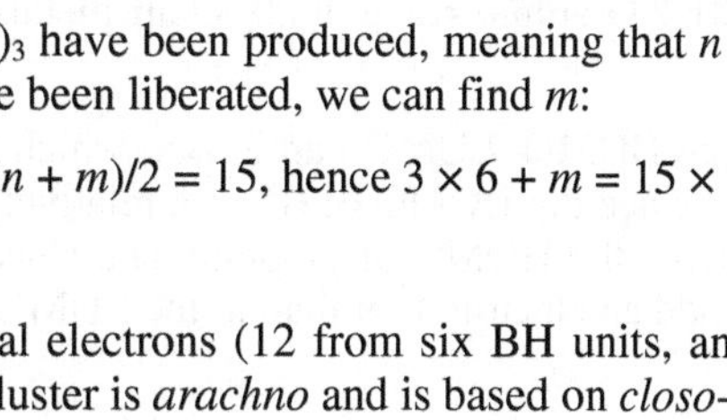

E13.25 **a)** The ^{11}B NMR spectrum of BH_3CO would show one quartet resonance due to ^{11}B being coupled to three equivalent H nuclei.

b) $[B_{12}H_{12}]^{2-}$ has two distinct ^{11}B environments in its structure: two apical B atoms and 10 equivalent B atoms in two five-membered rings. Thus the ^{11}B NMR should show two signals in relative intensity 1 : 5 both of which should be, if we assume no B–B coupling, doublets because each B atom in $[B_{12}H_{12}]^{2-}$ is bonded to only one H atom.

E13.26 Boron and aluminium have the same oxidation state +3 in their oxides. Boron radius is, however, much smaller than that of Al. This small radius is reducing boron's coordination number. As you have seen through the chapter, boron's compounds never exceed coordination number four, and coordination number 3 is also very frequent: recall the structures of boron halides and diboron tetrahalides. Aluminium, with its greater radius can accommodate higher coordination numbers such as six in Al_2O_3 and many others.

E13.27 **a)** The products are $B(OH)_3$ and HCl. See Section 13.7 *Boron Trihalides*.

b) BCl_3 is a Lewis acid while $S(CH_3)_3$ is a Lewis base, thus the product is a Lewis acid–base adduct, $BCl_3S(CH_3)_2$.

E13.28 Br is smaller than F, and accommodating four large Br anions around relatively small B(III) is sterically challenging. Thus, there is no reaction between CsBr and BBr_3.

E13.29 **a) The structures.** Both substances have layered structures. The planar sheets in boron nitride and in graphite consist of edge-shared hexagons such that each B or N atom in BN has three nearest neighbours that are the other type of atom, and each C atom in graphite has three nearest neighbour C atoms. The structure of graphite is shown in Figure 14.2. The B–N and C–C distances within the sheets, 1.45 Å and 1.42 Å, respectively, are much shorter than the perpendicular interplanar spacing, 3.33 Å and 3.35 Å, respectively. In BN, the B_3N_3 hexagonal rings are stacked directly over one another so that B and N atoms from alternating planes are 3.33 Å apart, whereas in graphite the C_6 hexagons are staggered (see Section 13.9) so that C atoms from alternating planes are either 3.35 Å or 3.64 Å apart (you should determine this yourself using trigonometry).

b) Their reactivity with Na and Br_2? Graphite reacts with alkali metals and with halogens. In contrast, boron nitride is quite unreactive.

c) Explain the differences? The large HOMO–LUMO gap in BN, which causes it to be an insulator, suggests an explanation for the lack of reactivity: since the HOMO of BN is a relatively low-energy orbital, it is more difficult to remove an electron from it than from the HOMO of graphite, and since the LUMO of BN is a relatively high energy orbital, it is more difficult to add an electron to it than to the LUMO of graphite.

E13.30 The hexagonal (layered) BN can be converted into the cubic (diamond-like) BN at high pressures and temperatures.

E13.31 **a) $Ph_3N_3B_3Cl_3$.** The reaction of a primary ammonium salt with boron trichloride yields *N*-substituted *B*-trichloroborazines:

$$3[PhNH_3]Cl + 3BCl_3 \rightarrow Ph_3N_3B_3Cl_3 + 9HCl.$$

b) $Me_3N_3B_3H_3$. First prepare $Me_3N_3B_3Cl_3$ using $[MeNH_3]Cl$ and the method described above, and then perform a Cl^-/H^- metathesis reaction using LiH (or $LiBH_4$) as the hydride source:

$$3MeNH_3^+Cl^- + 3BCl_3 \rightarrow Me_3N_3B_3Cl_3 + 9HCl$$

$$Me_3N_3B_3Cl_3 + 3LiH \rightarrow Me_3N_3B_3H_3 + 3LiCl.$$

The structures of $Ph_3N_3B_3Cl_3$ and $Me_3N_3B_3H_3$ are shown below:

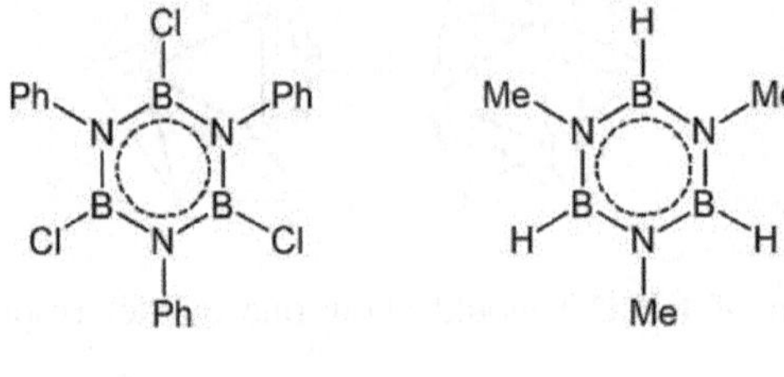

E13.32 Statement **(a)** is incorrect—in fact only boron has some non-metallic properties, all other elements are metallic.

Statement **(b)** is also incorrect—the hardness is decreasing down the group; in fact, of Group 13 elements only boron and aluminium (top of the group) show pronounced oxophilicity and boron fluorophilicity.

Statement **(c)** is only partially correct—while Lewis acidity of BX_3 does increase from F to Br, it is not due to stronger B–Br π bonding (as stated) but exactly due to the opposite trend: the π bonding is decreasing in the same order Lewis acidity is increasing, that is, π bonding in B–Br is weaker than in B–F (the latter explains the former).

Statement **(d)** is incorrect—*arachno* boron hydrides are less stable than *nido*-borohydrides, also, their skeletal electron count is $n + 3$ not $2(n + 3)$.

Statement **(e)** is correct.

Statement **(f)** is partially incorrect—while layered BN and graphite have similar structures, layered BN has a very large (not small as stated) HOMO–LUMO gap and is in fact an insulator (not a conductor as stated).

E13.30 The common oxidation states of indium are +1 and +3. The oxidation states of In in In_5Cl_9 must sum to +9 to maintain the electroneutrality of the compound. We can suggest having two of them in +3 oxidation state (giving +6) and the other three in +1 (for a total of +9). The formula is then rewritten as $In_3^{(I)}[In_2^{(III)}Cl_9]$. Because they have a similar chemical composition, the structures of $In_2Cl_9^{3-}$ and $Tl_2Cl_9^{3-}$ are the same and consist of two MCl_6 octahedra sharing a face (M = In or Tl):

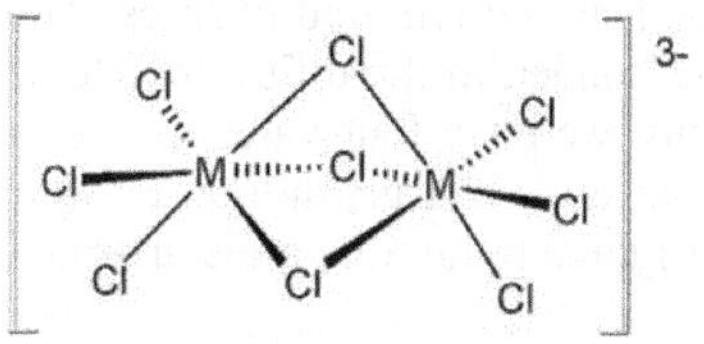

This structure can be proposed by a bit of "trial-and-error" with number of bridges and stoichiometry, the driving force behind the bridge formation is the Lewis acidity of Group 13 trihalides.

E13.33 Indium oxide is commonly doped with SnO_2 to give an n-type semiconductor. A possible dopant candidate for Al_2O_3, that would also produce an n-type semiconductor is SiO_2. The difference in oxidation states are the same, +3 (In, Al) and +4 (Sn, Si) and Al^{3+} commonly substitutes Si^{4+} in silicate minerals producing aluminosilicates.

E13.34 Table 13.5 provides the band gap energies for GaN, GaAs and GaSb. Since P lies between N and As in the periodic table and the band gap energies decrease going down Group 15, we can estimate the band gap for GaP as an average band gap energy of that for GaN and for GaAs:

$$(3.40 \text{ eV} + 1.35 \text{ eV})/2 = 2.37 \text{ eV}.$$

This corresponds to 3.79×10^{-19} J (1 eV = 1.60×10^{-19} J).

The energy is related to the wavelength:

$$E = \frac{hc}{\lambda} \Rightarrow \lambda = \frac{hc}{E}.$$

From here we have:

$$\lambda = \frac{6.626\times10^{-34}\,\text{Js}\times3\times10^{8}\,\text{m/s}}{3.79\times10^{-19}\,\text{J}} = 5.24\times10^{-7}\,\text{m} = 524\,\text{nm}.$$

The green light has a wavelength of about 550 nm. Thus, our estimate is consistent with the observation of green light emitted by GaP LEDs.

E13.35 The Pauling electronegativities (from Table 1.7) are 0.82 and 1.78 for Rb and In respectively. The difference is 0.96 and the mean value is 1.3. This would place Rb_2In_3 in the region of metallic bonding (see Ketalaar triangle in Figure 2.28).
Very likely structure of anion In_6^{4-} in Rb_4In_6 is octahedral with six indiums at the vertices. It resembles the structure of B_6^{2-} discussed in section 13.10 *Metal Borides*.

Guides to Selected Tutorial Problems

T13.1 The important comparisons are that of charge and ionic radii, consequently the charge density discussion is also relevant. Particularly interesting are biological activities of K^+ and Tl^+ with the difference in biological effect resulting from differences in Lewis acid hardness (soft vs. hard) and tendency to form complexes.

T13.7 Synthesis of GaN can be achieved using several methods. Simple internet search will provide basic results. The structure is that of wurtzite.

Chapter 14 The Group 14 Elements

Self-Tests

S14.1 Increase in pressure always favours a denser structure. In this case, that is the cubic structure. Thus, at higher pressures, the phase change can occur at even lower temperatures.

S14.2 It has to be noted that given formulas lack the required charges. This is easy to observe: in the first formula, we have germanium in oxidation state +2 bonded to three fluorine atoms in oxidation state -1, thus [Ge(II)F_3] should have overall negative one charge. Thus, we have four valence electrons from Ge, 21 valence electrons from three fluorine atoms plus one extra for the negative charge, in total 26 valence electrons to account for. We can suggest a tetrahedral electron geometry with trigonal pyramidal molecular geometry:

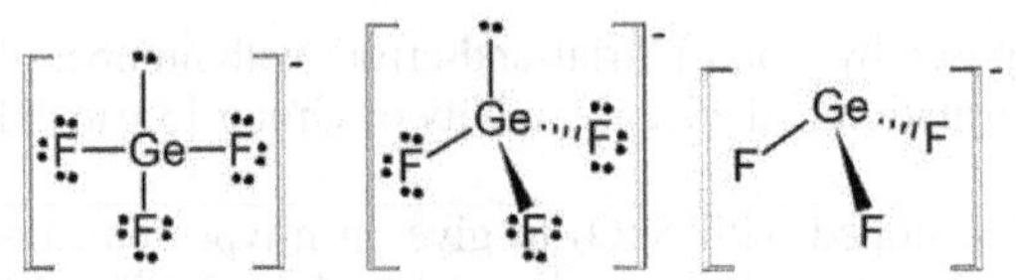

Similar analysis shows that [Ge(IV)F_6] has a 2- charge with a total valence electron count of 48 electrons. This composition and number of electrons result in an octahedral electron and molecular geometry:

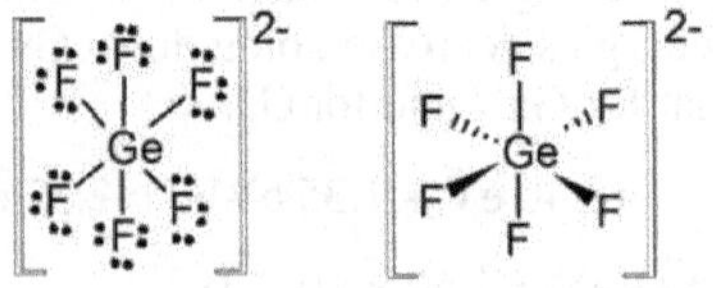

What might be somewhat confusing at the end are the charges: the sum of charges on the structural units is 3- and there are no additional cations, while the compound in question is neutral Ge_3F_8. The reason for this apparent discrepancy is that our focus in this self-test was on the geometries of the structural units *not* how the units interact in the structure. In the Ge_3F_8 structure the octahedra and trigonal pyramids are connected through bridging fluorides. This is also apparent when we look at the stoichiometry: the formula Ge_3F_8 is not the sum of its structural units 2 × [Ge(II)F_3] + [Ge(IV)F_6]—the sum has four more F atoms than the formula.

S14.3 **a)** The extended π system for each of the planes of graphite results in a band of π orbitals. The band is half filled in pure graphite; that is, all bonding MOs are filled and all antibonding MOs are empty. The HOMO–LUMO gap is ~0 eV, giving rise to the observed electrical conductivity of graphite. Chemical reductants, like potassium, can donate their electrons to the LUMOs (graphite π* MOs), resulting in a material with a higher conductivity.

b) Chemical oxidants, like bromine, can remove electrons from the HOMOs (π MOs) of graphite. This also results in a material with a higher conductivity.

S14.4 The enthalpy of formation is calculated as the difference between the bonds broken and the bonds formed in the reaction.

Four C–H bonds are formed in the formation of methane from graphite and hydrogen gas:

$$C(graphite) + 2H_2(g) \rightarrow CH_4(g).$$

$\Delta_f H(CH_4,g)$ = (bonds broken) – (bonds formed) = [715 + 2(436)] kJ mol^{-1} – [4(412)] kJ mol^{-1}

$\Delta_f H = -61$ kJ mol^{-1}

Four Si–H bonds are formed in the synthesis of silane from silicon with hydrogen gas:

$$Si(s) + 2H_2(g) \rightarrow SiH_4(g).$$

$\Delta_f H(SiH_4,g)$ = (bonds broken) – (bonds formed) = [439 + 2(436)] kJ mol^{-1} – [4(318)] kJ mol^{-1}

$\Delta_f H = +39$ kJ mol^{-1}

S14.5 The first step of synthesis would be the oxidation of ^{13}CO to $^{13}CO_2$. Carbon dioxide is Lewis acidic and can react with strong Lewis bases. Thus, we can treat $^{13}CO_2$ with a source of deuteride ion, D^-, a good Lewis base. A convenient source would be LiD (or NaD). The entire synthesis would be:

$$^{13}CO(g) + 2MnO_2(s) \rightarrow {}^{13}CO_2(g) + Mn_2O_3(s)$$

$$2Li(s) + D_2 \rightarrow 2LiD(s)$$

$$^{13}CO_2(g) + LiD(ether) \rightarrow Li^+D^{13}CO_2^-(ether)$$

S14.6 Oxidation of C_2^{2-} to C_2^- would result in removal of one of the electrons found in $2\sigma_g$ molecular orbital. This molecular orbital is bonding in character, and removal of one of the electrons from this orbital would give the $1\sigma_g^2$ $1\sigma_u^2$ $1\pi_u^4$ $2\sigma_g^1$ electronic configuration resulting in a decrease in bond order—from 3 to 2.5: $b = ½(2 - 2 + 4 + 1) = 2.5$. Therefore, C–C bond length in C_2^- would be longer than in C_2^{2-}.

Exercises

E14.1 Statement **(a)** is incorrect—tin and lead are metallic elements.

Statement **(b)** is correct.

Statement **(c)** is partially correct—whereas it is true that both CO_2 and CS_2 are weak Lewis acids, CS_2 is softer (not harder) than CO_2.

Statement **(d)** is incorrect—zeolites are framework aluminosilicates (not layered like mica) and they do contain metal cations in their composition to balance the negative charge of the aluminosilicate framework.

Statement **(e)** is correct—hydrolysis of CaC_2, an ionic compound, produces HC≡CH and $Ca(OH)_2$.

E14.2 **a)** Although two major allotropes of carbon are graphite and diamond, it makes sense to compare diamond with silicon because one of the silicon's solid phases shares the same structure with diamond. In the diamond-type structure, each atom forms single bonds with four neighbouring atoms located at the corners of a tetrahedron. Each atom in this structure can be seen as having sp^3-hybridization and a full octet achieved through electron sharing in covalent bonds. This structure results in a wide energy gap between filled and empty bands—a gap that is, however, smaller for Si(diamond) than for C(diamond). As a consequence, C(diamond) is generally considered an insulator, whereas Si(diamond) is a semiconductor.

b) Really striking differences between C and the rest of the elements of Group 14 are the properties and structures of the oxides. Carbon forms a rather stable monoxide, but silicon monoxide is an unstable species. Both carbon and silicon form very stable dioxides but the two compounds could hardly be more different. While CO_2 is gas and a linear molecule, SiO_2 forms a network solid with Si–O–Si bridges, with each Si atom tetrahedrally surrounded with four O atoms throughout the structure. As a consequence of this rigid structure, SiO_2 has high melting and boiling points.

c) Silicon tetrahalides are mild Lewis acids whereas analogous carbon compounds are not Lewis acidic. The reason for this difference in acidic properties is in the fact that silicon, unlike carbon, can extend its octet (i.e., it can form hypervalent compounds) allowing for formation of more than four bonds and thus bonding with Lewis bases.

E14.3 $SiCl_3F$, $SiCl_2F_2$, and $SiClF_3$ have tetrahedral structures as shown below. Each Si has an octet and each molecule conforms with VSEPR theory:

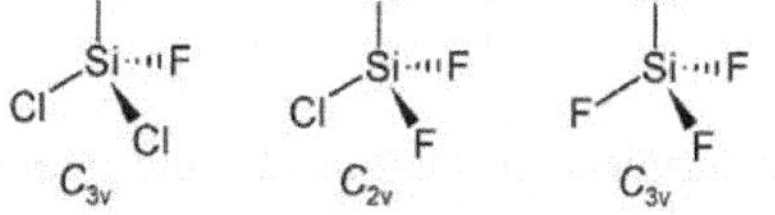

The point groups for each molecule are given below the structure. At this point in the course you should be very familiar in using the decision tree (Chapter 3, Figure 3.9) for identification of point groups.

E14.4 These unusual stoichiometries arise because Ge has two easily accessible oxidation states: +2 and +4. The fluorides mentioned in the exercises, like the fluoride from Example 14.2, contain germanium in both oxidation states. Ge_5F_{12} is $Ge(II)_4Ge(IV)F_{12}$ while Ge_7F_{16} is $Ge(II)_6Ge(IV)F_{16}$.

E14.5 The bond enthalpy of a C–F bond is higher than the bond enthalpy of a C–H bond, making C–F bond more difficult to break. Further, the combustion of CH_4 produces not only CO_2 but H_2O as well, a very stable compound. Formation of both CO_2 and H_2O thermodynamically drive the combustion of CH_4. There is no equivalent to the formation of H_2O in the combustion of CF_4.

E14.6 SiF_4 is a gas at standard conditions, while PbF_2 is a solid with two polymorphs: one has fluorite-type, cubic unit cell, the other orthorhombic unit cell. The electronegativity difference between Si and F is sufficiently large to indicate ionic bonding and thus a solid state structure. However, the atomic sizes and coordination in SiF_4 are the likely factors that prevent the lattice formation. Comparatively large and heavy Pb^{2+} can accommodate higher coordination numbers, up to nine in orthorhombic structure.

E14.7 The chemical equation for reaction of SiF_4 with $[(CH_3)_4N]F$:

$$SiF_4 + [(CH_3)_4N]F \rightarrow [(CH_3)_4N]^+[SiF_5]^-.$$

a) The cation is $[(CH_3)_4N]^+$. (At this point in the material, you should be very familiar with VSEPR theory so only a brief outline is given here!) The number of valence electrons present on central N atom is eight and the number of valence electron pairs is four. Therefore, VSEPR theory predicts that a species with four bonding pairs of electrons should be tetrahedral.

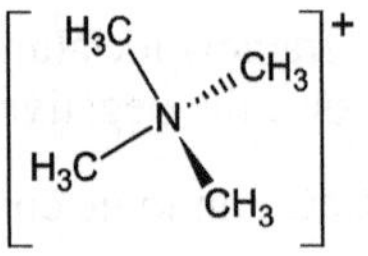

The anion is SiF_5^-. The number of valence electrons present on central Si atom is 10 and the number of valence electron pairs is five. Therefore, VSEPR theory predicts that a species with five pairs of electrons should be trigonal bipyramidal.

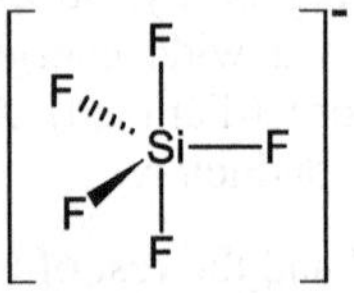

b) The ^{19}F NMR shows two fluorine environments because in the structure of the anion there are two different fluorine environments: axial and equatorial.

E14.8 The structure of cyclic anion $[Si_4O_{12}]^{n-}$:

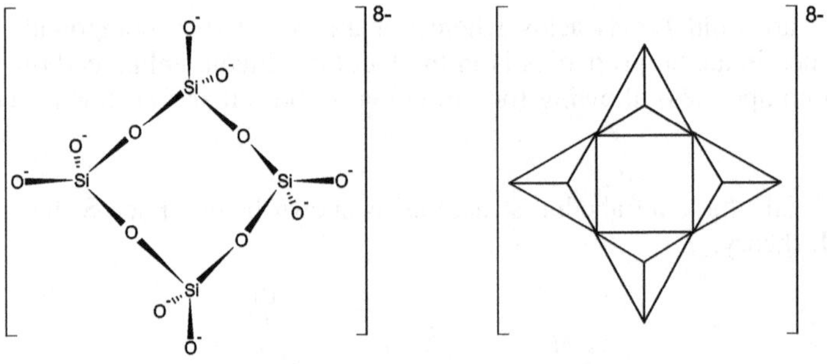

The charge of this anion can be determined if we consider the number of Si^{4+} cations and the number of O^{2-} anions. There are four Si^{4+} cations giving a total charge of $4 \times (+4) = +16$. There are 12 O^{2-} anions with total negative charge $12 \times (-2) = -24$. Thus, the total charge of the anion is the sum of total positive and total negative charges: $+16 + (-24) = -8$.

E14.9 Trimethylstannylphosphine is $(CH_3)_3SnPH_2$. The relevant NMR data can be found in Table 8.4: ^{119}Sn, abundance 8.58% with $I = ½$ and ^{31}P abundance 100% with $I = ½$.

a) For the ^{119}Sn NMR spectrum, we can ignore relatively weak, two bond $^{2}J(^{119}Sn\text{-}^{1}H)$ couplings or treat the spectrum as proton decoupled. In this case, the spectrum would show one doubled due to one bond coupling between ^{119}Sn and ^{31}P.

b) For the ^{31}P spectrum, we cannot ignore one bond couplings to two hydrogen nuclei directly bonded to P. This strong coupling would produce a triplet. This spectrum would also show satellites due to the one-bond coupling between ^{119}Sn and ^{31}P nuclei: a doublet of triplets lines would flank the central triplet.

E14.10 The enthalpy of hydrolysis of a compound can be calculated as the difference in energy between the bonds broken and the bonds formed in the hydrolysis reaction. The relevant equations for the hydrolysis of CCl_4 and CBr_4 are:

$$CCl_4(l) + 2H_2O(l) \rightarrow CO_2(g) + 4HCl(aq)$$

$$\Delta_h H^\circ(CCl_4) = [(2 \times 743\text{ kJ mol}^{-1}) + (4 \times 431\text{ kJ mol}^{-1})] - [(4 \times 322\text{ kJ mol}^{-1}) + 2(2 \times 463\text{ kJ mol}^{-1})]$$

$$= [1526\text{ kJ mol}^{-1} + 1724\text{ kJ mol}^{-1}] - [1288\text{ kJ mol}^{-1} + 2 \times 926\text{ kJ mol}^{-1}]$$

$$= 3250\text{ kJ mol}^{-1} - [1288\text{ kJ mol}^{-1} + 1852\text{ kJ mol}^{-1}]$$

$$= 3250\text{ kJ mol}^{-1} - 3140\text{ kJ mol}^{-1}$$

$$= 110\text{ kJ mol}^{-1}$$

$$CBr_4(l) + 2H_2O(l) \rightarrow CO_2(g) + 4HBr(aq)$$

$$\Delta_h H^\circ(CBr_4) = [(2 \times 743\text{ kJ mol}^{-1}) + (4 \times 366\text{ kJ mol}^{-1})] - [(4 \times 288\text{ kJ mol}^{-1}) + 2(2 \times 463\text{ kJ mol}^{-1})]$$

$$= [1526\text{ kJ mol}^{-1} + 1464\text{ kJ mol}^{-1}] - [1152\text{ kJ mol}^{-1} + 2 \times 926\text{ kJ mol}^{-1}]$$

$$= 3090\text{ kJ mol}^{-1} - [1152\text{ kJ mol}^{-1} + 1852\text{ kJ mol}^{-1}]$$

$$= 3090\text{ kJ mol}^{-1} - 3004\text{ kJ mol}^{-1}$$

$$= 86\text{ kJ mol}^{-1}$$

E14.11

$Si + 2Cl_2 \rightarrow SiCl_4$	**A** = $SiCl_4$
$SiCl_4 + RLi \rightarrow SiRCl_3 + LiCl$	**B** = $SiRCl_3$
$SiRCl_3 + 3H_2O \rightarrow RSi(OH)_3 + 3HCl$	**C** = $RSi(OH)_3$
$2RSi(OH)_3 \rightarrow RSiOSiR + H_2O$	**D** = RSiOSiR
$SiCl_4 + 4RMgBr \rightarrow SiR_4 + 4MgBrCl$	**E** = SiR_4
$SiCl_4 + 2H_2O \rightarrow SiO_2 + 4HCl$	**F** = SiO_2

Note that in the reaction with RLi the products can be $RSiCl_3$, R_2SiCl_2, R_3SiCl, and R_4Si depending on the stoichiometry used; thus, product **B** has more than one possibility considering that the stoichiometry is not clearly given. For the reaction with water, however, your product must contain Si–Cl bonds; this excludes R_4Si as a possible product. Considering that product **B** has more than one possibility, products **C** and **D** vary. For example, if **B** is R_2SiCl_2, then reaction with water would give $R_2Si(OH)_2$ as **C** that would further condense to cyclic $(R_2SiO)_3$ product **D**. Consult Section 14.16 for further details.

E14.12 **a)** +4 is the most stable oxidation state for the lighter elements, but +2 is the most stable oxidation state for Pb, the heaviest element in Group 14. Recall from Section 9.5 that the inert-pair effect is manifested in the relative stability of an oxidation state in which the oxidation number is 2 less than the group oxidation number. Pb therefore displays the inert-pair effect.

b) Table 14.1 shows that the most stable oxidation state for Sn is +4 and the most stable oxidation state for Pb is +2. Reactions (i) and (ii) illustrate Pb and Sn attaining their respective most stable oxidation states and are therefore both favourable reactions.

$$\text{(i) } Sn^{2+} + PbO_2 + 4H^+ \rightarrow Sn^{4+} + Pb^{2+} + 2H_2O$$

This reaction produces Pb^{2+}, which is more stable than starting Pb^{4+}, and Sn^{4+}, which is more stable than Sn^{2+}. Thus, the reaction fits the trend.

$$\text{(ii) } 2Sn^{2+} + O_2 + 4H^+ \rightarrow 2Sn^{4+} + 2H_2O.$$

Sn^{2+} is less stable than Sn^{4+}, thus Sn^{2+} should be easily oxidized. Again, the trend is observed.

E14.13 The most stable oxidation state of Pb is +2. Combining Pb^{4+}, an oxidizing agent, with a good reducing agent, I^-, will lead to a redox reaction and PbI_4 breaks down into PbI_2 and I_2. On the other hand, since F_2 is one of the strongest oxidizing agents, F^- has virtually no reducing ability. As such we can prepare PbF_4 as a stable compound. As a matter of fact, F_2 is such a strong oxidizing reagent that it can oxidize Pb straight to otherwise unstable Pb^{4+}.

E14.14 Consult Figure 9.9 in your textbook. Silicon is classified as a lithophile element according to the Goldschmidt's classification of elements. Lithophiles are hard cations that like to be associated with hard anions, such as O^{2-}. Lead on the other hand is a member of chalcophiles—elements that form soft cations and consequently prefer soft anions, such as S^{2-}. Hence, the major ore for Pb is a sulfide: galena or PbS.

E14.15 Silicon is recovered from silica (SiO_2) by reduction with carbon:

$$SiO_2(s) + C(s) \rightarrow Si(s) + CO_2(g) \quad \Delta_rH < 0.$$

The reaction takes place in an electric arc furnace because it has high activation energy due to the strong Si–O bonds.

Germanium is recovered from its oxide by reduction with hydrogen:

$$GeO_2(s) + 2H_2(g) \rightarrow Ge(s) + H_2O(g) \qquad \Delta_rH < 0.$$

GeO_2 can also be reduced by CO:

$$GeO_2(s) + 2CO(g) \rightarrow Ge(s) + 2CO_2(g).$$

E14.16 **a)** There is a decrease in band gap energy from carbon (diamond) to grey tin (See Section 14.1) as the s and p orbitals become larger (more extensive) down the group resulting in a better overlap and broader bands.

b) Silicon is a semiconductor. This type of material always experiences an increase in conductivity as the temperature is raised. Therefore, the conductivity of Si will be greater at 40°C than at 20°C.

E14.17 Consult Figure 14.13 and Section 14.13.

	Ionic carbides	Metallic carbides	Metalloid carbides
Group I elements	Li, Na, K, Rb, Cs		
Group II elements	Be, Mg, Ca, Sr, Ba		
Group 13 elements	Al		B
Group 14 elements			Si
3d elements		Sc, Ti, V, Cr, Mn, Fe, Co, Ni	
4d elements		Zr, Nb, Mo, Tc, Ru	
5d elements		La, Hf, Ta, W, Re, Os	
6d elements		Ac	
Lanthanides		Ce, Pr, Nd, Pm, Sm, Eu, Gd, Tb, Dy, Ho, Er, Tm, Yb, Lu	

E14.18 **a) KC_8?** This compound is formed by heating graphite with potassium vapour or by treating graphite with a solution of potassium in liquid ammonia. The potassium atoms are oxidized to K^+ ions; their electrons are added to the LUMO π^* orbitals of graphite. The K^+ ions intercalate between the planes of the reduced graphite so that there is a layered structure of alternating sp^2 carbon atoms and potassium ions. The structure of KC_8, which is an example of a saline carbide, is shown in Figures 14.4 and 14.13.

b) CaC_2? There are two ways of preparing calcium carbide, and both require very high temperatures (≥ 2000°C). The first is the direct reaction of the elements, and the second is the reaction of calcium oxide with carbon:

$$Ca(l) + 2C(s) \rightarrow CaC_2(s)$$

$$CaO(s) + 3C(s) \rightarrow CaC_2(s) + CO(g)$$

The structure is quite different from that of KC_8. Instead of every carbon atom bonded to three other carbon atoms, as in graphite and KC_8, calcium carbide contains discrete C_2^{2-} ions with carbon–carbon triple bonds. The structure of CaC_2 is similar to the NaCl structure, but because C_2^{2-} is a linear anion (not spherical like Cl^-) the unit cell of $CaCl_2$ is elongated in one direction making the unit cell tetragonal. This carbide is, hence, ionic.

c) K_3C_{60}? A solution of C_{60}, in a high boiling point solvent (such as toluene), can be treated with elemental potassium to form this compound. It is also ionic, and it contains discrete K^+ ions and C_{60}^{3-} ions (see Section 14.6(a) *Carbon clusters*). The ionic unit cell is based on a face-centred packing of C_{60}^{3-} anions with K^+ cations occupying every tetrahedral and octahedral hole (see Figure 14.5).

E14.19 Both compounds react with acid to produce the oxide, which for carbon is CO_2 and for silicon is SiO_2. The balanced equations are:

$$K_2CO_3(aq) + 2HCl(aq) \rightarrow 2KCl(aq) + CO_2(g) + H_2O(l)$$

$$Na_4SiO_4(aq) + 4HCl(aq) \rightarrow 4NaCl(aq) + SiO_2(s) + 2H_2O(l)$$

The second equation represents one of the ways that silica gel is produced. Note that in the second reaction the first product is silicic acid, H_4SiO_4. However, this acid rapidly polymerizes and loses H_2O, forming a network of very strong Si–O bonds.

E14.20 In both anions Si has a formal charge of 4+. If you recall from the chapter, Si in general prefers tetrahedral geometry (as is found in SiF_4 and SiO_4^{4-} structural units of silicates). However, F^- is small enough to allow an increase in coordination number at Si^{4+}. Chloride, however, has significantly larger ionic radius than F^-—Si^{4+} cannot accommodate six of them.

E14.21 In contrast with the extended three-dimensional structure of SiO_2, the structures of jadeite and kaolinite consist of extended one- and two-dimensional structures, respectively. The $[SiO_3^{2-}]_n$ ions in jadeite are a linear polymer of SiO_4 tetrahedra, each one sharing a bridging oxygen atom with the tetrahedron before it and the tetrahedron after it in the chain (see Structure **17** in the chapter). Each silicon atom has two bridging oxygen atoms and two terminal oxygen atoms. The two-dimensional aluminosilicate layers in kaolinite represent another way of connecting SiO_4 tetrahedra (see Figure 14.18). Each silicon atom has three oxygen atoms that bridge to other silicon atoms in the plane, and one oxygen atom that bridges to an aluminium atom.

E14.22 In linear $Si_3O_{10}^{8-}$ we have two different Si environments—one Si atom is in the centre of the chain with the remaining two Si atoms on each end. Thus, we can expect two signals in ^{29}Si-MASNMR with intensities 1 : 2 for central and terminal Si atoms. The cyclic anion $Si_3O_9^{6-}$ would show a single resonance. See also Self-Test 8.6.0.

E14.23 Both pyrophyllite, $Al_2Si_4O_{10}(OH)_2$, and muscovite, $KAl_2(Si_3Al)O_{10}(OH)_2$, belong to the phyllosilicate subclass of silicate class. The phyllosilicates are characterized with two-dimensional sheets composed of vertex-sharing SiO_4-tetrahedra; each SiO_4-tetrahedron is connected to three neighbouring SiO_4-tetrahedra via bridging O atoms. These sheets are negatively charged and held together via electrostatic interactions with cations sandwiched between the sheets. Because of this layered structure and relatively weak electrostatic interactions keeping them together, both pyrophyllite and muscovite have excellent cleavage parallel to the layers (just like graphite). The hardness of the two minerals is, however, somewhat different—muscovite is harder than pyrophyllite (although both are very soft and can be scratched with a nail). The reason for this is in the chemical composition. Pyrophyllite layers are composed of exclusively SiO_4 tetrahedra held together with sandwiched Al^{3+} cations located in octahedral holes. In muscovite structure, however, every fourth Si^{4+} in the layer is replaced with Al^{3+} (see the formula—Si_3Al). The replacement of one +4 charge with +3 decreases the overall charge of layers and an additional cation is required to balance the charge. Strictly speaking one Si^{4+} is replaced with the Al^{3+}/K^+ pair. Lower charge on layers and more cations in between layers hold the muscovite structure tighter than pyrophyllite, and the result is higher hardness in muscovite.

E14.24 Glasses are amorphous silicon-oxygen compounds. Starting from sand (which is mostly pure SiO_2) and other oxide additives, a wide variety of glasses with different properties can be obtained, used and seen every day.
Activated carbon (used as adsorbent), carbon fibres (used as additives in plastic to increase the strength), and carbon black (used as pigment) are a further three examples of amorphous and partially crystalline carbon.

E14.25 The silicate sheets are represented on Figures 14.17 and 14.18. An edge view of these sheets is sketched below where every tetrahedron represents one SiO_4 unit. There is a rather extensive substitution of Si^{4+} with Al^{3+} (every second Si^{4+} has been substituted with Al^{3+}, as can be seen from the mineral formula) in these sheets. Note that two tetrahedral AlSi sheets are oriented with respect to each other, in such a way that unshared oxygen atoms on each sheet are facing each other. It is within this space between the sheets that octahedral holes are formed—they are defined by six unshared oxygen atoms, three from the top and three from the bottom SiAl-sheet. The rest of Al^{3+} cations (those that are not substituted for Si^{4+}) occupy these holes. Ca^{2+} cations are, however, too large for the octahedral holes, and it is more likely that they will be found in larger holes with coordination number 12.

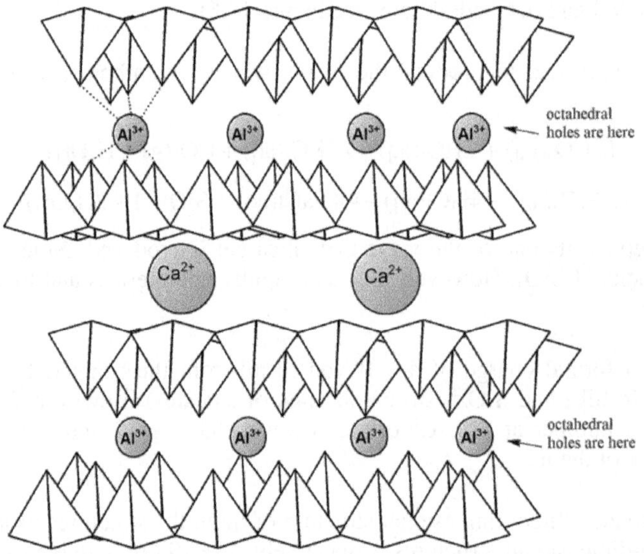

E14.26 Silly Putty is composed of siloxane polymers in the form of boric acid esters. The starting material for the manufacture of silly putty is dichlorodimethylsilane, $(CH_3)_2SiCl_2$. This compound is first hydrolysed with water to produce polydimethylsiloxane:

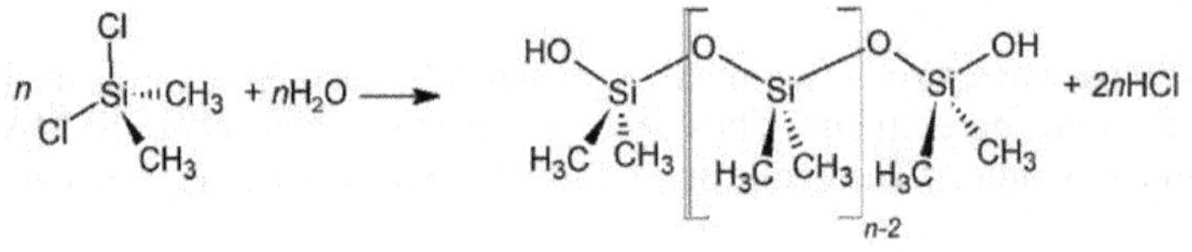

This is a linear chain and oily product. It is further cross-linked with boric acid to produce solid material that predominately gives silly putty its interesting properties:

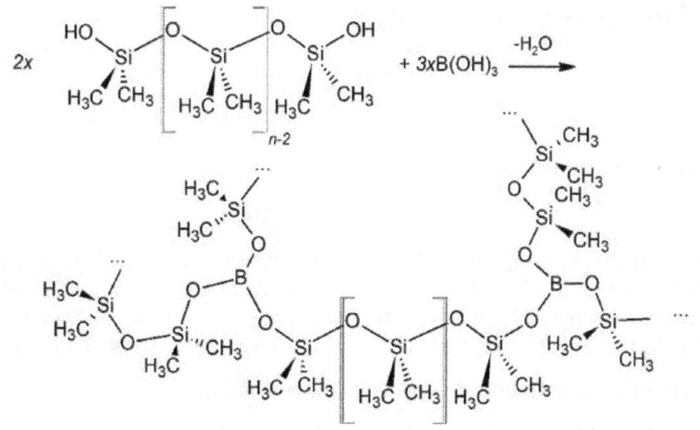

Guidelines for Selected Tutorial Problems

T14.2 You can use any mineralogy textbook as a source for this problem. Suggested text would be a classic mineralogy textbook *Dana's Manual of Mineralogy* by C. Klein and B. Dutrow, published by Wiley and Sons, currently in its 23rd edition.

T14.3 The obvious choice for silicon is, of course, the fact that it is in the same group as carbon. They also share the common highest (and most stable) valence and preferred geometry (tetrahedral). The fact that silicon is the second most abundant element in Earth's crust (as well as in crusts of the other rocky planets—at least as much as we know currently) also goes in favour of silicon-based life.
The major part of your arguments against should focus on explaining the reasons behind different catenation preferences for the two elements, lack of Si=Si bonds (or better said, their low stability), and silicon's high affinity toward oxygen (the most abundant element in Earth's crust—explain how this is an issue using thermodynamic parameters provided in the tables within this chapter).

T14.4 The text suggested for T14.2 also discusses the synthetic diamonds and provides further sources on this topic, thus it is a good starting point for this tutorial problem as well.

T14.10 The structure of Cu_2ZnSnS_4 has appeared several times in the literature, most recently in Ritscher *et al. Journal of Solid State Chemistry* (2016) **238**, p. 68. The structure is tetragonal and can be approximately described as two cubic close packed unit cells composed of S^{2-} stacked one on top of the other with metal cations occupying half of tetrahedral holes. The introduction section of this article also contains information and sources relevant to other questions of this tutorial.

Chapter 15 The Group 15 Elements

Self-Tests

S15.1 The combustion reaction is:

$$P_4 + 5O_2 \rightarrow P_4O_{10}.$$

P_4 molecule is tetrahedral (structure **1** in your textbook) with four P–P single bonds. Structure **6** in the textbook shows one molecule of P_4O_{10}: this molecule has four P=O double bonds and 10 P–O single bonds. This gives us the summary of bonds broken and bonds formed: four P–P single bonds and five O=O double bonds have to be broken and four P=O double bonds and 10 P–O single bonds have to be formed:

$$\Delta_f H = 4 \times B(\text{P–P}) + 5 \times B(\text{O=O}) - 4 \times B(\text{P=O}) - 10 \times B(\text{P–O})$$

$$= 4 \times 201 \text{ kJ mol}^{-1} + 5 \times 497 \text{ kJ mol}^{-1} - 4 \times 560 \text{ kJ mol}^{-1} - 10 \times 407 \text{ kJ mol}^{-1}$$

$$= -3021 \text{ kJ mol}^{-1}.$$

This very negative value shows how stable P_4O_{10} is and why P_4 spontaneously bursts into flames when exposed to air. The actual value for $\Delta_f H(P_4O_{10})$ is –2986 kJ mol^{-1}, very close to our calculated one. The discrepancy lies in the sublimation enthalpies of P_4 and P_4O_{10}—under standard conditions both are solids. P_4 has to be converted into gas phase before P–P bonds can be broken, and this process requires energy. The product is converted from the gas phase into solid and the sublimation enthalpy is released.

S15.2 **a)** N_2O is linear molecule because it is resonance stabilized:

b) NO_2^- anion has a bent geometry (with trigonal planar electron geometry):

S15.3 **a)** Yes, the structure is consistent with the VSEPR model. Considering the three nearest neighbours only, the structure around each Bi atom is trigonal pyramidal, like NH_3. VSEPR theory predicts this geometry for an atom that has a Lewis structure with three bonding pairs and a lone pair. A side view of the structure of bismuth is shown below.

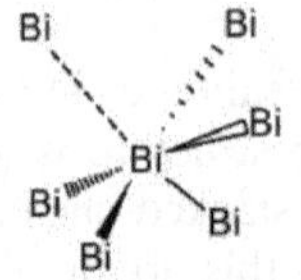

b) Two nitrogen atoms in N_2 molecule have 10 valence electrons (each contributing five electrons). If we make only one bond between N atoms, then we have a sextet around each N atom (as shown in Figure S15.3(A)). Making the double bond gives each N atom an additional electron pair and a septet at each N (Figure S15.3(B)). Only if we make a triple bond between two N atoms (Figure S15.3(C)) can each nitrogen achieve an octet giving also a final Lewis structure for N_2.

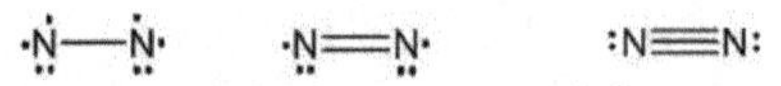

S15.3(A) S15.3(B) S15.3(C)

Nitrogen is a gas under standard contitions as can be expected from the final structure: non-polar, low-molecular weight molecule. It is also diamagnetic as can be expected from the fact that all electrons are paired with an octet. N_2 could also react as a Lewis base: each N atom still has a lone electron pair, but it is likely to be a weak one due to the fact that N atoms have high electronegativity and the electrons have low energy. N_2 is also a very unreactive gas as can be expected from a triple bond between its atoms.

S15.4 **a)** As we descend Group 15, the element–hydrogen bonds become weaker. This is also the trend in basicity decrease: weak element–hydrogen bonds make the formation of fourth unlikely particularly if proton has to be transferred from a weak acid. For example, H_2O can protonate ammonia but stronger acids (such as HCl are required to protonate PH_3 and AsH_3. Formation of the fourth As–H bond (to make AsH_4^-) is not sufficient driving force to allow for heterolytical cleavage of H–O bond in H_2O.

b) We have to estimate the enthalpies of formation for NF_3 and NCl_3 based on the following equations:

$$N_2(g) + 3F_2(g) \rightarrow 2NF_3$$

$$N_2(g) + 3Cl_2(g) \rightarrow 2NCl_3.$$

The estimate can be made following our established route: looking at the change in bond enthalpies before and after the reaction (if you need more details, see Self-Test 15.1). From Table 2.7 the relevant bond enthalpies are:

$B(N{\equiv}N) = 946\ \text{kJ mol}^{-1}$,

$B(F–F) = 155\ \text{kJ mol}^{-1}$,

$B(Cl–Cl) = 242\ \text{kJ mol}^{-1}$,

$B(N–F) = 270\ \text{kJ mol}^{-1}$ and

$B(N–Cl) = 200\ \text{kJ mol}^{-1}$.

First, we consider the formation of NF_3 and its $\Delta_f H$:

$$\Delta_f H(NF_3) = B(N{\equiv}N) + 3 \times B(F–F) - 6 \times B(N–F)$$

$$\Delta_f H(NF_3) = 946\ \text{kJ mol}^{-1} + 3 \times 155\ \text{kJ mol}^{-1} - 6 \times 270\ \text{kJ mol}^{-1}$$

$$\Delta_f H(NF_3) = -209\ \text{kJ mol}^{-1}.$$

Now we consider the formation of NCl_3 and its $\Delta_f H$:

$$\Delta_f H(NCl_3) = B(N{\equiv}N) + 3 \times B(Cl–Cl) - 6 \times B(N–Cl)$$

$$\Delta_f H(NCl_3) = 946\ \text{kJ mol}^{-1} + 3 \times 242\ \text{kJ mol}^{-1} - 6 \times 200\ \text{kJ mol}^{-1}$$

$$\Delta_f H(NCl_3) = 472\ \text{kJ mol}^{-1}.$$

We can see that the synthesis of NF_3 from elements is a thermodynamically favourable process and this fluoride is stable, but that the synthesis of NCl_3 from elements is not a thermodynamically favourable process and NCl_3 is unstable. The reasons for the instability (and reactivity) of NCl_3 is strong Cl–Cl bond (in comparison to F–F bond) and weak N–Cl bond (in comparison to N–F).

S15.5 Dimethylhydrazine ignites spontaneously; unlike H_2 and low hydrocarbons, it is not a gas and does not need to be liquified making its handling easier and more energy efficient. In comparison to hydrocarbons, it produces less CO_2.

S15.6 Sulfur is on the right-hand side of phosphorus in the same period (period 3) in the periodic table. Two trends are important for this pair: 1) moving from left to right the non-metallic character of the elements is increasing and 2) the electronegativity values are increasing. High electronegativity and strong non-metallic character are typical of good oxidizing agents. Thus, because sulfur is on the right-hand side, it is more non-metallic than phosphorus and has higher electronegativity, making it a better oxidizing agent than phosphorus. This becomes more qualitative if we look at the reduction potentials for phosphorus and sulfur. From the table of standard reduction potentials in Resource Section 3 you can obtain the following information:

$$P(s) + 3e^- + 3H^+(aq) \rightarrow PH_3(aq) \qquad E^\circ = -0.063\ \text{V}$$

$$S(s) + 2e^- + 2H^+(aq) \rightarrow H_2S(aq) \qquad E^\circ = 0.144\ \text{V}.$$

Since $\Delta G = -\nu FE^\circ$, sulfur is more readily reduced and is therefore the better oxidizing agent. This is in harmony with the higher electronegativity of sulfur ($\chi = 2.58$) than phosphorus ($\chi = 2.19$).

S15.7 **a)** Brown, gaseous NO_2 is not easily oxidized. Since it contains N in an intermediate oxidation state (+4), it is prone to disproportionation in aqueous solutions. Unlike NO_2, the colourless NO is rapidly oxidized by O_2 in air producing NO_2. N_2O is unreactive for purely kinetic reasons.
b) The reactions that are employed to synthesize hydrazine are as follows:

$$NH_3 + ClO^- + H^+ \rightarrow [H_3N—Cl\text{-}O]^- + H^+ \rightarrow H_2NCl + H_2O$$

$$H_2NCl + NH_3 \rightarrow H_2NNH_2 + HCl.$$

Both can be thought of as redox reactions because the formal oxidation state of the N atoms changes (from −3 to −1 in the first reaction and from −1 and −3 to −2 and −2 in the second reaction). Mechanistically, both reactions appear to involve nucleophilic attack by NH_3 (a Lewis base) on either ClO^- or NH_2Cl (acting as Lewis acids).

The reaction employed to synthesize hydroxylamine, which produces the intermediate $N(OH)(SO_3)_2$, probably involves the attack of HSO_3^- (acting as a Lewis base) on NO_2^- (acting as a Lewis acid), although, because the formal oxidation state of the N atom changes from +3 in NO_2^- to +1 in $N(OH)(SO_3)_2$, this can also be seen as a redox reaction. Whether one considers reactions such as these to be redox reactions or nucleophilic substitutions may depend on the context in which the reactions are being discussed. It is easy to see that these reactions do not involve simple electron transfer, such as in $2Cu^+(aq) \rightarrow Cu^0(s) + Cu^{2+}(aq)$.

S15.8 The strongly acidic OH groups are titrated by the first 30.4 cm^3 and the two terminal –OH groups are titrated by the remaining 45.6 cm^3 – 30.4 cm^3 = 15.2 cm^3. The concentrations of analyte and titrant are such that each OH group requires 15.2 cm^3/2 = 7.6 cm^3 of NaOH. There are therefore (30.4 cm^3)/(7.6 cm^3) = 4 strongly acidic OH groups per molecule. A molecule with two terminal –OH groups and four central –OH groups is a tetrapolyphosphate.

Exercises

E15.1 The answers are summarized in the table below:

	Type of element	Diatomic gas?	Achieves maximum oxidation state?	Displays inert pair effect?
N	nonmetal	yes	yes	no
P	nonmetal	no	yes	no
As	nonmetal	no	yes	no
Sb	metalloid	no	yes	no
Bi	metalloid	no	yes	yes

E15.2 For this exercise, recollect your knowledge from Chapters 1 and 9 and combine with the group trends discussed here.

Moscovium, being at the bottom of the group, is expected to be a metallic element. Its atomic radius should be about 190–200 pm. First ionization energy probably between 500 to 600 kJ mol^{-1}. Pauling's electronegativity 2.0–1.80. Due to the inert pair effect, moscovium's preferred oxidation state would be +3, with +5 unstable and very good to excellent oxidizing agent (more so than Bi(V)). Its hydride, if stable at all, should have a formula McH_3, and (if stable at all) might be close to saline hydride (in terms of classification) and the most hydridic hydride in the group. McH_5 is not expected to exist at all. The stable oxide is Mc_2O_3 with basic to amphoteric properties. Mc_2O_5 is a very strong oxidizing agent, with acidic properties. It might be possible to stabilize Mc(V) oxo anions in very basic solutions (pH = 14). In terms of halides, all halides for Mc(III) should be expected with varying degree of covalency (depending on the halide). McF_5 would be the only example of Mc(V) halides.

E15.3 The bond enthalpy of N≡N bond is very large (946 kJ mol^{-1}, Table 2.7) making N_2 molecule a preferred form for nitrogen. Any other allotrope would be based on N–N and N=N bonds, which are significantly weaker (163 kJ mol^{-1} and 946 kJ mol^{-1} respectively). This difference in bond enthalpies would make any nitrogen allotrope unstable (even explosively so).
P–P single bond is stable enough (bond enthalpy 201 kJ mol^{-1}) to allow for many different combinations and phosphorus exists in several allotropes.

E15.4 The cubic BN has a diamond-type structure with directional covalent bonds. The electrons in these bonds are very localized and cannot hop easily. The fact that boron and nitrogen are both rather electronegative elements further localize electrons around the bonds and atoms. The bonds in GaN are more polar, less localized and in combination with higher energies of Ga valence orbitals (and smaller energy difference between the orbitals in the valance shell in comparison to boron) the band gap decreases. As we combine Ga with heavier Group 15 elements, the valence orbitals have better overlaps, are closer in energy and the band gap decreases. Decrease in electronegativity down the group also helps increase mobility of electrons and expansion of bands in the structure.

E15.5 **a) High-purity phosphoric acid.** The starting point is hydroxyapatite, $Ca_5(PO_4)_3OH$, which is converted to crude $Ca_3(PO_4)_2$. This compound is treated with carbon to reduce phosphorus from P(V) in PO_4^{3-} to P(0) in P_4 and with silica, SiO_2, to keep the calcium-containing products molten for easy removal from the furnace. The impure P_4 is purified by sublimation and then oxidized with O_2 to form P_4O_{10}, which is hydrated to form pure H_3PO_4.

$$2Ca_3(PO_4)_2 + 10C + 6SiO_2 \rightarrow P_4 + 10CO + 6CaSiO_3$$

$$P_4(\text{pure}) + 5O_2 \rightarrow P_4O_{10}$$

$$P_4O_{10} + 6H_2O \rightarrow 4H_3PO_4 \text{ (pure)}.$$

b) Fertilizer-grade H_3PO_4. In this case, hydroxyapatite is treated with sulfuric acid, producing phosphoric acid that contains small amounts of impurities.

$$Ca_5(PO_4)_3OH + 5H_2SO_4 \rightarrow 3H_3PO_4(\text{impure}) + 5CaSO_4 + H_2O$$

c) Account for the difference in cost. The synthesis of fertilizer-grade (i.e., impure) phosphoric acid involves a single step and gives a product that requires little or no purification for further use. In contrast, the synthesis of pure phosphoric acid involves several synthetic steps and a time-consuming and expensive purification step, the sublimation of white phosphorus, P_4.

E15.6 **a) Hydrolysis of Li_3N.** Lithium nitride can be prepared directly from the elements. Since it contains N^{3-} ions, which are very basic, Li_3N can be hydrolysed with water to produce ammonia:

$$6Li(s) + N_2(g) \rightarrow 2Li_3N(s)$$

$$Li_3N(s) + 3H_2O(l) \rightarrow NH_3(g) + 3LiOH(aq).$$

b) Reduction of N_2 by H_2. This reaction requires high temperatures and pressures as well as a presence of the solid-state catalyst to both proceed at an appreciable rate and provide sufficient conversion:

$$N_2(g) + 3H_2(g) \rightarrow 2NH_3(g).$$

c) Account for the difference in cost. The second process is considerably cheaper than the first, although the second process must be carried out at high temperature and pressure. This is because the cost of using lithium metal is higher than using $N_2(g)/H_2(g)$ from the air and water.

E15.7 NH_4NO_3 in water dissociates to give ammonium cation and nitrate anion:

$$NH_4NO_3(s) + H_2O(l) \rightarrow NH_4^+(aq) + NO_3^-(aq).$$

The ammonium ion is the conjugate acid of the weak base, ammonia, and hence expected to be moderately acidic (relative to water). Accordingly, the equation below explains the observed acidity of ammonium ions.

$$NH_4^+(aq) + H_2O(l) \rightarrow NH_3(aq) + H_3O^+(aq)$$

Note that NO_3^- is the conjugate base of the strong acid HNO_3 and therefore itself is an extremely weak base.

E15.8 CO forms a complex with haemoglobin in the blood; this complex is more stable than oxy-haemoglobin. This prevents the haemoglobin in the red blood cells from carrying oxygen around the body. This causes an oxygen deficiency, leading to unconsciousness and then death. CO is a polar molecule and binds to iron in haemoglobin through a lone pair on C atom.

N_2 itself, with a triple bond between the two atoms, is strikingly unreactive. In comparison to CO, N_2 is a non-polar molecule that has lone electron pairs on nitrogen atoms. The general unreactivity of N_2 molecule, its non-polarity, and lone pairs on more electronegative nitrogen (in comparison to carbon of CO) make N_2 a very weak ligand that does not bind to haemoglobin. Consequently, N_2 molecule is not toxic.

E15.9 Inductive effects of very electronegative fluorine atoms make the lone pair on nitrogen atom unavailable for bonding: partial positive charge on N atom, $N^{\delta+}$–$F^{\delta-}$, stabilizes the lone pair. PF_3 also has stabilized lone pair on P atom but larger size of phosphorus atom is better at reducing charge density and making this lone pair somewhat more available. However, more important for the formation of the metal complexes is PF_3's ability to behave as a π acceptor. Its empty antibonding MOs are a good match for the d orbitals on metal centres in complexes. Thus, while PF_3 is a weak σ donor, it is a very good π acceptor and can form a strong bond by accepting electronic density from the metals. Similar MOs exist in NF_3 but their energy is too low for an overlap with d orbitals on the metal. Consequently, NF_3 cannot act as a ligand.

Since NF_5 does not exist, the existence of NOF_3 (another N(V)species) might be surprising. However, oxygen is very good at stabilizing high oxidation states—if you do not recall this fact from before, just look at the N(V) species covered in this chapter: they all contain NO bonds. Each bonded O atom "neutralizes" two positive charges, in comparison to one for each fluorine. Thus, while it is difficult to make five F–N(V) covalent bonds (there is not enough room around N(V) for them) and have a trigonal pyramidal geometry, it is feasible to have coordination four and tetrahedral arrangement around N(V).

E15.10 The only isolable nitrogen chloride is NCl_3, and it is thermodynamically unstable with respect to its constituent elements (i.e., it is endoergic, see Self-Test 15.4). The compound NCl_5 is unknown. In contrast, both PCl_3 and PCl_5 are stable and can be prepared directly from phosphorus and chlorine. The nitrogen equivalent of $POCl_3$ does not exist either.

E15.11 The Lewis structures and molecular geometries are summarized in the table below.

a) PCl_4^+: Four bonding pairs of electrons around the central phosphorus atom, and no lone electron pairs, suggest tetrahedral molecular geometry.

b) PCl_4^-: Four bonding pairs of electrons and one lone pair of electrons around the central phosphorus atom, the molecular structure is a see-saw (trigonal bipyramidal electron pair geometry with the lone pair in the equatorial plane).

c) $AsCl_5$: Five bonding pairs of electrons around the central arsenic atom, the structure is a trigonal bipyramid (both the electron pair and molecular geometries).

d) SbF_5^{2-}: Five bonding and one lone electron pair around Sb give a tetragonal pyramidal geometry.

e) SbF_6^-: Six bonding pairs around Sb result in an octahedral geometry.

	PCl_4^+	PCl_4^-	$AsCl_5$	SbF_5^{2-}	SbF_6^-
Lewis structure					
Molecular geometry					

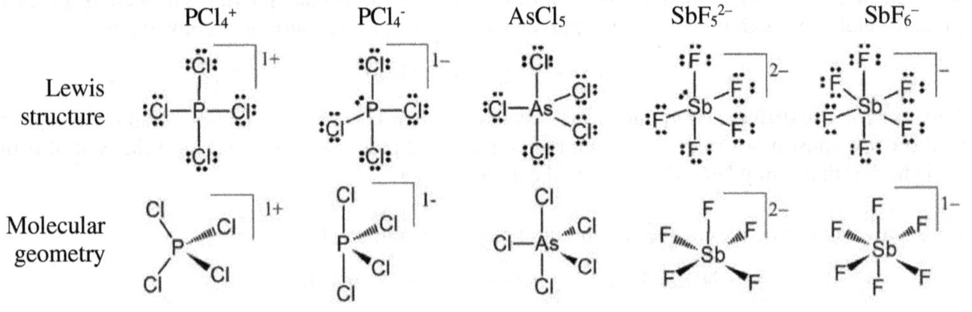

E15.12 Large Bi(III) can accommodate higher coordination numbers. But the presence of an inert electron pair distorts the geometries from ideal due to strong repulsion with bonding electron pairs (see also Self-Test 15.3(a)).

E15.13 a) Oxidation of P_4 with excess O_2. The balanced equation is:

$$P_4 + 5O_2 \rightarrow P_4O_{10}.$$

b) Reaction of the product from part (a) with excess H_2O. The balanced equation is:

$$P_4O_{10} + 6H_2O \rightarrow 4H_3PO_4.$$

c) Reaction of the product from part (b) with solution of $CaCl_2$. The products of this reaction would be calcium phosphate and a solution of hydrochloric acid. The balanced equation is:

$$2H_3PO_4(l) + 3CaCl_2(aq) \rightarrow Ca_3(PO_4)_2(s) + 6HCl(aq).$$

E15.14 a) Synthesis of HNO_3. The synthesis of nitric acid, an example of the most oxidized form of nitrogen, starts with ammonia, the most reduced form:

$$4NH_3(aq) + 7O_2(g) \rightarrow 6H_2O(g) + 4NO_2(g).$$

High temperatures are obviously necessary for this reaction to proceed at a reasonable rate: ammonia is a flammable gas, but at room temperature it does not react rapidly with air. The second step in the formation of nitric acid is the high-temperature disproportionation of NO_2 in water:

$$3NO_2(aq) + H_2O(l) \rightarrow 2HNO_3(aq) + NO(g).$$

b) Synthesis of NO_2^-. Prepare NO_2 from ammonia as above. Whereas the disproportionation of NO_2 in acidic solution yields NO_3^- and NO (see part **(a)**), in basic solution nitrite ion is formed:

$$2NO_2(aq) + 2OH^-(aq) \rightarrow NO_2^-(aq) + NO_3^-(aq) + H_2O(l).$$

c) Synthesis of NH_2OH. The protonated form of hydroxylamine is formed in an unusual reaction between nitrite ion, prepared following a) and b) procedures and bisulfite ion in cold aqueous acidic solution (the NH_3OH^+ ion can be deprotonated with base):

$$NO_2^-(aq) + 2HSO_3^-(aq) + H_2O(l) \rightarrow NH_3OH^+(aq) + 2SO_4^{2-}(aq).$$

d) Synthesis of N_3^-. The azide ion can be prepared from anhydrous molten sodium amide (m.p. ~200°C) and either nitrate ion or nitrous oxide at elevated temperatures:

$$3NaNH_2(l) + NaNO_3 \rightarrow NaN_3 + 3NaOH + NH_3(g)$$

$$2NaNH_2(l) + N_2O \rightarrow NaN_3 + NaOH + NH_3.$$

E15.15 Phosphorus(V) oxide, P_4O_{10}, is formed by the complete combustion of white phosphorus:

$$P_4(s) + 5O_2(g) \rightarrow P_4O_{10}(s).$$

Where P_4 is white phosphorus. As discussed in Section 15.1, white phosphorus (P_4) is adopted as the reference phase for thermodynamic calculations although it is not the most stable phase of elemental phosphorus–in this sense using P_4 as a reference state differs from the usual practice.

E15.16 The main points you will want to remember are (i) Bi(III) is much more stable than Bi(V), and (ii) P(III) and P(V) are both about equally stable (i.e., Bi(V) is a strong oxidant but P(V) is not). This suggests that the point for Bi(III) on the Frost diagram lies *below* the line connecting Bi(0) and Bi(V), whereas the point for P(III) lies very close to the line connecting P(0) and P(V). The essential parts of the Frost diagrams for these two elements are shown below (see Figure 15.6 for the complete Frost diagrams).

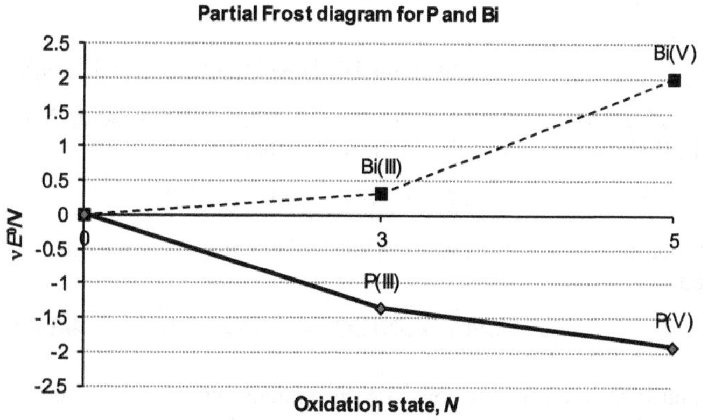

E15.17 The rates of reactions in which nitrite ion is reduced (i.e., in which it acts as an oxidizing agent) are increased as the pH is lowered. That is, acid enhances the rate of oxidations by NO_2^-. The reason is that NO_2^- is converted to the nitrosonium ion, NO^+, in strong acid:

$$HNO_2(aq) + H^+(aq) \rightarrow NO^+(aq) + H_2O(l).$$

This cationic Lewis acid can form complexes with the Lewis bases undergoing oxidation (species that can be oxidized are frequently electron rich and hence are Lewis bases). Therefore, at low pH the oxidant (NO^+) is a different chemical species than at higher pH (NO_2^-) (see Section 15.13(b) *Nitrogen(IV) and nitrogen(III) oxides and oxoanions*).

E15.18 The given observations suggest that more than one NO molecule is involved in the slow step of the reaction mechanism because the rate of reaction is dependent on the concentration of NO. Therefore, the rate law must be more than first order in NO concentration. It turns out to be second order: an equilibrium between NO and its dimer N_2O_2 precedes the rate-determining reaction with oxygen. At high concentrations of NO, the concentration of the dimer is higher (see Section 15.13(c) *Nitrogen(II) oxide*).

E15.19 The problem provides us with $\Delta_rG = -26.5$ kJ mol^{-1} for the following reaction:

$$\tfrac{1}{2}N_2(g) + 3/2H_2(g) \rightarrow NH_3(aq).$$

This is a redox reaction, which can be broken into two half-reactions:

Reduction half-reaction: $\tfrac{1}{2}N_2(g) + 3e^- + 3H^+ \rightarrow NH_3(aq)$ $\quad E_{red} = ?$

Oxidation half-reaction: $3/2H_2 \rightarrow H^+(aq) + 3e^-$ $\quad E_{ox} = -0.059V \times pH$

We are looking for the standard reduction potential in basic aqueous solution, so the potential of standard hydrogen electrode has to be modified for the pH of the solution (pH = 14).

From the thermodynamic data, we can calculate the potential of the overall reaction from $\Delta_rG = \nu FE°$:

$$E^o = -\frac{\Delta_f G}{\nu F} = -\frac{-26.5\,\text{kJ mol}^{-1}}{3\times 96.48\,\text{kC mol}^{-1}} = +0.09\,\text{V}.$$

At pH = 14 the potential of H^+/H_2 electrode is:

$$E_{ox} = -0.059\ \text{V} \times 14 = -0.826\ \text{V}.$$

Finally, the unknown potential is:

$$E = +0.09\ \text{V} = E_{red} - (-0.826\ \text{V})$$

$$E_{red} = -0.736\ \text{V}.$$

E15.20 **a) H_2O (1 : 1).** In this case, water will form a strong P=O double bond, releasing two equivalents of HCl and forming tetrahedral $POCl_3$.

$$PCl_5 + H_2O \rightarrow POCl_3 + 2HCl$$

b) H_2O in excess. When water is in excess, all the P–Cl bonds will be hydrolysed, and phosphoric acid will result:

$$PCl_5(g) + 4H_2O(l) \rightarrow H_3PO_4(aq) + 5HCl(aq).$$

c) $AlCl_3$. By analogy to the reaction with BCl_3, a Group 13 Lewis acid, PCl_5 will donate a chloride ion and a salt will result. Both the cation and the anion are tetrahedral.

$$PCl_5 + AlCl_3 \rightarrow [PCl_4]^+[AlCl_4]^-.$$

d) NH_4Cl. A phosphazene, either cyclic or linear chain polymers of (=P(Cl)$_2$–N=P(Cl)$_2$–N=P(Cl)$_2$–N=) bonds will be formed:

$$nPCl_5 + nNH_4Cl \rightarrow -[(N{=}P(Cl)_2)_n]- + 4nHCl.$$

E15.21 In both PF_3 and POF_3 there is only one P environment and we can expect only one signal in ^{31}P NMR. For both compounds this signal is going to appear as a quartet because in each case P atom is bonded to three F atoms with $I = \tfrac{1}{2}$. Thus, splitting pattern will not be helpful. In POF_3, however, P atom is in higher oxidation state (+5) and is bonded to one more electronegative atom (oxygen) in comparison to PF_3 (P oxidation state +3). Therefore, P atom in POF_3 is more deshielded and its resonance will appear downfield in comparison to the P resonance in PF_3.

E15.22 From the standard potentials for the following two half-reactions, the standard potential for the net reaction can be calculated:

$$H_3PO_2(aq) + H_2O(l) \rightarrow H_3PO_3(aq) + 2e^- + 2H^+(aq) \qquad E° = 0.499\ \text{V}$$

$$Cu^{2+}(aq) + 2e^- \rightarrow Cu^0(s) \qquad E° = 0.340\ \text{V}$$

Therefore, the net reaction is:

$$H_3PO_2(aq) + H_2O(l) + Cu^{2+}(aq) \rightarrow H_3PO_3(aq) + Cu^0(s) + 2H^+(aq)$$

and the net standard potential is $E° = +0.839$ V. To determine whether HPO_3^{2-} and $H_2PO_2^-$ are useful as oxidizing or reducing agents, you must compare E values for their oxidations and reductions at pH 14 (these potentials are listed in Resource Section 3):

$$HPO_3^{2-}(aq) + 3OH^-(aq) \rightarrow PO_4^{3-}(aq) + 2e^- + 2H_2O(l) \qquad E(pH14) = 1.12\ V$$

$$HPO_3^{2-}(aq) + 2e^- + 2H_2O(l) \rightarrow H_2PO_2^-(aq) + 3OH^-(aq) \qquad E(pH14) = -1.57\ V$$

$$H_2PO_2^-(aq) + e^- \rightarrow P(s) + 2OH^-(aq) \qquad E(pH14) = -2.05\ V$$

Since a positive potential will give a negative value of ΔG, the oxidations of HPO_3^{2-} and $H_2PO_2^-$ are much more favourable than their reductions, so these ions will be much better reducing agents than oxidizing agents.

E15.23 In P_4 structure, each P atom contributes three valence electrons to the skeletal valence electron count. Since there are four P atoms, there are 12 valence skeletal electrons or six skeletal electron pairs. So P_4 would be n + 2 cluster (n = 4) making the structure *nido* type. Formally, its structure then should be derived from a trigonal bipyramid. This is indeed a reasonable solution because removal of one vertex from a trigonal bipyramid leaves a tetrahedral structure.

E15.24 $As + 3/2Cl_2 \rightarrow AsCl_3$ — **A** = $AsCl_3$

$AsCl_3 + Cl_2/h\nu \rightarrow AsCl_5$ — **B** = $AsCl_5$

$AsCl_3 + 3RMgBr \rightarrow AsR_3 + 3MgClBr$ — **C** = AsR_3

$AsCl_3 + LiAlH_4 \rightarrow AsH_3$ — **D** = AsH_3

E15.25 The two possible isomers of $[AsF_4Cl_2]^-$ are shown below. The chlorines are adjacent in the *cis* form and are opposite in the *trans* form, leading to two fluorine environments in the *cis* form and one in the *trans*. The *cis* isomer gives two ^{19}F signals (both will appear as doublets) and the *trans* isomer gives one signal (a singlet).

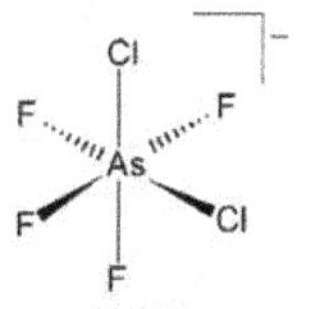

cis-$[AsF_4Cl_2]^-$

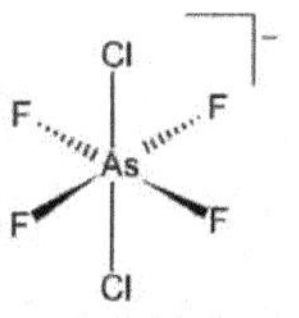

trans-$[AsF_4Cl_2]^-$

E15.26 $NO + O_2 \rightarrow NO_2$ — **A** = NO_2

$H_2O + 3NO_2 \rightarrow 2HNO_3 + NO$ — **B** = HNO_3; **C** = NO

$2NO_2 \rightarrow N_2O_4$ — **D** = N_2O_4

$2HNO_3 + Cu + 2H^+ \rightarrow 2NO_2 + Cu^{2+} + 2H_2O$ — **E**= NO_2

$HNO_3 + 4Zn + 9H^+ \rightarrow NH_4^+ + 3H_2O + 4Zn^{2+}$ — **F** = NH_4^+

Note: The reaction (b) initially produces HNO_3 and HNO_2 as first products of NO_2 disproportionation in water. However, HNO_2 further disproportionates to produce HNO_3 and NO. Thus, the final products of reaction between NO_2 and water are HNO_3 and NO.

E15.27 Recall that if the potential to the right is more positive than the potential to the left, the species is thermodynamically unstable and likely to disproportionate. However, no kinetic information (i.e., how fast a species would disproportionate) can be deduced. The species of N and P that disproportionate are N_2O_4, NO, N_2O, NH_3OH^+, $H_4P_2O_6$, and P.

Guidelines for Selected Tutorial Problems

T15.4 Note that traditionally phosphates have been removed from sewage waters by precipitation. Recently, however, new methods have been developed, including biological treatment of waters. Try to analyse the chemical precipitation reactions, look at the metals used and issues related to each (including cost and potential side-effects). Any textbook covering water treatment and analysis will have both chemical and engineering sides of the process covered.

Chapter 16 The Group 16 Elements

Self-Tests

S16.1 The principles of molecular orbital theory are described in *Molecular Orbital Theory* part of Chapter 2, in sections 2.8 *Homonuclear diatomic molecules* and 2.10 *Bond Properties*.

Considering that oxygen and sulfur are in the same group of the periodic table of the elements, it might be tempting to take the molecular orbitals for dioxygen (O_2) and apply it to S_2. Since S is below oxygen, the energy difference between 3s and 3p orbitals is much smaller than the energy difference between 2s and 2p orbitals in O. Consequently, the order of molecular orbitals in S_2 would be closer to that in C_2 (see Figure 2.17). Another difference is the number of shells in S atom vs. in O atom: three and two respectively. This means we have to have one more set of σ and π molecular orbitals in S_2 molecule (and by extension S_2^{2-} anion). Summing this up, we have the following molecular orbitals for S_2 molecule/S_2^{2-} dianion:

$$1\sigma_g 1\sigma_u 2\sigma_g 2\sigma_u\, 1\pi_u 3\sigma_g 1\pi_g 3\sigma_u 4\sigma_g 4\sigma_u 2\pi_u 5\sigma_g 2\pi_u 5\sigma_u.$$

We have to ill these orbitals with 34 electrons (16 from each S atom and two additional for minus two charge) keeping in mind that s molecular orbitals are singly degenerate and take two electrons while p molecular orbitals are doubly degenerate and can accommodate up to four electrons:

$$1\sigma_g^2 1\sigma_u^2 2\sigma_g^2 2\sigma_u^2 1\pi_u^4 3\sigma_g^2 1\pi_g^4 3\sigma_u^2 4\sigma_g^2 4\sigma_u^2 2\pi_u^4 5\sigma_g^2 2\pi_g^4 5\sigma_u^0.$$

The bond order can be calculated as half of the difference between the number of electrons in bonding and electrons in antibonding molecular orbitals. Also recall that σ_g and π_u are bonding while σ_u and π_g are antibonding orbitals:

$$b = ½(2 - 2 + 2 - 2 + 4 + 2 - 4 - 2 + 2 - 2 + 4 + 2 - 4) = 1.$$

The bond order between two sulfur atoms in S_2^{2-} is one.

S16.2 In order to behave as a catalyst, Br^- must react with H_2O_2 (reaction 1, below) and then be regenerated (reaction 2):

1) $2Br^-(aq) + H_2O_2(aq) + 2H^+(aq) \rightarrow Br_2 + 2H_2O(l);$

2) $Br_2(aq) + H_2O_2(aq) \rightarrow 2Br^-(aq) + O_2(g) + 2H^+(aq).$

The net reaction is simply the decomposition of H_2O_2:

3) $2H_2O_2 \rightarrow O_2(g) + 2H_2O(l).$

The first reaction is the difference between the half-reactions:

$H_2O_2(aq) + 2H^+(aq) + 2e^- \rightarrow 2H_2O(l)$ $E^\ominus = +1.77$ V

$Br_2 + 2\,e^- \rightarrow 2\,Br^-$ $E^\ominus = +1.06$ V.

And therefore $E^\ominus_{cell} = +\,0.71$ V. This reaction is spontaneous because $E_{cell} > 0$.

The second reaction is the difference between the half-reactions:

$Br_2 + 2e^- \rightarrow 2Br^-$ $E^\ominus = +1.06$ V

$O_2(g) + 2\,H^+ + 2e^- \rightarrow H_2O_2$ $E^\ominus = +0.70$ V.

And therefore $E^\ominus_{cell} = +0.36$ V. This reaction is also spontaneous.

Because both reactions are spontaneous, the decomposition of H_2O_2 is thermodynamically favoured in the presence of Br^-.

Similar analysis applies to Cl^-. Starting from the analogous reactions 1) and 2) we have:

1) $2Cl^-(aq) + H_2O_2(aq) + 2H^+(aq) \rightarrow Cl_2 + 2H_2O(l);$

2) Cl_2 (aq) + H_2O_2 (aq) → $2Cl^-$(aq) + O_2(g) + $2H^+$(aq).

The net reaction is again simply the decomposition of H_2O_2:

3) $2H_2O_2 \rightarrow O_2(g) + 2H_2O(l)$

The first reaction is the difference between the half-reactions

$H_2O_2(aq) + 2H^+(aq) + 2e^- \rightarrow 2\ H_2O(l)$ $E^\circ = +1.77$ V

$Cl_2 + 2e^- \rightarrow 2Cl^-$ $E^\circ = +1.36$ V.

And therefore $E^\circ_{cell} = +0.41$ V. This reaction is spontaneous.

The second reaction is the difference between the half-reactions

$Cl_2 + 2e^- \rightarrow 2Cl^-$ $E^\circ = +1.36$ V

$O_2(g) + 2H^+ + 2e^- \rightarrow H_2O_2$ $E^\circ = +0.70$ V.

And therefore $E^\circ_{cell} = +0.66$ V. This reaction is spontaneous.

Because both reactions are spontaneous, the decomposition of H_2O_2 is thermodynamically favoured in presence of Cl^-.

S16.3 The Lewis structure of gaseous SO_3 is shown below. Note that SO_3 is a hybrid structure of three resonance structures. The SO_3 molecule is planar and belongs to the D_{3h} point group.

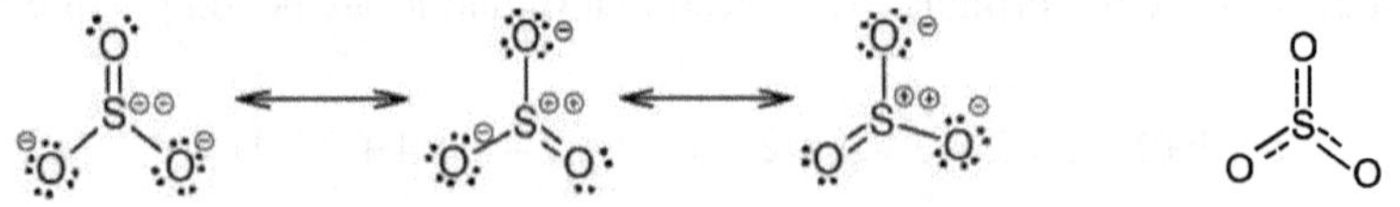

The Lewis structure of SO_3F^- anion is shown below. The molecular geometry is tetrahedral, but the point group is C_{3v}.

S16.4 It is reasonable to make the same assumption as in the Example: each S and N have a lone electron pair and form two bonds in the cyclic S_4N_4 molecule. That leaves four S atoms each still carrying an electron pair available for π bonding. This gives eight electrons from S atoms that could contribute to the π bonding. Four N atoms have one electron each, thus N atoms combined contribute four electrons. In total, there are 8 + 4 = 12 electrons in the system that could be used for π bonding. Consequently, S_4N_4 is aromatic because the $2n + 2$ rule is satisfied for n = 5: 2 × 5 + 2 = 12.

Exercises

E16.1 Remember that in general (with some exceptions) the oxides of non-metals are acidic and of metals are basic. Thus, CO_2, SO_3, and P_2O_5, are acidic, Al_2O_3 can react with either acids or bases, so it is considered amphoteric (an exception). CO is neutral (again an exception); and MgO and K_2O are basic. Review Section 5.4 if you had problems with this exercise.

E16.2 Refer to the Figures 2.12 and 2.17 for the distribution of molecular orbitals in O_2 molecule. From four valence orbitals on each oxygen atom we can form eight molecular orbitals; in order of increasing energy they are: $1\sigma_g$, $1\sigma_u$, $2\sigma_g$, $1\pi_u$, $1\pi_g$, $2\sigma_u$ (note that all π molecular orbitals are doubly degenerate). The bonding orbitals are $1\sigma_g$, $2\sigma_g$, and $1\pi_u$, whereas the rest are antibonding molecular orbitals. Two oxygen atoms in total have 12 valence electrons which, in O_2 molecule, can be distributed in the molecular orbitals as follows (obeying the Hund's rules and Pauli's exclusion principle): $1\sigma_g^2 1\sigma_u^2 2\sigma_g^2 1\pi_u^4 1\pi_g^2$. The bond order in O_2 molecule is ½[2 – 2 +2 + 4 – 2] = 2, and it corresponds to the bond length of 121 pm.

The cation O_2^+ is obtained when one electron is removed from the O_2 molecule. Thus, O_2^+ has the electronic configuration $1\sigma_g^2 1\sigma_u^2 2\sigma_g^2 1\pi_u^4\ 1\pi_g^1$—note that the electron is lost from the $1\pi_g$ antibonding orbital. The bond order in O_2^+ cation is ½[2 – 2 + 2 + 4 – 1] = 2.5, higher than in O_2 molecule. This higher bond order is reflected in a shorter bond length of 112 pm.

If we add two electrons to O_2 molecule we obtain the peroxide anion, O_2^{2-}, with the electronic configuration $1\sigma_g^2 1\sigma_u^2 2\sigma_g^2 1\pi_u^4 1\pi_g^4$—note that this time the electrons are added to the $1\pi_g$ antibonding orbital. The bond order for O_2^{2-} is ½[2 – 2 + 2 + 4 – 4] = 1. Of the three dioxygen species, O_2^{2-} has the lowest bond order and the longest bond: 149 pm.

E16.3 **a) The disproportionation of H_2O_2.** To calculate the standard potential for a disproportionation reaction, you must sum the potentials for the oxidation and reduction of the species in question. For hydrogen peroxide in acid solution, the oxidation and reduction are:

$$H_2O_2(aq) \rightarrow O_2(g) + 2e^- + 2H^+(aq) \quad E° = -0.695\ V$$

$$H_2O_2(aq) + 2e^- + 2H^+(aq) \rightarrow 2H_2O(l) \qquad E° = +1.763\ V.$$

Therefore, the standard potential for the net reaction $2H_2O_2(aq) \rightarrow O_2(g) + 2H_2O(1)$ is (–0.695 V) + (+1.763 V) = +1.068 V.

b) Catalysis by Cr^{2+}. Cr^{2+} can act as a catalyst for the decomposition of hydrogen peroxide if the Cr^{3+}/Cr^{2+} reduction potential falls between the values for the reduction of O_2 to H_2O_2 (+0.695 V) and the reduction of H_2O_2 to H_2O (+1.76 V). Reference to Resource Section 3 reveals that the Cr^{3+}/Cr^{2+} reduction potential is –0.424 V, so Cr^{2+} is *not* capable of decomposing H_2O_2. See also Example and Self-Test 16.1.

c) The disproportionation of HO_2. The oxidation and reduction of superoxide ion (O_2^-) in acid solution are:

$$HO_2(aq) \rightarrow O_2(g) + e^- + H^+(aq) \qquad E° = +0.125\ V$$

$$HO_2(aq) + e^- + H^+(aq) \rightarrow H_2O_2(aq) E° = +1.51\ V$$

Therefore, the standard potential for the net reaction $2HO_2(aq) \rightarrow O_2(g) + H_2O_2(aq)$ is (+0.125 V) + (+1.51 V) = +1.63 V. $\Delta_r G° = -\nu F E_{cell} = -1 \times 96500\ C\ mol^{-1} \times 1.63\ V = 157\ kJ\ mol^{-1}$. Similarly, we can obtain the Gibbs energy for the disproportionation of H_2O_2 (part **(a)**), $\Delta_r G° = 103\ kJ\ mol^{-1}$.

E16.4 Statement **(a)** is incorrect—this group exhibits alternation effect. Selenium, one of the group's middle elements, is not easy to oxidize to Se(VI) in contrast to both sulfur and tellurium.

Statement **(b)** is partially correct–the ground state of O_2 is a triplet state but it does not react as an electrophile in Diels–Alder reactions. The triplet ground state is $^3\Sigma_g^-$ with two unpaired electrons occupying different π^* ($1\pi_u$) molecular orbitals. The electrophilic O_2 molecule is one of the singlet states, the one that has two antiparallel electrons in one π^* molecular orbital, $^1\Delta_g$ singlet state. The triplet state reacts as a radical.

Statement **(c)** is partially correct—while ozone is considered a pollutant, it is very unlikely that the ozone pollution is caused by the diffusion of this gas from the stratosphere. Rather, this ozone is formed from other atmospheric pollutants (such as nitrogen oxides) in gas-phase atmospheric reactions.

E16.5 Hydrogen bonds are stronger in molecules where hydrogen is bonded directly to a highly electronegative atom such as O, N, or F. Since S has smaller electronegativity compared to oxygen, we expect O–H···S hydrogen bonds to be stronger.

E16.6 Using bond enthalpies from Table 2.7, we can estimate the enthalpies of formation for each compound. The relevant bond enthalpies are:

B(S–S) = 264 kJ mol^{-1},

B(F–F) = 155 kJ mol^{-1},

B(O=O) = 497 kJ mol^{-1},

B(H–H) = 436 kJ mol^{-1},

B(S–F) = 343 kJ mol^{-1},

B(O–F) = 185 kJ mol^{-1} and

B(S–H) = 338 kJ mol^{-1}.

The synthesis of SF_6 from the elements follows the equation:

$$1/8S_8 + 3F_2 \rightarrow SF_6.$$

Sulfur is taken in its most stable form, orthorhombic structure based on S_8 molecule. This cyclic molecule has eight S–S bonds. In one eighth of S_8 we have to break one S–S bond. We also have to break three F–F bonds but six S–F bonds are formed:

$$\Delta_f H(SF_6) = B(S\text{–}S) + 3 \times B(F\text{–}F) - 6 \times (S\text{–}F)$$

$$= 264 \text{ kJ mol}^{-1} + 3 \times 155 \text{ kJ mol}^{-1} - 6 \times 343 \text{ kJ mol}^{-1}$$

$$= -1329 \text{ kJ mol}^{-1}.$$

This negative, large $\Delta_f H$ indicates that SF_6 is a very stable compound.

The reaction for the synthesis of hypothetical OF_6 would follow the following equation:

$$1/2O_2 + 3F_2 \rightarrow OF_6$$

Similar analysis to the one above reveals that we have to break half a mole of O=O bond (note: this *does not* mean we are breaking a single O–O bond!), and three F–F bonds but we form six O–F bonds:

$$\Delta_f H(OF_6) = ½B(O\text{=}O) + 3 \times B(F\text{–}F) - 6 \times B(O\text{–}F)$$

$$= ½ \times 497 \text{ kJ mol}^{-1} + 3 \times 155 \text{ kJ mol}^{-1} - 6 \times 185 \text{ kJ mol}^{-1}$$

$$= -397 \text{ kJ mol}^{-1}.$$

For SH_6 we use the same approach as for SF_6, but replace F–F and S–F with H–H and S–H data:

$$\Delta_f H(SH_6) = B(S\text{–}S) + 3 \times B(H\text{–}H) - 6 \times B(S\text{–}H)$$

$$= 264 \text{ kJ mol}^{-1} + 3 \times 436 \text{ kJ mol}^{-1} - 6 \times 338 \text{ kJ mol}^{-1}$$

$$= -456 \text{ kJ mol}^{-1}.$$

The above calculations show that SF_6 is thermodynamically more stable than either OF_6 or SH_6 since it has a very negative $\Delta_f H$. The same calculations, however, indicate that both OF_6 and SH_6 are stable. But neither actually exists or can be prepared for reasons other than thermodynamics. In OF_6, oxygen would have an oxidation state of +6 which we know this element cannot achieve. Even if it could, its radius would be so small that fitting six F atoms around it would be sterically impossible. Although sulfur is sufficiently large to accommodate six H atoms around it, it still is very unlikely that it would be found in -6 oxidation state.

E16.7 Anionic species such as S_4^{2-} and Te_3^{2-} are intrinsically basic species that cannot be studied in solvents that are Lewis acids because complex formation (the Lewis acid–base pair) will destroy the independent identity of the anion. Since the basic solvent ethylenediamine will not react with Na_2S_4 or with K_2Te_3, it is a better solvent choice than sulfur dioxide.

E16.8 The three values of E^o for the $S_2O_8^{2-}/SO_4^{2-}$, the SO_4^{2-}/SO_3^{2-}, and the $SO_3^{2-}/S_2O_3^{2-}$ couples are +1.96 V, +0.158 V, and +0.400 V, respectively. Peroxydisulfate, $S_2O_8^{2-}$, is very easily reduced, so it is the strongest oxidizing agent. Sulfate dianion, SO_4^{2-}, is neither strongly oxidizing nor strongly reducing. Sulfite dianion, SO_3^{2-}, on the other hand, is relatively readily oxidized to sulfate, and hence is a moderate reducing agent.

E16.9 The reduction potential of sulfite ions in basic solution = –0.576 V. Using the data in Resource Section 3 for Mn we can construct the following table:

Oxidation states of Mn	Standard reduction potential of Mn couples	$E^\ominus_{cell} = E^\ominus_{red} - E^\ominus_{ox}$
7	+7 to +6 = +0.56 V	+1.136 V
6	+6 to +5 = +0.27 V	+0.847 V
5	+5 to +4 = +0.93 V	+1.506 V
4	+4 to +3 = +0.15 V	+0.726 V
3	+3 to +2 = –0.25 V	+0.326 V
2	+2 to 0 = –1.56 V	–0.984 V

From the table we can see that except the +2 oxidation state of Mn, all other oxidation states of Mn will be reduced by sulfite ions in basic solution because in each case we can obtain $E_{cell} > 0$, indicating spontaneous reaction.

E16.10 **a)** The formulas for these ions in acidic solution are $H_5TeO_6^-$ and HSO_4^- (the parent acids are H_6TeO_6, or $Te(OH)_6$, and H_2SO_4).

b) Tellurium(VI) is a much larger centre than sulfur(VI) and can readily increase its coordination number (increase from 4 in S(VI) to 6 in Te(VI)). This trend is found in other groups in the p-block.

E16.11 If the potential to the right is more positive than the potential to the left, the species is unstable and likely to disproportionate. $S_2O_6^{2-}$ and $S_2O_3^{2-}$ are unstable with respect to disproportionation.

E16.12 In general, from Latimer diagrams the stability of a given species against disproportionation depends on whether the right-hand side potential is greater or smaller than the left-hand side potential. If the right-hand side potential is greater than the left-hand side potential, the species is inherently unstable against disproportionation. Now looking at the Latimer diagram for SeO_3^{2-} in both acidic and basic solutions, the right-hand side potential is smaller than the left-hand side potential indicating that SeO_3^{2-} is stable both in acidic as well as basic solution. However, if one wants to estimate the relative stability then it would be instructive to calculate the overall potentials for disproportionation reaction.

$$E^\ominus = E^\ominus_{reduction} - E^\ominus_{oxidation}.$$

For basic solution:

Reduction potential for SeO_3^{2-}/Se = –0.36 V

Reduction potential for SeO_4^{2-}/SeO_3^{2-} = +0.03 V

$$E^\ominus_{selenate\ ion} = -0.36\ V - (+0.03\ V)$$

$$= -0.39\ V.$$

For acidic solution:

Reduction potential for SeO_3^{2-}/Se = +0.74 V

Reduction potential for SeO_4^{2-}/SeO_3^{2-} = +1.15 V

$$E^\ominus_{selenate\ ion} = +0.74\ V - (+1.15\ V)$$

$$= -0.41\ V$$

From this it can be concluded that the SeO_3^{2-} is marginally more stable in acid solutions because $E^\ominus$ is more positive.

E16.13 The half-potentials for the reduction of VO^{2+}, Fe^{3+}, and Co^{3+} are +1.00 V, +0.771 V, and +1.92 V, respectively. All will be reduced as their potentials are greater than the potential of thiosulfate.

E16.14 SF_3^+ has a trigonal pyramidal molecular geometry with tetrahedral electron pair geometry, BF_4^- tetrahedral (both molecular and electron pair geometry).

E16.15 The Frost diagram reveals that the stability of the lowest oxidation state is decreasing descending the group. Oxygen is very stable in oxidation state 2- and species containing oxygen in this state will be prevalent. Looking at the diagram overall, the same can be said for the other Group 16 elements: negative two oxidation state is the lowest on the diagram (besides the elemental form for all) and sulfides, selenides and tellurides should be the most stable compounds of these elements. Descending the group (starting with S already) positive oxidation states are accessible. All have good oxidizing properties. In the case of tellurium +4 and -2 have about the same stability. This is a consequence of the increased stability of higher oxidation states as we descend a group.

E16.16 **a)** $SF_4 + (CH_3)_4NF \rightarrow [(CH_3)_4N]^+[SF_5]^-$

The above reaction is balanced once the stoichiometry of the reaction is determined: convert the masses of the reagents to moles and see that they react in 1 : 1 molar ratio.

b) Square pyramidal structure (see structure below)

c) Two F environments (basal and axial, F_b and F_a respectively on the structure below) would result in two signals in the ^{19}F NMR. The F_a resonance, with intensity 1, would show as a quintet due to the coupling to four equivalent basal F atoms. The F_b resonance, with intensity 4, would appear as a doublet due to the coupling to one axial F atom.

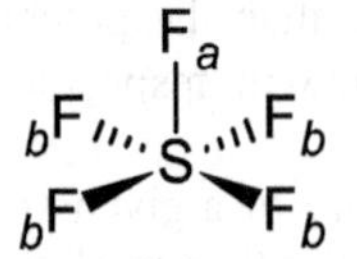

E16.17

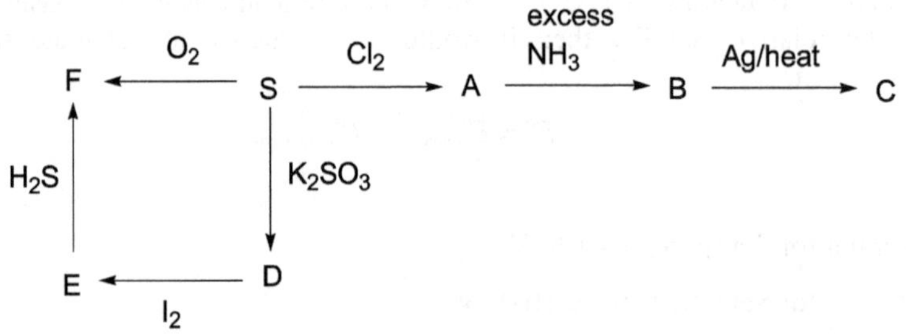

A = S_2Cl_2, **B** = S_4N_4, **C** = S_2N_2, **D** = $K_2S_2O_3$, **E** = $S_2O_6^{2-}$, **F** = SO_2

Note: Sulfur reacts with Cl_2 to produce both S_2Cl_2 and SCl_2. However, of the two possible chlorides, only S_2Cl_2 can be taken in further reactions with NH_3 (to give S_4N_4) and Ag (to produce S_2N_2 from S_4N_4). Thus, the acceptable product **A** is S_2Cl_2.

E16.18 **a)** $S_3N_3^-$ has three electron pairs from S atoms, three electrons from N atoms, and one electron from the negative charge available for π bonding. In total, there are 10 electrons available for π bonding, and this anion is conforming to the $2n + 2$ rule with $n = 4$ ($2 \times 4 + 2 = 10$) and it is expected to exhibit inorganic aromaticity.

b) The $S_4N_3^+$ cation has, in total, 10 electrons available for π bonding: eight from electron pairs on four S atoms, three electrons from three N atoms, and minus one for positive charge. It is expected to be aromatic because it obeys the $2n + 2$ rule with $n = 4$.

c) S_5N_5 has 15 electrons available for π bonding: 10 from five sulfur atoms and five from five N atoms. It is not expected to be aromatic because it does not obey the $2n + 2$ rule.

E16.19 Sulfuric and selenic acids are rather similar. When pure, both are hygroscopic, very viscous liquids. Both form crystalline hydrates. Sulfuric and selenic acids react with their anhydrides, SO_3 and SeO_3, to produce trisulfuric and triselenic acids respectively. Much like sulfuric acid, selenic acid is a strong acid in the first dissociation step, and the acids have the same order of magnitude for Ka_2 (~10^{-2}). Both produce two types of salts: hydrogensulfates/sulfates and hydrogenselentates/selenates. The salts also have similar solubilities. The most important difference between the two acids is their oxidizing power: H_2SeO_4 is a very strong oxidizing agent capable even of oxidizing gold and palladium.

Telluric acid is remarkably different from sulfuric and selenic acid. When pure, it is a solid composed of discrete $Te(OH)_6$ octahedra. This structure is retained in the solution as well. As it can be expected, it forms more than two types of salts: $H_5TeO_6^-$, $H_4TeO_6^{2-}$, $H_2TeO_4^{4-}$, and TeO_6^{6-}. It is a weak acid even in the first dissociation step. At elevated temperatures it loses water molecules to produce various polytelluric acids. What is similar to sulfuric and selenic acid is that $Te(OH)_6$ can produce well-defined, crystalline hydrates. Telluric acid is a weaker oxidizing agent than selenic acid in both acidic and alkaline medium, but is a stronger oxidant than sulfuric acid in either medium.

Chapter 17 The Group 17 Elements

Self-Tests

S17.1 A quick glance at the Latimer diagram for chlorine in acidic solution (Resource Section 3) shows that $HClO_2$ can disproportionate to ClO_3^- and HClO:

$$2HClO_2(aq) \rightarrow ClO_3^-(aq) + HClO(aq).$$

S17.2 The anion Cl_3^- has 2 valence electrons to account for: 3 × 7 = 21 for Cl atoms and one more for the negative charge. This anion, thus, as a trigonal bipyramidal electron geometry but the shape is linear:

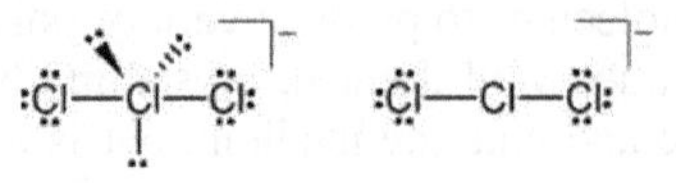

S17.3 Since the IO_3^-/I_2 standard reduction potential is +1.19 V (see Resource Section 3), many species can be used as reducing agents, including $SO_2(aq)$ and $Sn^{2+}(aq)$: the HSO_4^-/H_2SO_3 and Sn^{4+}/Sn^{2+} reduction potentials are +0.158 V and +0.15 V, respectively. The balanced equations would be as follows:

$$2IO_3^- + 5H_2SO_3 \rightarrow I_2 + 5HSO_4^- + H_2O + 3H^+$$

$$2IO_3^- + 5Sn^{2+} + 12H^+ \rightarrow I_2 + 5Sn^{4+} + 6H_2O.$$

The reduction of iodate with aqueous sulfur dioxide would be far cheaper than reduction with Sn^{2+} because sulfur dioxide costs less than tin. One reason this is so is because sulfuric acid, for which SO_2 is an intermediate, is prepared worldwide on an enormous scale.

S17.4 The ordering of molecular orbitals in F_2 is the same as in Cl_2. The difluorine molecule has 14 valence electrons and the following electronic configuration:

$$1\sigma_g^2 1\sigma_u^2 2\sigma_g^2 1\pi_u^4 1\pi_g^4.$$

The bond order is:

$$b = ½(8 - 6) = 1.$$

And just like in Cl_2 molecule from Example 17.4, F–F is a single bond.

The anion F_2^- has one more valence electron—15 in total—and the electronic configuration would be:

$$1\sigma_g^2 1\sigma_u^2 2\sigma_g^2 1\pi_u^4 1\pi_g^4 2\sigma_u^1.$$

The bond order is:

$$b = ½(8 - 7) = 0.5.$$

Since F_2^- has a lower bond order than F_2, the F–F bond is longer in the anion, F_2^-.

S17.5 Iodine is the central atom in the molecule. Seven valence electrons form bonds with seven fluorine atoms, giving a total of seven bonding electron pairs, and structure is pentagonal bipyramid.

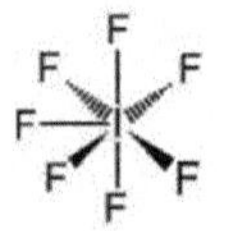

The ^{19}F NMR spectrum of IF_7 contains two resonances. One is a sextet for two axial F atoms (coupled to five equatorial F atoms), and another one is a triplet due to the five equatorial F atoms (coupled to two axial F atoms). The intensity ratios of the two resonances is 1 : 5.

S17.6 The reaction is:

$$ClF_3 + SbF_5 \rightarrow [ClF_2][SbF_6].$$

The shape of the reactants:

ClF_3: This is an XY_3 interhalogen. The central Cl atom is surrounded by five electron pairs: two are lone electron pairs and three are boding pairs resulting in a trigonal bipyramidal electron geometry and a distorted T-shaped molecular geometry.

SbF_5: There are five electron pairs around the central atom, Sb, all of which are bonding pairs resulting in a trigonal bipyramidal geometry.

The shape of the products:

$[ClF_2]^+$: After the removal of one F^- during the reaction, the central Cl atom is now surrounded by four electron pairs: two bonding and two lone electron pairs. This results in a tetrahedral electron group geometry and a bent molecular shape.

$[SbF_6]^-$: After bonding to one F^- the central Sb now has six electron pairs around it, all of which are bonding. This results in an octahedral molecular shape.

The structures of all four are shown below:

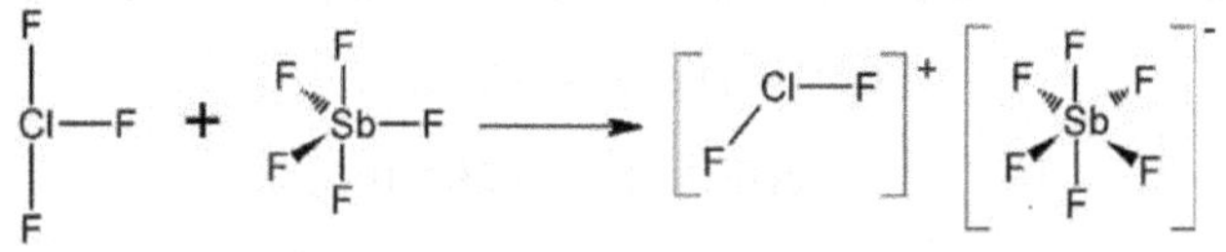

S17.7 Examples include I_3^-, IBr_2^-, ICl_2^-, and IF_2^-. One way of describing the bonding in these species is based on Lewis acids and bases. In all cases, consider that the central I^+ species has a vacant orbital (and is a Lewis acid) that interacts with two electron pair donors (Lewis bases): pyridine (C_5H_5N) in the case of $[py–I–py]^+$ or I^- in the case of I_3^-.

In terms of VSEPR and molecular orbital theories, the same analysis applies as in Exercise 17.7. The listed species have 10 valence electrons at central I+. This results in trigonal bipyramidal electron geometry with linear shapes for the complexes. The four electrons in two bonds are also placed in two molecular orbitals, one bonding, and one nonbonding. The antibonding orbital remains empty.

S17.8 **a)** Bromine species that will disproportionate in a basic solution are Br_2 and BrO^-:

$$Br_2 + 2OH^- \rightarrow Br^- + BrO^- + H_2O$$

$$BrO^- + 2OH^- \rightarrow Br^- + BrO_3^- + H_2O$$

Note that the disproportionation of Br_2 is likely going to produce Br^- and BrO_3^- because BrO^-, the initial product, is also prone to disproportionation.

b) Iodine species that will disproportionate in a basic solution are: I_2 and IO^-:

$$I_2 + 2OH^- \rightarrow I^- + IO^- + H_2O$$

$$IO^- + 2OH^- \rightarrow I^- + IO_3^- + H_2O$$

with the similar note pertaining to the I_2 disproportionation product as in **(a)** for bromine above.

Exercises

E17.1 The table summarizes the facts:

		Physical state	Electronegativity	Hardness of halide ion	Colour
Fluorine	F_2	gas	highest (4.0)	hardest	light yellow
Chlorine	Cl_2	gas	lower	softer	yellow-green
Bromine	Br_2	liquid	lower	softer	dark red-brown
Iodine	I_2	solid	lowest	softest	dark violet

E17.2 Since chlorine and fluorine have strongly positive reduction potentials, the oxidation of the chloride and fluoride anions require very strong oxidizing agents. As a result, only electrolytic oxidation of Cl^- and F^- is commercially reasonable.

The principal source of fluorine is CaF_2. It is converted to HF by treating it with a strong acid such as sulfuric acid. Liquid HF is electrolysed to H_2 and F_2 in an anhydrous cell, with KF as the electrolyte:

$$CaF_2 + H_2SO_4 \rightarrow CaSO_4 + 2HF;$$

$$2HF + 2KF \rightarrow 2K^+HF_2^-;$$

$$2K^+HF_2^- + \text{electricity} \rightarrow F_2 + H_2 + 2KF.$$

The principal source of the other halides is seawater (natural brines). Chlorine is liberated by the electrolysis of aqueous NaCl (the chloralkali process):

$$2Cl^- + 2H_2O + \text{electricity} \rightarrow Cl_2 + H_2 + 2OH^-.$$

Bromine and iodine are prepared by treating aqueous solutions of the halides with chlorine:

$$2X^- + Cl_2 \rightarrow X_2 + 2Cl^- \quad (X^- = Br^-, I^-).$$

Cl_2 is a stronger oxidizing reagent than Br_2 and I_2, so it can be used as an oxidant to oxidize Br^- and I^-.

E17.3 Follow the procedure described in Exercise 17.4 and its accompanying Self-Test. The Br_2 molecule has 14 valence electrons (just like Cl_2 and F_2) and the following simplified electronic configuration:

$$1\sigma_g^2 1\sigma_u^2 2\sigma_g^2 1\pi_u^4 1\pi_g^4.$$

This gives bond order 1 and a single Br–Br bond.

Br_2^+ has one valence electron less (13 in total) and the following simplified electronic configuration:

$$1\sigma_g^2 1\sigma_u^2 2\sigma_g^2 1\pi_u^4 1\pi_g^3.$$

This electronic configuration results in bond order 1.5. Since the cation has a higher bond order, Br–Br bond in Br_2^+ is shorter than in Br_2.

E17.4 A drawing of the cell is shown in Figure 17.4. Note that Cl_2 is liberated at the anode and H_2 is liberated at the cathode, according to the following half-reactions:

$$\text{anode: } 2Cl^-(aq) \rightarrow Cl_2(g) + 2e^-$$

$$\text{cathode: } 2H_2O(l) + 2e^- \rightarrow 2OH^-(aq) + H_2(g).$$

To maintain electroneutrality, Na^+ ions diffuse through the membrane. Because of the chemical properties of the membrane, anions such as Cl^- and OH^- cannot diffuse through it. If OH^- ions did diffuse through the membrane, they would react with Cl_2 and spoil the yield of the electrolysis because Cl_2 easily disproportionates in a basic medium:

$$2OH^-(aq) + Cl_2(aq) \rightarrow ClO^-(aq) + Cl^-(aq) + H_2O(l).$$

E17.5 According to Figure 17.5, the vacant anti-bonding orbital of a halogen molecule is $2\sigma_u$* and is composed primarily of halogen atomic p orbitals; recall that the *ns*–*np* gap is larger toward the right-hand side of the periodic table, and the larger the gap, the smaller the amount of s–p mixing. A sketch of the empty $2\sigma_u$* antibonding orbital is shown below:

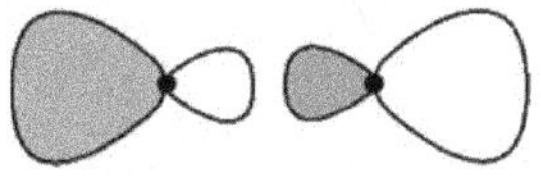

$2\sigma_u$* (antibonding)

Since the $2\sigma_u$* antibonding orbital is empty (hence it is LUMO for an X_2 molecule), it is the orbital that accepts the pair of electrons from a Lewis base B when a dative $B{:}\rightarrow X_2$ bond is formed. From the shape of the LUMO, we can conclude that the B–X–X unit should be linear.

17.6 The standard reduction potentials for different dihalogens and H_2O are:

$I_2(s) + 2e^- \rightarrow 2I^-(aq)$ $\quad E^\circ = +0.53$ V

$Br_2(l) + 2e^- \rightarrow 2Br^-(aq)$ $\quad E^\circ = +1.06$ V

$Cl_2(g) + 2e^- \rightarrow 2Cl^-(aq)$ $\quad E^\circ = +1.36$ V

$F_2(g) + 2e^- \rightarrow 2F^-(aq)$ $\quad E^\circ = +2.85$ V

$O_2(g) + 2H^+ + 4e^- \rightarrow 2H_2O(l)$ $\quad E^\circ = +1.23$ V

For a halogen to be thermodynamically capable of oxidizing water, the following reaction

$$X_2 + H_2O(l) \rightarrow 2HX(aq) + \tfrac{1}{2}O_2(g)$$

should have $E^\circ_{cell} = E^\circ_{red} - E^\circ_{ox} > 0$. From the above given data, the reduction potentials of I_2 and Br_2 are lower than the reduction potential of H_2O; thus, for these two halogens $E^\circ_{cell} < 0$. The reduction potentials of Cl_2 and F_2 are higher than the reduction potential of H_2O, and for them $E^\circ_{cell} > 0$. Therefore, Cl_2 and F_2 are *thermodynamically* capable of oxidizing H_2O to O_2.

E17.7 **a) The difference in volatility.** Ammonia is one of many substances that exhibit very strong hydrogen bonding. The very strong $N–H^{\delta+}\cdots N^{\delta-}$ intermolecular interactions lead to a relatively large enthalpy of vaporization and a relatively high boiling point for NH_3 (–33°C). In contrast, the intermolecular forces in liquid NF_3 are relatively weak dipole–dipole forces. In comparison to hydrogen bonding, dipole–dipole forces require less energy to be broken, resulting in a smaller enthalpy of vaporization and a low boiling point (–129°C) for NF_3.

b) Explain the difference in basicity. The strong electron-withdrawing effect of the three fluorine atoms in NF_3 lowers the energy of the nitrogen atom lone pair. This lowering of energy has the effect of reducing the electron-donating ability of the nitrogen atom in NF_3, reducing the basicity.

E17.8 **a) The reaction of $(CN)_2$ with NaOH.** When a halogen such as chlorine is treated with an aqueous base, it undergoes disproportionation to yield chloride ion and hypochlorite ion, as follows:

$$Cl_2(aq) + 2OH^-(aq) \rightarrow Cl^-(aq) + ClO^-(aq) + H_2O(l).$$

The analogous reaction of cyanogen with base is:

$$(CN)_2(aq) + 2OH^-(aq) \rightarrow CN^-(aq) + NCO^-(aq) + H_2O(l).$$

The linear NCO^- ion is called cyanate.

b) The reaction of SCN^- with MnO_2 in aqueous acid. If MnO_2 is an oxidizing agent, then the probable reaction is oxidation of thiocyanate ion to thiocyanogen, $(SCN)_2$, coupled with reduction of MnO_2 to Mn^{2+} (if you do not recall that Mn^{3+} is unstable to disproportionation, you will discover it when you refer to the Latimer diagram for manganese in Resource Section 3). The balanced equation is:

$$2SCN^-(aq) + MnO_2(s) + 4H^+(aq) \rightarrow (SCN)_2(aq) + Mn^{2+}(aq) + 2H_2O(l).$$

c) The structure of trimethylsilyl cyanide. Just as halides form Si–X single bonds, trimethylsilyl cyanide contains an Si–CN single bond. Its structure is shown below:

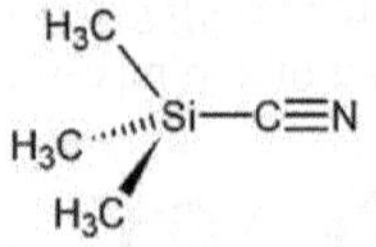

E17.9 **a)** IF_3 (1.84 g, 10 mmol) + $[(CH_3)_4N]F$ (0.93 g, 10 mmol) $\longrightarrow$ $[N(CH_3)_4][IF_4]$ (10 mmol)

Note that IF_3 and $[(CH_3)_4N]F$ react in a 1 : 1 ratio and that only one product X is formed; therefore the product X is $[N(CH_3)_4][IF_4]$.

b) A reliable way to predict the structure of IF_3 is to draw its Lewis structure and then apply VSEPR theory. In the Lewis structure for IF_3, iodine is the central atom. Three bonding pairs and two lone pairs on the iodine yield trigonal bipyramidal electron geometry and T-shaped molecular geometry, as shown below.

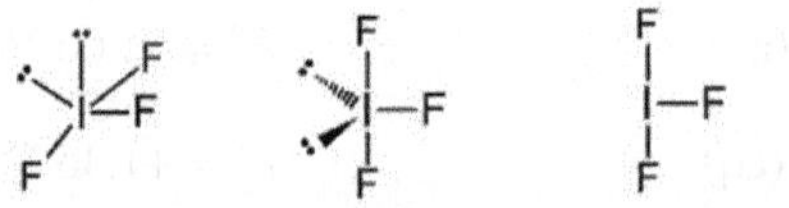

The shape of anion IF_4^- is square planar (see the structure below). The iodine atom is at the centre of the ion. Four electron pairs form bonds to F atoms, and two electron pairs do not take part in bonding.

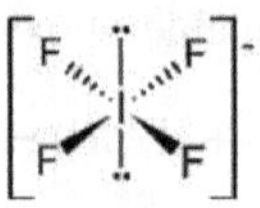

The shape of cation $(CH_3)_4N^+$ is tetrahedral. The nitrogen atom is at the centre of the cation. Nitrogen has five valence electrons. Three electrons form bonds to CH_3, plus one lone pair forms coordinate covalent bond with CH_3^+.

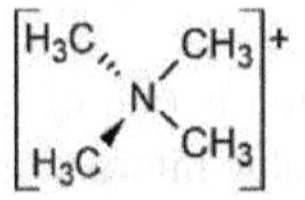

c) The ^{19}F NMR spectrum of IF_3 contains two resonances: one doublet is for axial F atoms and another triplet is for an equatorial F atom.

The ^{19}F NMR spectrum of IF_4^- contains one resonance, a singlet.

E17.10 As a stronger oxidizing agent, ozone is going to oxidize Br_2. The product of oxidation is a bromine oxide that can be identified based on the colour as BrO_2, or more specifically a mixed Br(I)–Br(VII) oxide, $BrOBrO_3$:

$$Br_2 + 4O_3 \rightarrow BrOBrO_3 + 4O_2.$$

The reaction with the equimolar NaOH produces two anions:

$$6BrOBrO_3 + 6OH^- \rightarrow 5BrO_3^- + Br^- + 3H_2O.$$

Note that this reaction can be seen as a disproportionation reaction since the average oxidation state of Br in $BrOBrO_3$ is +4; after the reaction with OH^- we have two Br species: one has Br in oxidation state +5, the other -1.

E17.11 $SbCl_5$ is trigonal bipyramidal, and $FClO_3$ is tetrahedral.

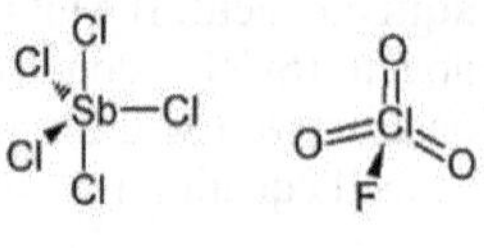

E17.12 SbF_5 is a strong Lewis acid and extracts a fluoride ion from the interhalogen compound ClF_5, leading to the formation of $[ClF_4]^+[SbF_6]^-$. The structures are shown below:

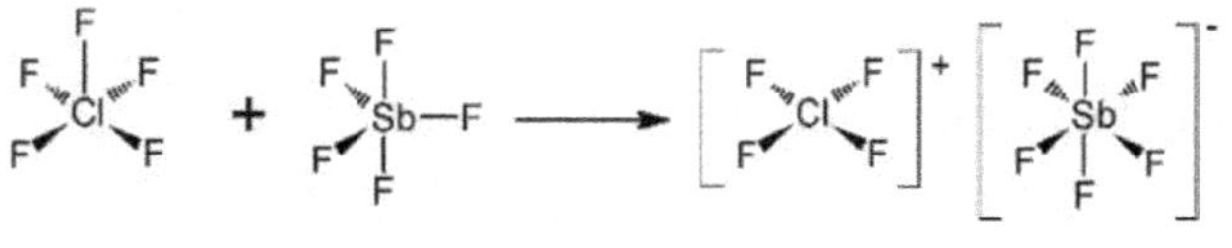

E17.13 The complex MCl_4F_2 can have two isomers: *cis* and *trans* (see below). Both isomers have one resonance in ^{19}F NMR. MCl_3F_3 also has two isomers: *fac* and *mer*. The *fac* isomer has one resonance and the *mer* isomer has two resonances—one will appear as a doublet for two mutually *trans* F atoms coupled to the *cis* F atom, and one triplet for the *cis* F atom coupled to two mutually *trans* F atoms.

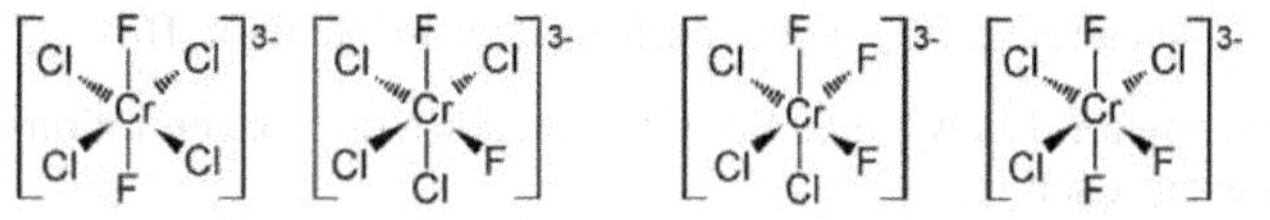

trans-$[CrCl_4F_2]^{3-}$ *cis*-$[CrCl_4F_2]^{3-}$ *fac*-$[CrCl_3F_3]^{3-}$ *mer*-$[CrCl_3F_3]^{3-}$

17.14 **a) The structures of $[IF_6]^+$ and IF_7.** The structures are shown below. The structures represent the Lewis structures as well, except that the three lone pairs of electrons on each fluorine atom have been omitted. VSEPR theory predicts that a species with six bonding pairs of electrons, such as IF_6^+, should be octahedral. Possible structures of species with seven bonding pairs of electrons, like IF_7, were not covered in Section 2.3, but a reasonable and a symmetrical structure would be a pentagonal bipyramid.

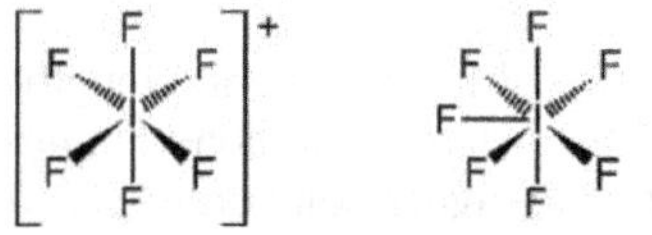

b) The preparation of $[IF_6][SbF_6]$. Judging from the structures shown above, it should be possible to abstract an F^- ion from IF_7 to produce IF_6^+. The strong Lewis acid SbF_5 should be used for the fluoride abstraction so that the salt $[IF_6][SbF_6]$ will result:

$$IF_7 + SbF_5 \rightarrow [IF_6][SbF_6].$$

17.15 I_2Cl_6 is the dimer of ICl_3 with two Cl atoms acting as bridges. Therefore, each iodine atom contains 12 valence electrons, or four bonding pairs and two non-bonding pairs. VSEPR theory predicts that the electron geometry around each iodine atom with six pairs of electrons should be octahedral, and with four bonding and two non-bonding electron pairs, the molecular geometry should be square planar for each iodine atom. In this structure four Cl atoms occupy the corners of the square plane. In the solid state, the compound is a planar dimer (I_2Cl_6) with bridging Cl atoms.

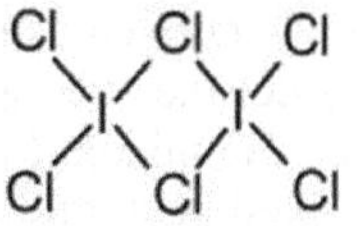

The point group of I_2Cl_6 is D_{2h}.

17.16 The Lewis structure for ClO_2F is shown below. (Note that chlorine is the central atom, not oxygen or fluorine. The heavier, less electronegative element is always the central atom in interhalogen compounds and in compounds containing two different halogens and oxygen). Three bonding pairs and one lone pair yield a trigonal pyramidal molecular geometry, also shown below. This structure possesses only a single symmetry element, a mirror plane that bisects the O–Cl–O angle and contains the Cl and F atoms; therefore, ClO_2F has a C_s symmetry.

The Lewis structure of ClO_2F:

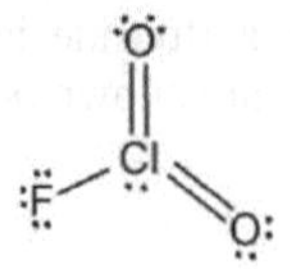

The trigonal pyramidal structure of ClO_2F:

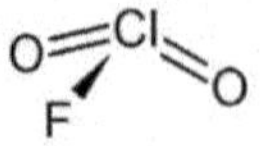

E17.17 The acid/base properties of liquid BrF_3 result from the following autoionization reaction:

$$2BrF_3 \rightarrow BrF_2^+ + BrF_4^-.$$

The acidic species is the cation BrF_2^+ and the basic species is the anion BrF_4^-.

a) SbF_5? Adding the powerful Lewis acid SbF_5 will increase the concentration of BrF_2^+, thus increasing the acidity of BrF_3 by the following reaction:

$$BrF_3 + SbF_5 \rightarrow BrF_2^+ + SbF_6^-.$$

b) SF_6? Adding SF_6 will have no effect on the acidity or basicity of BrF_3 because SF_6 is neither Lewis acidic nor Lewis basic.

c) CsF? Adding CsF, which contains the strong Lewis base F^-, will increase the concentration of BrF_4^-, thus increasing the basicity of BrF_3, by the following reaction:

$$BrF_3 + F^- \rightarrow BrF_4^-.$$

E17.18 The structure of I_5^+ cation reveals two different I–I bond lengths: longer central I–I distances (290 pm) and shorter terminal I–I distances (268 pm). The difference in bond lengths can be explained by different bond order—recall that shorter bond lengths correspond to higher bond orders. You can also note that the shorter bond distances in I_5^+ are identical to I–I distances in I_3^+ (structure **4**) and that the bond angles are about the same. This gives us one description of I_5^+ structure as a hybrid of two structures:

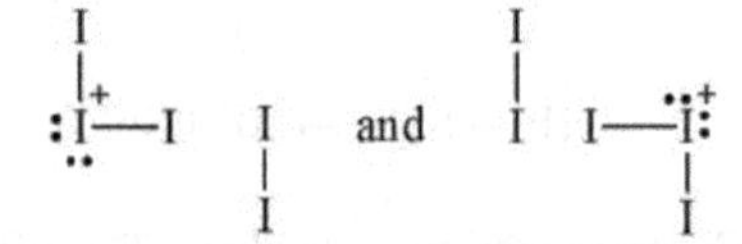

This description uses two-centre bonding with the overall hybrid structure having bond order lower than one for the three central I atoms accounting for the longer bond length between these atoms. Also note that in each structure the positively charged central I atom has four electron pairs—two bonding and two lone pairs—which results in overall bent geometry at those I atoms. This, combined with the relatively large size of an iodine atom, explains the fact that the bond angles are around 90°. Consequently, the structure conforms to VSEPR theory.

A bit more rigorous bonding analysis can be derived using simplified molecular orbital approach. If we place the z-axis along I–I–I bonds then we can consider the overlap of three p_z atomic orbitals on three I atoms. In this case we are going to place a positive charge on the middle I atom and make the empty p orbital on this I^+ a p_z atomic orbital. Two I atoms on either side have filled p_z atomic orbitals. Starting from these three atomic orbitals we can produce three molecular orbitals: a bonding, a non-bonding, and an antibonding molecular orbital. The bonding molecular orbital is going to be filled and will have electron density between all three I nuclei giving a three-centre bonding description. Non-bonding orbital will also be filled but would be mostly located on I atoms on either side of the central cation, thus contributing little to bonding (a more rigorous computational analysis can reveal that this orbital is actually slightly bonding in character). The non-bonding molecular orbital is empty. The central I–I bonds are longer because the bonding orbital contains only two electrons but spans three nuclei. The terminal I–I bonds are shorter because they are typical two-centre, two-electron bonds.

E17.19 Iodine is the central atom in IF_5. Five unpaired electrons form bonds with five fluorine atoms, plus one lone electron, giving a total of six electron pairs. Therefore, the electronic geometry is octahedral with one position occupied by a lone pair giving a square pyramidal arrangement of atoms.

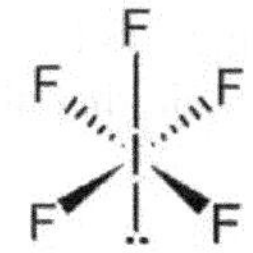

The structure of the cation IF_5^+ is similar—removal of one electron from IF_5 converts a lone electron pair into an unpaired electron. There will be less repulsion between a lone electron and bonding electron pairs in I–F bonds but the arrangement of atoms would remain square pyramidal. Thus, the ^{19}F NMR spectrum of IF_5^+ contains two resonances—one is a doublet for four basal F atoms coupled to the axial F, and another one is a quintet for the axial F atom coupled to the four basal F atoms.

E17.20 **a) SbF_5.** All interhalogens, including BrF_3, are strong oxidizing agents. The antimony atom in SbF_5 is already in its highest oxidation state, and the fluorine cannot be chemically oxidized (fluorine is the most electronegative element). Since SbF_5 cannot be oxidized, it will not form an explosive mixture with BrF_3.

b) CH_3OH. Methanol, being an organic compound, is readily oxidized by strong oxidants. Therefore, you should expect it to form an explosive mixture with BrF_3.

c) F_2. Both F_2 and BrF_3 are strong oxidizing agents. However, F_2 can oxidize Br from +3 in BrF_3 and produce BrF_5 with Br in oxidation state +5. The two will not form an explosive mixture but will react.

d) S_2Cl_2. In this compound, sulfur is in the +1 oxidation state. Recall that SF_4 and SF_6 are stable compounds. Therefore, you should expect S_2Cl_2 to be oxidized to higher valent sulfur fluorides. The mixture $S_2Cl_2 + BrF_3$ will be an explosion hazard.

E17.21 Since Br_3^- is only moderately stable it is likely going to produce Br_2 and Br^- in solution. IBr is, however, fairly stable. Hence, the probable reaction will be:

$$Br_3^- + I_2 \rightarrow 2IBr + Br^-.$$

The IBr may associate to some extent with Br^- to produce IBr_2^-.

E17.22 This is an example of an important principle of inorganic chemistry that you have encountered before: large cations stabilize large anions. A detailed analysis of this situation is as follows. The enthalpy change for the reaction

$$NaI_3(s) \rightarrow NaI(s) + I_2(s)$$

is negative, which is just another way of stating that NaI_3 is not stable. This enthalpy change for the above reaction is composed of four terms:

$$\Delta_{rxn}H = \Delta_{latt}H(NaI_3) + B([I–I_2]^-) - \Delta_{latt}H\,(I_2(s)) - \Delta_{latt}H(NaI(s)).$$

The fourth term, the lattice enthalpy for NaI(s), is larger than the first term, the lattice enthalpy for $NaI_3(s)$ because I^- is smaller than I_3^-. This difference provides the driving force for the reaction to occur as written. If you substitute Cs^+ for Na^+, the two middle terms will remain constant. Now, the fourth term is still larger than the first term, *but by a significantly smaller amount.* The net result is that $\Delta_{rxn}H$ for CsI_3 is positive. In the limit where the cation becomes infinitely large, the difference between the first and fourth terms becomes negligible.

E17.23 **a) ClO_2.** The Lewis structure and the predicted shape of ClO_2 are shown below. The bent shape is a consequence of repulsions between the bonding and non-bonding electrons. With three non-bonding electrons, the O–Cl–O angle in ClO_2 is 118°. With four non-bonding electrons, as in ClO_2^-, the repulsions are greater and the O–Cl–O angle is only 111°. The point group is C_{2v}.

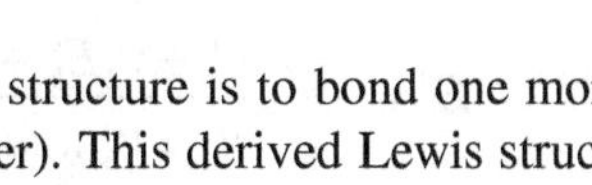

b) I_2O_6. The easiest way to determine this structure is to bond one more oxygen atom to any iodine atoms in the structure of I_2O_5 (structure 15 in this chapter). This derived Lewis structure and the predicted molecular geometry of I_2O_6 are shown below. The central O atom has two bonding pairs of electrons and two lone pairs, like H_2O, so it should be no surprise that the I–O–I bond angle is less than 180°. One iodine atom is trigonal pyramidal because it has three bonding pairs and one lone pair; while the other iodine atom, with four bonding pairs, is tetrahedral. The point group is C_s for the geometry shown.

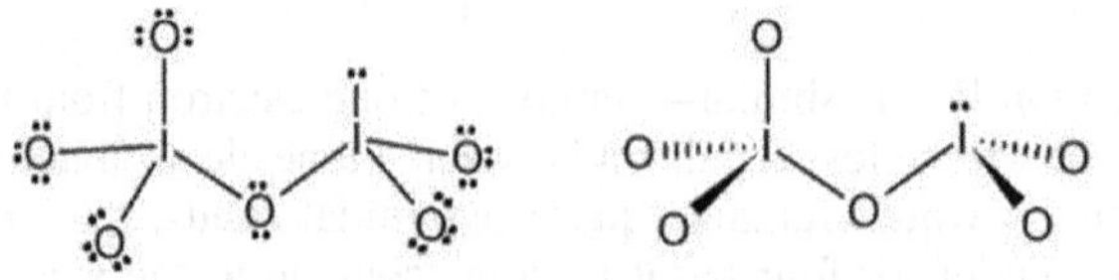

E17.24 **a) Formulas and acidities.** The formulas are $HBrO_4$ and H_5IO_6. The difference lies in iodine's ability to expand its coordination shell, a direct consequence of its large size.

Recall from Section 5.3 that the relative strength of an oxoacid can be estimated from the *q*/*p* ratio (*p* is the number of oxo groups and *q* is the number of OH groups attached to the central atom). A high value of *p*/*q* correlates with strong acidity, whereas a low value correlates with weak acidity. For $HBrO_4$, $p/q = 3/1$, so it is a strong acid. For H_5IO_6, $p/q = 1/5$, so it is a weak acid.

b) Relative stabilities. Periodic acid is thermodynamically more stable with respect to reduction than perbromic acid. Bromine, like its period 4 neighbours arsenic and selenium, is more oxidizing in its highest oxidation state than the members of the group immediately above and below. $HBrO_4$ also releases O_2 when heated (leaving $HBrO_3$ behind). H_5IO_6 loses water and condenses to polyperiodic acids.

E17.25 **a) The expected trend.** Looking at Figure 17.14, we observe that *E* decreases as the pH increases.

b) *E* at pH 0 and pH 7 for ClO_4^-. The balanced equation for the reduction of ClO_4^- is:

$$ClO_4^-(aq) + 2H^+(aq) + 2e^- \rightarrow ClO_3^-(aq) + H_2O(l).$$

The value of $E°$ for this reaction at pH = 0 is +1.201 V. The potential at any $[H^+]$, given by the Nernst equation, is

$$E = E° - (0.059\ V/2)(\log([ClO_3^-]/[ClO_4^-][H^+]^2)).$$

At pH 7, $[H^+] = 10^{-7}$ M and if both perchlorate and chlorate ions are present at unit activity, the reduction potential is

$$E = +1.201\ V - (0.0295\ V)(\log 10^{14}) = +1.201\ V - 0.413\ V = 0.788\ V.$$

Thus, the potential at pH = 7 is less positive than at pH = 0 resulting in lower oxidizing power of ClO_4^- as we would predict from part (a).

E17.26 The disproportionation of ClO_3^-, as an example of an oxoanion, can be broken down into a reduction and an oxidation, as follows:

$$\text{Reduction: } ClO_3^-(aq) + 6H^+(aq) + 6e^- \rightarrow Cl^-(aq) + 3H_2O(l)$$

$$\text{Oxidation: } 3ClO_3^-(aq) + 3H_2O(l) \rightarrow 3ClO_4^-(aq) + 6H^+(aq) + 6e^-.$$

Any effect that changing the pH has on the potential for the reduction reaction will be counteracted by an equal but opposite change on the potential for the oxidation reaction. In other words, the net reaction does not include H^+, so the net potential for the disproportionation cannot be pH dependent. Therefore, the promotion of disproportionation reactions of some oxoanions at low pH *cannot* be a thermodynamic promotion. Low pH results in a kinetic promotion. Protonation of an *oxo* group that occurs at low pH values aids oxygen–halogen bond scission (see Section 17.12). The disproportionation reactions have the same driving force at high pH and at low pH, but they are much faster at low pH.

E17.27 Recall from Chapter 5 that the reduction potential of a non-adjacent couple in the Latimer diagram can be calculated using the formula:

$$E^o = \frac{\nu_a \times E_a^o + \nu_b E_b^o + \ldots}{\nu_a + \nu_b \ldots}.$$

From the Frost diagram, the potential can be calculated as a slope of the line connecting the points of the two species in question. Since the reduction potential from the Latimer diagrams are easily accessible (i.e., do not require geometrical constructions), they are a better choice for calculations such as the ones in this exercise.

a) ClO_4^-/ClO^- potential. The relevant part of the Latimer diagram for chlorine is shown below. The numbers above the arrows connecting two species are the standard reduction potential (as is written per convention) for that redox couple. The numbers in the brackets represent the number of electrons received (ν in the previous equation).

$$ClO_4^- \xrightarrow{+0.374\,V(\nu=2)} ClO_3^- \xrightarrow{+0.295\,V(\nu=2)} ClO_2^- \xrightarrow{+0.681\,V(\nu=2)} ClO^-$$

Substituting the values from the diagram into the equation we have:

$$E^o_{ClO_4^-/ClO^-} = \frac{2\times(+0.374\text{ V})+2\times(+0.295\text{V})+2\times(+0.681\text{V})}{2+2+2} = +0.45\text{V}.$$

The same procedure is used to calculate the potentials in (b) and (c).

b) BrO_4^-/BrO^- potential.

$$E^o_{BrO_4^-/BrO^-} = \frac{2\times(+1.025\text{V})+4\times(+0.492\text{V})}{2+4} = +0.67\text{ V}.$$

c) IO_4^-/IO^- potential.

$$E^o_{IO_4^-/IO^-} = \frac{2\times(+0.65\text{ V})+4\times(+0.15\text{ V})}{2+4} = +0.317\text{ V}.$$

E17.28 The Frost diagram for chlorine is shown in Figure 17.14. If you connect the points for Cl^- and ClO_4^- with a line, you will see that the intermediate oxidation state species HClO, $HClO_2$, and ClO_3^- lie above that line. Therefore, they are unstable with respect to disproportionation. As discussed in Section 17.13, the redox reactions of halogen *oxo* anions become progressively faster as the oxidation number of the halogen *decreases*. Therefore, the rates of disproportionation are probably HClO > $HClO_2$ > ClO_3^-. Note that ClO_4^- cannot undergo disproportionation because there are no species with a higher oxidation number.

E17.29 The key to answering this question is to decide if the perchlorate ion, which is a very strong oxidant, is present along with a species that can be oxidized. If so, the compound *does* represent an explosion hazard.

a) NH_4ClO_4. Ammonium perchlorate is a dangerous compound because the N atom of the NH_4^+ ion is in its lowest oxidation state (–3) and can be oxidized.

b) $Mg(ClO_4)_2$. Since Mg^{2+} cannot be oxidized to a higher oxidation state, magnesium perchlorate is a stable compound and is not an explosion hazard.

c) $NaClO_4$. The same answer applies here as above for magnesium perchlorate. Sodium has only one common oxidation state. Even the strongest oxidants, such as F_2, FOOF, and ClF_3, cannot oxidize Na^+ to Na^{2+}.

d) $[Fe(H_2O)_6][ClO_4]_2$. Although the H_2O ligands cannot be oxidized, the metal ion can. This compound presents an explosion hazard because Fe(II) can be oxidized to Fe(III) by a strong oxidant such as perchlorate ion.

E17.30 **a) Cr^{3+}.** Yes: $Cr^{3+}/Cr_2O_7^{2-}$ potential is +1.38V; less positive than Cl_2/ClO^- (+1.659V).

b) V^{3+}. Yes $E^0(V^{3+}/VO_2^+ = +0.668\text{V})$.

c) Fe^{2+}. Yes, $E^0(Fe^{2+}/Fe^{3+}) = +\ 0.77\text{V}$.

d) Co^{2+}. No. The reduction potential for Co^{2+}/Co^{3+} is +1.92 V, which is more positive than Cl_2/ClO^- potential.

E17.31

A = ClF	$F_2(g) + Cl_2(g) \rightarrow 2ClF(g)$
E = $CsClF_2$	$ClF + CsF \rightarrow CsClF_2$
B = OF_2	$2\ F_2(g) + 2\ OH^-(aq) \rightarrow OF_2(g) + H_2O(l) + 2\ F^-(aq)$
C = HF	$F_2(g) + H_2O\ (l) \rightarrow 2\ HF(aq) + ½\ O_2(g)$

D = SiF_4	$SiO_2(s) + 2\,F_2(g) \rightarrow SiF_4(g) + O_2(g)$
G = CaF_2	$CaF_2(s) + H_2SO_4 \rightarrow 2\,HF(g) + CaSO_4(s)$

E17.32 **a)** Unlike ClO_4^- and IO_4^-, BrO_4^- is capable of oxidizing water. This makes preparation and isolation of $KBrO_4$ very difficult.

b) All compounds are kinetically reasonably stable to be isolated.

c) Unlike its Cl and Br analogues, IO_2^- ion is too unstable for its salts to be isolated; even the acid HIO_2 has been identified only as a transient species.

d) BrO^- and IO^- are disproportionating too rapidly to be isolated.

E17.32 Statement **(a)** is correct.

Statement **(b)** is incorrect—the two anions do not have identical structures.

Statement **(c)** is incorrect—it is the nucleophilicity of the Cl atom in ClO^- that is crucial for the oxidation mechanism.

Statement **(d)** is correct.

Chapter 18 The Group 18 Elements

Self-Tests

S18.1 The XeF^+ cation has 14 electrons: eight from Xe plus seven from F and minus one electron to account for a positive charge. The electronic configuration is:

$$1\sigma_g^2 1\sigma_u^2 2\sigma_g^2 1\pi_u^4 1\pi_g^4 2\sigma_u^0.$$

Bond order is:

$$b = ½(8 - 6) = 1.$$

S18.2 The reaction we are looking at:

$$Xe(g) + 3F_2(g) \rightarrow XeF_6(s).$$

From the reaction, three F–F bonds have to be broken while 6 Xe–F bonds are formed:

$$\begin{aligned}\Delta_f H &\approx 3 \times B(\text{F–F}) - 6 \times B(\text{Xe–F})\\ &= 3 \times 155\ \text{kJ mol}^{-1} - 6 \times 144\ \text{kJ mol}^{-1}\\ &= -408\ \text{kJ mol}^{-1}.\end{aligned}$$

This is only an estimate because our $\Delta_f H$ does not consider the sublimation (or lattice) enthalpy of XeF_6: $XeF_6(g) \rightarrow XeF_6(s)$.

S18.3 As the self-test states, xenate ($HXeO_4^-$) decomposes to perxenate (XeO_6^{4-}), xenon, and oxygen, so the equation that must be balanced is:

$$HXeO_4^- \rightarrow XeO_6^{4-} + Xe + O_2 \ \text{(not balanced)}.$$

Since the reaction occurs in a basic solution, we can use OH^- and H_2O to balance the equation, if necessary. You notice immediately that there is no species containing hydrogen on the right-hand side of the equation, so H_2O should go on the right and OH^- should go on the left. The balanced equation is

$$2HXeO_4^-(aq) + 2OH^-(aq) \rightarrow XeO_6^{4-}(aq) + Xe(g) + O_2(g) + 2H_2O(l).$$

Since the products are perxenate (an Xe(VIII) species) and elemental xenon, this reaction is a disproportionation of the Xe(VI) species $HXeO_4^-$. Oxygen is produced from a thermodynamically unstable intermediate with Xe(IV), possibly as follows:

$$HXeO_3(aq) \rightarrow Xe(g) + O_2(g) + OH^-(aq)$$

Exercises

E18.1 All the original He and H_2 present in the Earth's atmosphere when our planet originally formed has been lost. Earth's gravitational field is not strong enough to hold these light gases, and they eventually diffuse away into space. The small amount of helium that is present in today's atmosphere is the product of ongoing radioactive decay.

E18.2 **a) The lowest-temperature refrigerant.** The noble gas with the lowest boiling point would best serve as the lowest-temperature refrigerant. Helium, with a boiling point of 4.2 K, is the refrigerant of choice for very low-temperature applications, such as cooling superconducting magnets in modern NMR spectrometers. The boiling points of the noble gases are directly related to their atomic volume. The larger atoms Kr and Xe are more polarizable than He or Ne, so Kr and Xe experience larger van der Waals attractions and have higher boiling points.

b) An electric discharge light source requiring a safe gas with the lowest ionization energy. The noble gases are a safe choice due to their low reactivity. Among them, the largest one, Xe, has the lowest ionization potential (see Table 18.1). Radon is even larger than Xe and has even lower ionization potential, but because it is radioactive it is not safe.

c) The least expensive inert atmosphere. Ar is the choice here because it is, relative to the other noble gases, very abundant in the atmosphere and is generally cheaper than helium, which is rare in the atmosphere. However, a great deal of helium is collected as a by-product of natural gas production. In places where large amounts of natural gas are produced, such as the United States, helium may be marginally less expensive than argon.

E18.3 **a) Synthesis of XeF_2.** XeF_2 can be prepared in two different ways. First, a mixture of Xe and F_2 that contains an excess of Xe is heated to 400°C. The excess Xe prevents the formation of XeF_4 and XeF_6. The second way is to photolyse a mixture of Xe and F_2 at room temperature. For either method of synthesis, the balanced equation is:

$$Xe(g) + F_2(g) \rightarrow XeF_2(s).$$

At 400°C the product is a gas, but at room temperature it is a solid.

b) Synthesis of XeF_6. For the synthesis of this compound you would want also to use a high temperature, but unlike the synthesis of XeF_2, you want to have a *large* excess of F_2:

$$Xe(g) + 3F_2(g) \rightarrow XeF_6(s).$$

As with XeF_2, xenon hexafluoride is a solid at room temperature.

c) Synthesis of XeO_3. This compound is endergonic, so it cannot be prepared directly from the elements. However, a sample of XeF_6 can be carefully hydrolysed to form the desired product:

$$XeF_6(s) + 3H_2O(l) \rightarrow XeO_3(s) + 6HF(g).$$

If a large excess of water is used, an aqueous solution of XeO_3 is formed instead.

E18.4 The reaction we are looking at:

$$Xe(g) + F_2(g) \rightarrow XeF_2(s).$$

From the reaction, one F–F bond has to be broken while two Xe–F bonds are formed:

$$\Delta_f H \approx B(\text{F–F}) - 2 \times B(\text{Xe–F})$$

$$= 155 \text{ kJ mol}^{-1} - 2 \times 144 \text{ kJ mol}^{-1}$$

$$= -133 \text{ kJ mol}^{-1}.$$

This is only an estimate because our $\Delta_f H$ does not consider the sublimation (or lattice) enthalpy of XeF_2: $XeF_2(g) \rightarrow XeF_2(s)$ which should be subtracted from the final value.

E18.5 The Lewis structures of $XeOF_4$, XeO_2F_2, and XeO_6^{4-} are shown below.

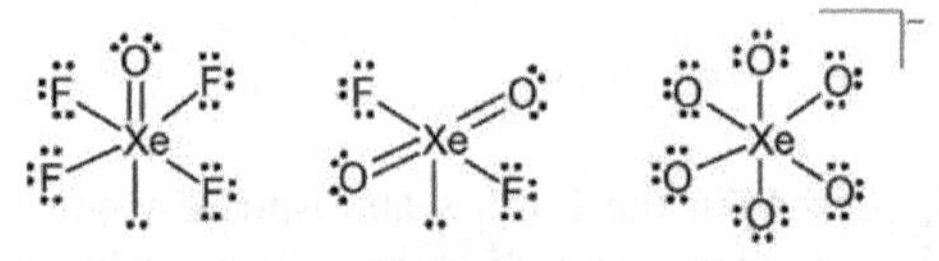

$XeOF_4$ XeO_2F_2 $[XeO_6]^{4-}$

E18.6

a) Isostructural with ICl_4^-. The Lewis structure of ICl_4^- has four bonding pairs and two lone pairs of electrons around the central iodine atom. Therefore, the anion has a square planar geometry. The noble gas compound that is isostructural is XeF_4.

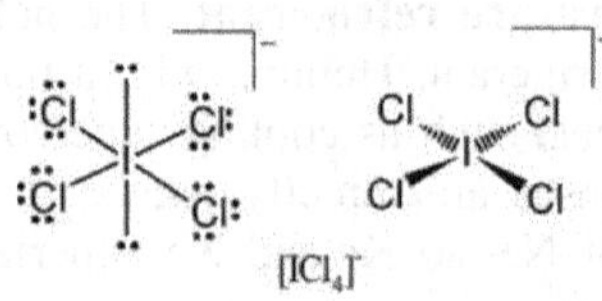

$[ICl_4]^-$

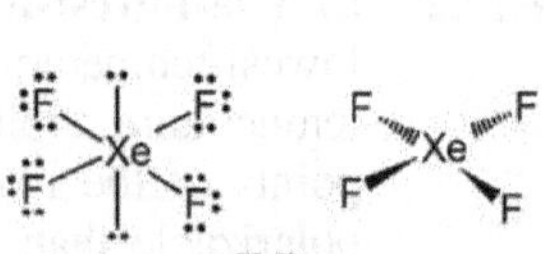

XeF_4

b) Isostructural with IBr_2^-. The Lewis structure of IBr_2^- anion has two bonding pairs and three lone pairs of electrons around the central iodine atom. Therefore, the anion has a linear geometry.

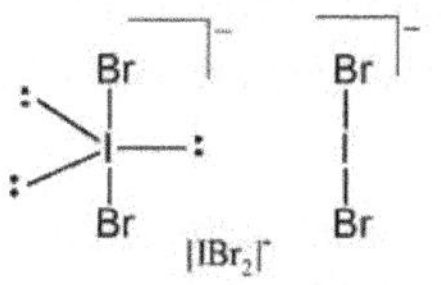

$[IBr_2]^-$

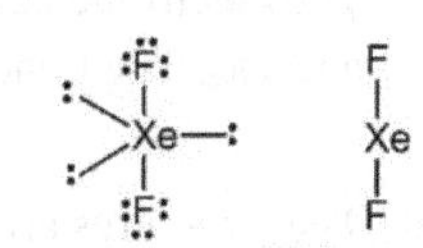

XeF_2

c) Isostructural with BrO_3^-. The Lewis structure of BrO_3^- anion has three bonding pairs and one lone pair of electrons around the central bromine atom. Therefore, the anion has a trigonal pyramidal geometry. Note that it is also a resonance stabilized anion. The noble gas compound that is isostructural is XeO_3.

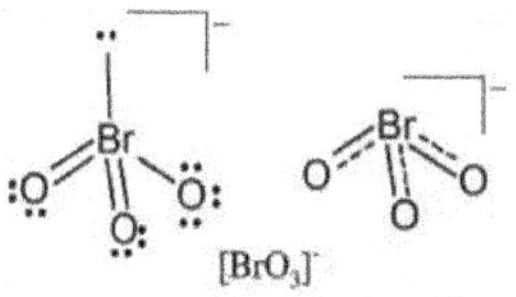

$[BrO_3]^-$

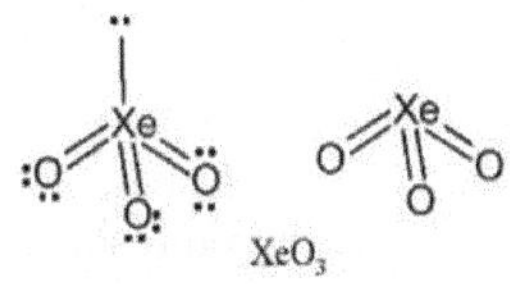

XeO_3

d) Isostructural with ClF. The ClF diatomic molecule is isostructural with the cation XeF^+.

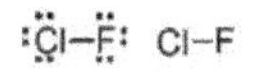

ClF

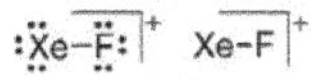

XeF^+

E18.7 a) The Lewis structure for XeF_7^- is shown below.

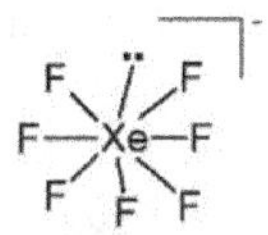

b) The structure of XeF_5^-, with seven pairs of electrons, is based on a pentagonal bipyramidal array of electron pairs with two stereochemically active lone pairs (see Structure **9**) resulting in overall pentagonal planar molecular geometry. The ion XeF_8^{2-}, on the other hand, has nine pairs of electrons, but only eight are stereochemically active (see Structure **10**) giving a square-antiprism molecular geometry. It is possible that the structure of XeF_7^-, with eight pairs, would be a pentagonal bipyramid, with one stereochemically inactive lone pair.

The real structure of this anion is capped octahedral with a very long "capping" Xe–F bond, as shown below, confirming that the lone pair is indeed stereochemically inert.

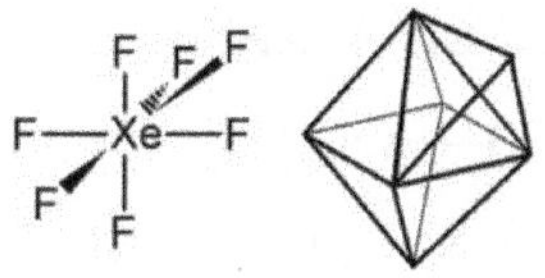

E18.8 $[XeF_5][RuF_6]$ contains $[XeF_5]^+$ cation and $[RuF_6]^-$ anion. The cation has 42 valence electrons. This gives octahedral electron pair geometry and square pyramidal molecular geometry:

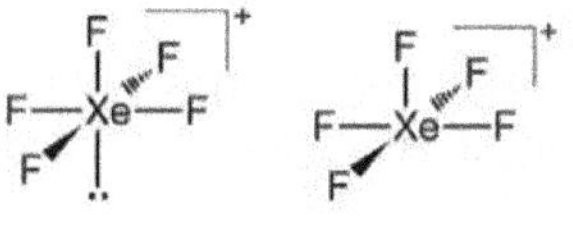

The anion is a complex anion, with Ru as a central metal cation and six fluoride ligands. The geometry is octahedral:

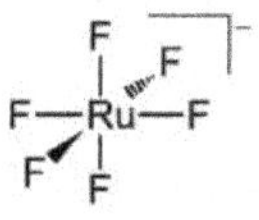

E18.9 According to molecular orbital theory, the electronic configuration of He_2^+ is $1\sigma_g^2 1\sigma_u^1$. There are two electrons in bonding σ ($1\sigma_g$) orbital and one electron in an antibonding σ* orbital ($1\sigma_u$), therefore the bond order is ½(2 – 1) = 0.5.

The electronic configuration of Ne_2^+ is $1\sigma_g^2 1\sigma_u^2 2\sigma_g^2 1\pi_u^4 1\pi_g^4 2\sigma_u^1$. Recall that the bonding orbitals are $1\sigma_g$, $2\sigma_g$, $1\pi_u$, and $2\pi_u$. The bond order is therefore ½(8 – 7) = 0.5.

E18.10 The structures are summarized in the table below. Since the lone pairs on F atoms do not influence the geometry, they have been omitted for clarity.

	(a) XeF_3^+	(b) XeF_3^-	(c) XeF_5^+	(d) XeF_5^-
Lewis structures	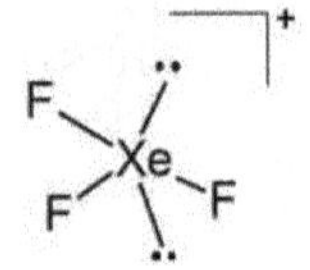	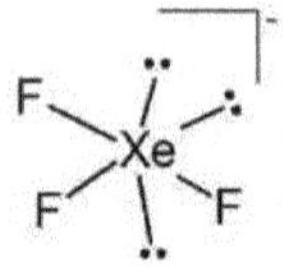	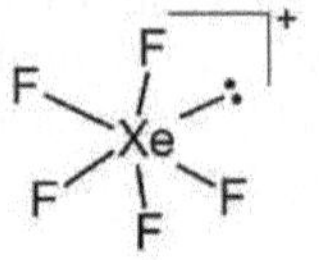	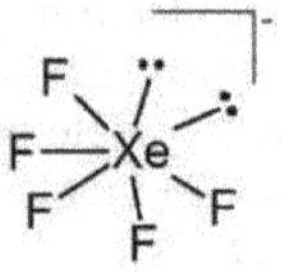
Electron pair geometry	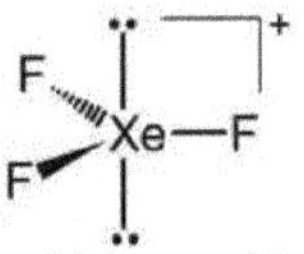	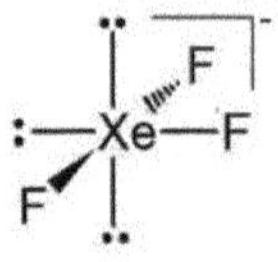	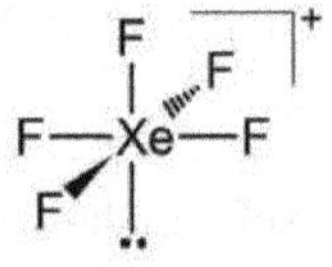	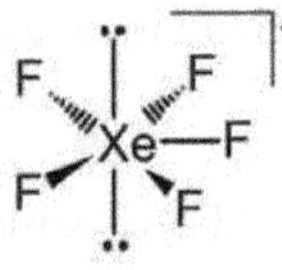
	Trigonal bipyramid	Octahedral	Octahedral	Pentagonal bipyramid
Molecular geometry	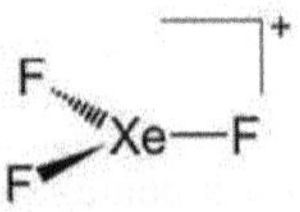	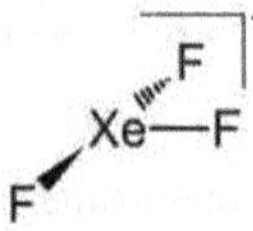	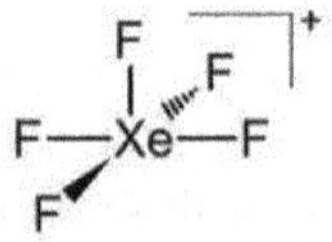	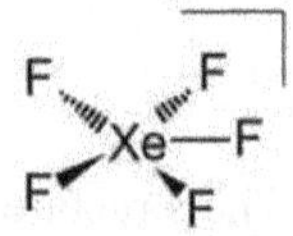
	Trigonal planar	T-shaped	Square pyramidal	Pentagonal planar

E18.11 **A** = $XeF_2(g)$ $Xe + F_2 \rightarrow XeF_2$

B = $[XeF]^+[MeBF_3]^-$ $XeF_2(g) + MeBF_2 \rightarrow [XeF]^+[MeBF_3]^-$

C = XeF_6 $Xe + F_2(\text{excess}) \rightarrow XeF_6$

D = XeO_3 $XeF_6 + H_2O \rightarrow XeO_3$

E = XeF_4 (g) $Xe + 2F_2 \rightarrow XeF_4(g)$

E18.12 The structure of this square planar compound has been covered before (see for example, Exercise 18.6(a)). The ^{19}F NMR would consist of a single resonance flanked by a doublet due to the coupling to ^{129}Xe isotope (spin = ½; abundance 26.4%).

E18.13 The structure of $XeOF_3^+$ is shown below. The ^{129}Xe NMR would consist of one resonance split into a 1 : 3 : 3 : 1 quartet due to the coupling to three equivalent ^{19}F nuclei.

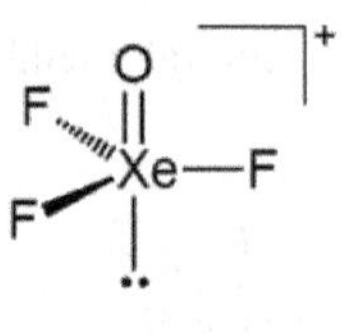

E18.14 The molecular structure of $XeOF_4$ is a square pyramid with the O-atom residing at the apical position (see structure **8**). The four fluorine atoms are located at the equatorial positions in a basal plane. Because of the equivalency of the fluorine atoms, only one resonance is expected. Because of the interactions between fluorine and the isotope of Xe ^{129}Xe (spin = ½; abundance 26.4%), we do expect an additional weak feature. The predicted spectra will consist of a central line and two other lines, called satellites, symmetrically distributed around this central line. These are satellites due to the coupling between ^{19}F and ^{129}Xe nuclei and would be a doublet.

Chapter 19 The d-block Elements

Self-Tests

S19.1 The Latimer diagram for vanadium in basic solution is provided in Resource Section 3. The reduction potential for O_2/H_2O couple in alkaline solution is:

$$E = +1.229\text{ V} - 0.059\text{ V} \times \text{pH} = +1.229\text{ V} - 0.059\text{ V} \times 14 = +0.403\text{ V}.$$

The reduction potential for VO/V_2O_3 couple in basic solution is –0.486 V. The potential difference for oxidation of VO to V_2O_3 with oxygen is

$$E_{\text{cell}} = +0.403\text{ V} - (-0.486\text{ V}) = +0.889\text{ V}.$$

Thus, VO will be oxidized to V_2O_3 by oxygen in air. The next oxidation step is oxidation from V_2O_3 to $HV_2O_5^-$ with reduction potential +0.542V. This reaction has the following potential difference:

$$E_{\text{cell}} = +0.403\text{ V} - (+0.542\text{ V}) = -0.139\text{ V}.$$

This step has a negative potential and is not a thermodynamically spontaneous process. This means that a basic solution oxidation of VO to V_2O_3 is a thermodynamically favourable process, and V_2O_3 is more stable than VO in presence of oxygen.

S19.2 Re_3Cl_9 is a trimer (see Structure 19). When ligands are added, such as PPh_3, discrete molecular species such as $Re_3Cl_9(PPh_3)_3$ are formed (see below). Sterically, the most favourable place for each bulky triphenylphosphine ligand to go is in the terminal position in the Re_3 plane.

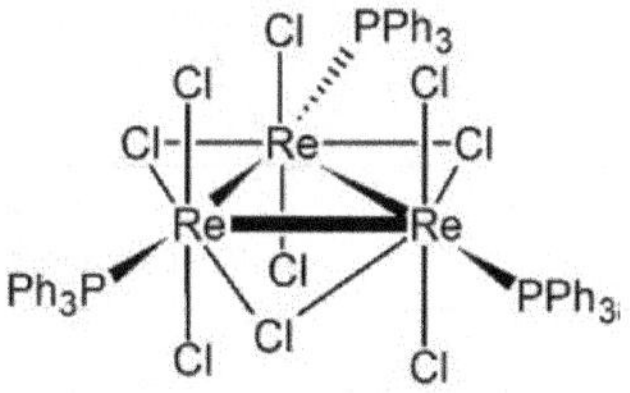

Exercises

E19.1 The table below summarizes the highest oxidation states of the first-row transition metals, gives examples of oxo species containing metals in the highest oxidation state, and contrasts them with the second- and third-row elements.

Group	**3**	**4**	**5**	**6**	**7**	**8**	**9**	**10**	**11**	**12**
1st row	Sc	Ti	V	Cr	Mn	Fe	Co	Ni	Cu	Zn
Max. OS	3	4	5	6	7	6	4	4	2	2
Max. group OS?	*Yes*	*Yes*	*Yes*	*Yes*	*Yes*	*No*	*No*	*No*	*No*	*No*
Example*	Sc_2O_3	TiO_4	VO_2^+	CrO_3	MnO_4^-	FeO_4^{2-}	Co_2O_3	NiO	CuO	ZnO
2nd/3rd row	Y/La	Zr/Hf	Nb/Ta	Mo/W	Tc/Re	Ru/Os	Rh/Ir	Pd/Pt	Ag/Au	Cd/Hg
Max. OS	3	4	5	6	7	8	6	4/6	3/5	2
Example	Y_2O_3 La_2O_3	ZrO_2 HfO_2	Nb_2O_5 Ta_2O_5	MoO_3 WO_3	Tc_2O_7 ReO_4^{3-}	RuO_4 OsO_4	RhO_2 IrO_2	PdO_2 PtO_2	AgO Au_2O_3	CdO HgO

*Note that the highest oxidation state oxo species do not exist for all elements, particularly late transition metals. For example, the highest well-defined oxidation state for Rh and Ir is +6, with representative compound being hexafluoride of two metals. However, the maximum oxidation state in which oxo species occur is +4 in RhO_2 and IrO_2.

It is clear from the table that the maximum oxidation state of the group increases from left to right, peaks at Group 7 for the first row and at Group 8 for the second- and third-row transition metals, and then steadily decreases toward the end of the transition metal series. Within a group, the heavier elements can achieve higher oxidation states than first-row members (see, for example, Group 8).

E19.2 Recall that the Frost diagrams can tell us a lot about the thermodynamic stability of an oxidation state but under strict conditions. In this case we are constructing the diagram from the *standard* reduction potentials, thus the diagram below is strictly correct for pH = 0 and unit activity of all species involved. Nevertheless, the conclusions hold rather generally. Note that a completely correct Frost diagram must clearly indicate chemical species associated with each data point. In some cases, such as for Cr +5 and +4, the chemical species are not well-defined. In these instances, it is sufficient to indicate the element's oxidation state (i.e., Cr(V) and Cr(IV)).

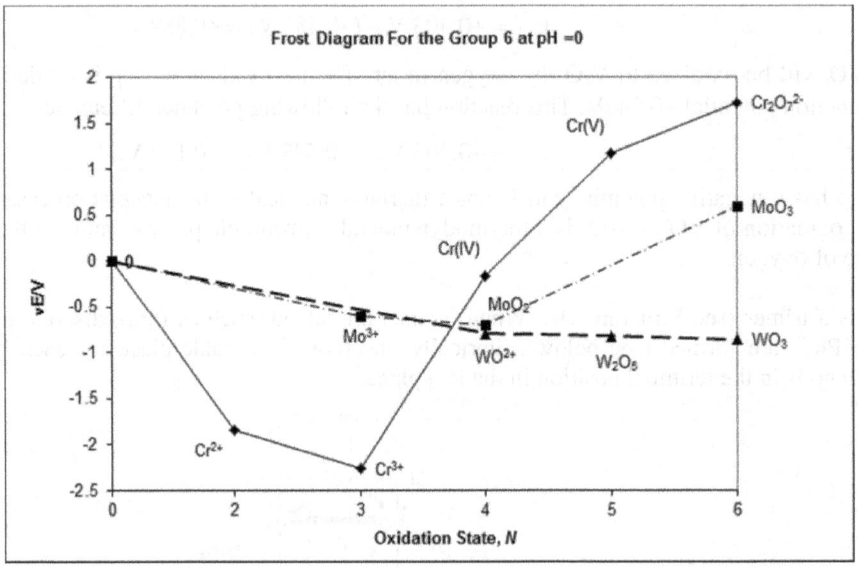

(a) The most oxidizing state for Cr and Mo is +6. Tungsten has no species with oxidizing properties.

(b) Cr, Cr(IV) and Cr(V) are susceptible to disproportionation. Other elements do not have species that would be expected to disproportionate.

E19.3 **(a) $Cr_2O_7^{2-}$**

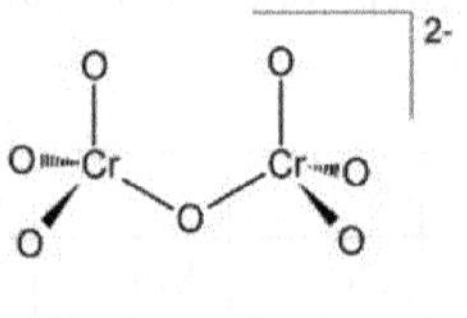

(b) VO^{2+}

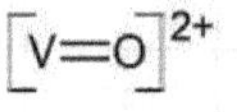

(c) VO_4^{3-}

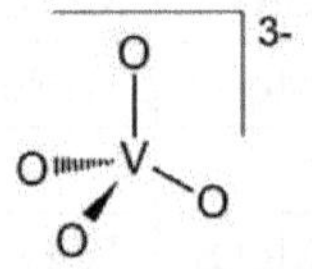

(d) MnO_4^{2-}

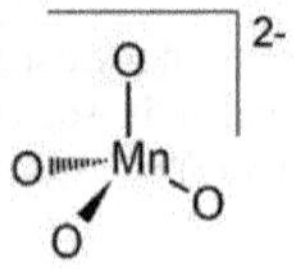

Note: In structures (a), (b), and (d) the terminal metal–oxygen bond order is higher than 1 but lower than 2 due to the resonance effects; these given structures depict the geometry rather than exact bond orders.

E19.4 The oxides FeO_4 and Co_2O_9 contain iron and cobalt in oxidation states +8 and +9 respectively, which are not stable for either element. The massive amount of energy required to ionize either Fe to +8 or Co to +9 cannot be offset by the lattice energy of the oxides, and the two compounds do not exist. Fe(+8) and Co(+9) would be a sufficiently strong oxidizing agent to oxidize some of the O^{2-} anions back to O_2.

E19.5 Statement **(a)** is correct.

Statement **(b)** is also correct.

Statement **(c)** is incorrect: the atomic volumes of the transition metals decrease steadily from left to right; the lanthanide contraction also affects the atomic volumes of third-row transition elements so that their atomic volumes are almost identical to those found for the second-row transition metals directly above. Consequently, transition metals are very dense. For example, osmium with density of 22.6 g/cm^3 is one of the densest materials known.

Statement **(d)** is correct.

E19.6 The data required are summarized in the table below:

Electron gain enthalpy (kJ mol^{-1})*		**First ionization energies (kJ mol^{-1})**	
Cu	119.24	**Li**	513
Ag	125.86	**Na**	495
Au	222.75	**K**	419
		Rb	403
		Cs	375

*Data from Greenwood, N.N. and Earnshaw, A. (1997). *Chemistry of the Elements,* 2nd ed. Elsevier. (p. 1176). The first ionization energies of the Group 1 metals have been taken from Chapter 11 of the textbook.

Analysing the data from the table we can conclude that most likely compounds of general formula MM′ are rubidium and caesium aurides, RbAu and CsAu, containing Au^-. This is because gold has the highest electron gain enthalpy which is large enough to, in combination with other effects such as lattice energy and ion solvation energy, offset the energy required for ionization of Group 1 elements. Indeed, the evidence points to the formation of Au^- in ammonia solutions containing K, Rb, and Cs. Also, CsAu can be isolated and has a salt-like character.

E19.7 HfO_2 and ZrO_2 are isostructural and, because ionic radii of Hf^{4+} and Zr^{4+} are almost identical, they likely have very similar unit cell parameters and consequently unit cell volume. Thus, the difference in density can be explained by difference in atomic mass for Hf and Zr. Since, the atomic mass of Hf (and consequently molecular mass of HfO_2) is higher than Zr (and ZrO_2 has lower molecular weight than HfO_2) more mass is "packed" in about the same volume. Consider also that the ratio of molecular masses is very similar to the ratio of the densities: $Mr(HfO_2)/Mr(ZrO_2) = 210.49/123.22 = 1.72$ vs. $\rho(HfO_2)/\rho(ZrO_2) = 9.68/5.73 = 1.69$.

E19.8 The Frost diagram for mercury is shown below:

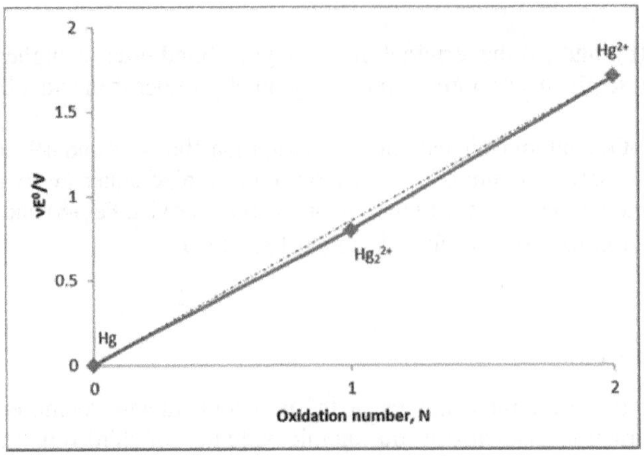

Considering that the point corresponding to Hg_2^{2+} lies below the line connecting Hg and Hg^{2+} (dashed line), Hg_2^{2+} will not disproportionate. Rather, Hg and Hg^{2+} would comproportionate to Hg_2^{2+}.

E19.9 Pigments should be permanent and stable. They should be as inert as possible to the atmosphere—they should be stable toward oxidation with O_2, should not react with moisture, and should not be light-sensitive. Considering the increased atmospheric pollution, they should be as resistant as possible to the effects of gases such as H_2S, CO_2, SO_2, etc. They should also be non-toxic and have low vapour pressure. A good pigment would have very good coverage ability; that is, small amount of pigment should cover a large surface area. If the material is natural in origin, then its extraction and storage should be easy. If the pigment is synthetic, then the manufacturing process should be cheap.

Guidelines for Selected Tutorial Problems

T19.3 Very good overview of both processes can be found in Greenwood, N.N. and Earnshaw, A. (1997). *Chemistry of the Elements,* 2nd ed. Elsevier, in chapter on the Group 4 elements.

T19.4 After the Earth's atmosphere became oxidizing, the availability of many biologically important elements changed: Fe and N became very difficult to extract whereas Cu became available. For Fe availability, start by considering in which compounds iron frequently occurs in nature. Then compare the solubilities, solution properties, and movability of Fe^{2+} and Fe^{3+} analogues. Also, look up biologically important compounds known as siderophores. Compare their structures, coordination properties, and stabilities of Fe-siderophore complexes with the man-made chelating reagents such as edta. Pay attention to the process of Fe release from a siderophore.

T19.7 A good starting point for this tutorial assignment is the text Schmid, G. (ed.). (2010). *Nanoparticles: From Theory to Application*, 2nd ed. Wiley-VCH. The text gives a general overview of nanoparticles, but has a significant part devoted to gold nanoparticles.

T19.9 Protiodide was a medical trade name for mercury(I) iodide, Hg_2I_2. Various sources can be used to summarize the side-effects and overdose risks, such as MSDS sheets. Sweetman, S.C. (ed.). *Martindale: The Complete Drug Reference*, published by Pharmaceutical Press, could also be a good source for the toxicity and effects of inorganic mercury salts.

Chapter 20 d-Metal Complexes: Electronic Structure and Properties

Self-Tests

S20.1 A high-spin d^7 configuration is $t_{2g}^5e_g^2$. To calculate the ligand field stabilization energy LFSE we should note that each electron in the t_{2g} stabilizes the CFSE for –0.4 units and destabilizes by 0.6 units when placed in the e_g level. Accordingly, for the high spin d^7 configuration, we should expect $5 \times (-0.4\Delta_o) + 2 \times 0.6\Delta_o$ or $-0.8\Delta_o$. The LFSE of a low-spin d^7 with a configuration of $t_{2g}^6e_g^1$, is expected to have $-1.8\Delta_o + P[6 \times (-0.4\ \Delta_o) + 0.6\ \Delta_o = -1.8\Delta_o]$ where P is electron-pairing energy. Note that although high-spin d^7 configuration has two electron pairs in t_{2g} level, they do not contribute to LFSE because they would be paired in a spherical field as well. The low-spin configuration adds one more electron pair (a pair that did not exist in a spherical field before the orbital splitting), and this pair has to be accounted for.

S20.2 Since each isothiocyanate ligand has a single negative charge, the oxidation state of the manganese ion is +2. Mn(II) is d^5, and there are two possibilities for an octahedral complex, a low-spin (t_{2g}^5), with one unpaired electron, or a high-spin ($t_{2g}^3e_g^2$), with five unpaired electrons. The observed magnetic moment of 6.06 μ_B is close to the spin-only value for five unpaired electrons, $(5 \times 7)^{1/2} = 5.92\ \mu_B$ (see Table 20.3). Therefore, this complex is high spin and has a $t_{2g}^3e_g^2$ configuration.

S20.3 If it were not for LFSE, MF_2 lattice enthalpies would increase from Mn(II) to Zn(II). This is because the decreasing ionic radius, which is due to the increasing Z_{eff} as you cross through the d-block from left to right, leads to decreasing M–F separations (straight line on diagram below). Therefore, you expect that ΔH_{latt} for MnF_2 (2 780 kJ mol^{-1}) would be smaller than ΔH_{latt} for ZnF_2 (2985 kJ mol^{-1}). In addition, as discussed for aqua ions and oxides, we expect additional LFSE for these compounds. From Table 20.2, you see that LFSE = 0 for Mn(II), $-0.4\Delta_o$ for Fe(II), $-0.8\Delta_o$ for Co(II), $-1.2\Delta_o$ for Ni(II), and 0 for Zn(II). The deviations of the observed values from the straight line connecting Mn(II) and Zn(II) are not quite in the ratio 0.4 : 0.8 : 1.2, but note that the deviation for Fe(II) is smaller than that for Ni(II).

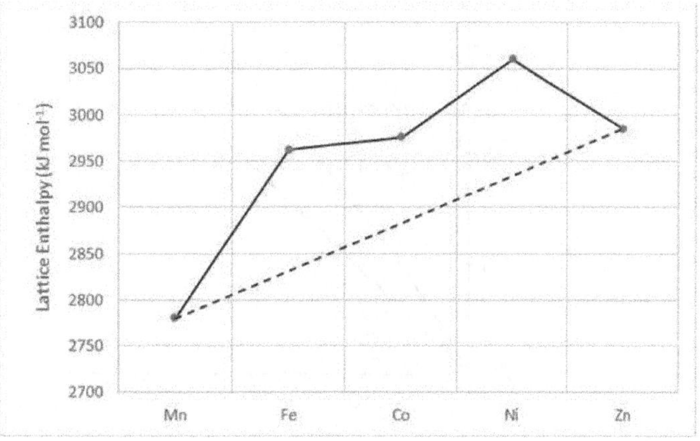

S20.4 In the spectrum of $[Mo(CO)_6]$ (Figure 20.17), the ionization energy around 8 eV was attributed to the t_{2g} electrons that are largely metal-based (see Example 20.4). The biggest difference between the photoelectron spectra of ferrocene and magnesocene is in the 6–8 eV region. Fe(II) has six d electrons (like Mo(0)), while Mg(II) has no d electrons. Therefore, the differences in the 6–8 eV region can be attributed to the lack of d electrons in Mg(II) complex. The group of energies around 14 eV and 18 eV in both compounds can be attributed to M–Cp orbitals and ionization of Cp itself, respectively.

S20.5 A p^1d^1 configuration will have $L = 1 + 2 = 3$ (an electron in a p orbital has $l = 1$, whereas an electron in a d orbital has $l = 2$). This value of L corresponds to an F term. Since the electrons may have antiparallel ($S = -½ + ½ = 0$, multiplicity $2S + 1 = 2 \times 0 + 1 = 1$) or parallel spins ($S = +½ + ½ = 1$, multiplicity $2S + 1 = 2 \times 1 + 1 = 3$), both ^{1}F and ^{3}F terms are possible.

S20.6 **a) $2p^2$ ground term**. Two electrons in a p subshell can occupy separate p orbitals and have parallel spins ($S = +½ + ½ = 1$), so the maximum multiplicity $2S + 1 = 2 \times 1 + 1 = 3$. The maximum value of M_L is $1 + 0 = 1$, which corresponds to a P term (according to the Pauli principle, m_l cannot be +1 for both electrons if the spins are parallel, thus one electron has $m_l = +1$ and the other $m_l = 0$). Thus, the ground term is ^3P (called a "triplet P" term).

b) $3d^9$ ground term. The largest value of M_S for nine electrons in a d subshell is 1/2 (eight electrons are paired in four of the five orbitals, while the ninth electron has $m_s = \pm 1/2$). Thus $S = 1/2$ and the multiplicity $2S + 1 = 2$. Notice that the largest value of M_S is the same for one electron in a d subshell. Similarly, the largest value of M_L is 2, which results from the following nine values of m_l: +2, +2, +1, +1, 0, 0, –1, –1, –2. Notice also that $M_L = 2$ corresponds to one electron in a d subshell. Thus, the ground term for a d^9 or a d^1 configuration is ^2D (called a "doublet D" term).

S20.7 An F term arising from a d^n configuration correlates with $T_{1g} + T_{2g} + A_{2g}$ terms in an O_h complex. The multiplicity is unchanged by the correlation, so the terms in O_h symmetry are $^3T_{1g}$, $^3T_{2g}$, and $^3A_{2g}$. Similarly, a D term arising from a d^n configuration correlates with $T_{2g} + E_g$ terms in an O_h complex. The O_h terms retain the singlet character of the ^1D free ion term, and so are $^1T_{2g}$ and ^1E_g.

S20.8 The two transitions in question are $^4T_2 \leftarrow {}^4A_2$ and $^4T_1 \leftarrow {}^4A_2$ (the first one has a lower energy and hence a lower wave number). Since $\Delta_O = 17\,600\ cm^{-1}$ and $B = 700\ cm^{-1}$, you must find the point 17 600/700 = 25.1 on the *x*-axis of the Tanabe–Sugano diagram. Then, the ratios *E/B* for the two bands are the *y* values of the points on the $^4T_2$ and $^4T_1$ lines, 25 and 32, respectively, as shown below on a partial Tanabe–Sugano diagram for the d^3 configuration. Therefore, the two lowest-energy spin-allowed bands in the spectrum of $[Cr(H_2O)_6]^{3+}$ will be found at $(700\ cm^{-1})(25) = 17\,500\ cm^{-1}$ and $(700\ cm^{-1})(32) = 22\,400\ cm^{-1}$.

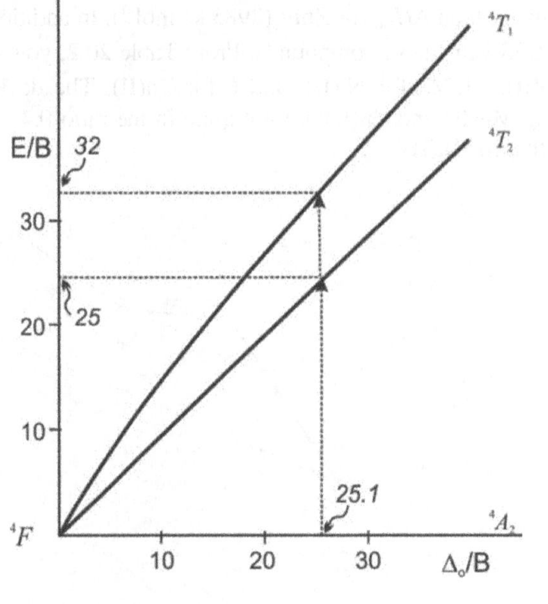

S20.9 This six-coordinate d^3 complex undoubtedly has an O_h symmetry, so the general features of its spectrum will resemble the spectrum of $[Cr(NH_3)_6]^{3+}$, shown in Figure 20.23. The very low intensity of the band at 16 000 cm^{-1} is a clue that it is a spin-forbidden transition, probably $^2E_g \leftarrow {}^4A_{2g}$. The spin-allowed but Laporte-forbidden bands typically have $\varepsilon \sim 100\ M^{-1}\ cm^{-1}$, so it is likely that the bands at 17 700 cm^{-1} and 23 800 cm^{-1} are of this type (they correspond to the $^4T_{2g} \leftarrow {}^4A_{2g}$ and $^4T_{1g} \leftarrow {}^4A_{2g}$ transitions, respectively). The band at 32 400 cm^{-1} is probably a charge transfer band because its intensity is too high to be a ligand field (d–d) band. Since you are provided with the hint that the NCS^- ligands have low-lying π* orbitals, it is reasonable to conclude that this band corresponds to a MLCT transition. Notice that the two spin-allowed ligand field transitions of $[Cr(NCS)_6]^{3-}$ are at lower energy than those of $[Cr(NH_3)_6]^{3+}$, showing that NCS^- induces a smaller Δ_o on Cr^{3+} than does NH_3. Also notice that $[Cr(NH_3)_6]^{3+}$ lacks an intense MLCT band at ~30 000–40 000 cm^{-1}, showing that NH_3 does not have low-lying empty molecular orbitals.

Exercises

E20.1 **a) $[Co(NH_3)_6]^{3+}$.** Since the NH_3 ligands are neutral, the cobalt ion in this octahedral complex is Co^{3+}, which is a d^6 metal ion. Ammonia is in the middle of the spectrochemical series but because the cobalt ion has a 3+ charge, this is a strong field complex and hence is a low spin complex, with $S = 0$ and has no unpaired electrons (the configuration is t_{2g}^6). The LFSE is $6(-0.4\Delta_o) = -2.4\Delta_o + 2P$.

b) $[Fe(OH_2)_6]^{2+}$. The iron ion in this octahedral complex, which contains only neutral water molecules as ligands, is Fe^{2+}, a d^6-metal ion. Since water is lower in the spectrochemical series than NH_3 (i.e., it is a weaker field ligand than NH_3) *and* since the charge on the metal ion is only 2+, this is a weak field complex and hence is high spin, with $S = 2$ and four unpaired d electrons (the configuration is $t_{2g}^4e_g^2$). The LFSE is $4(-0.4\Delta_o) + 2(0.6\Delta_o) = -0.4\Delta_o$. Compare this small value to the large value for the low spin d^6 complex in part **(a)** above.

c) $[Fe(CN)_6]^{3-}$. The iron ion in this octahedral complex, which contains six negatively charged CN^- ligands, is Fe^{3+}, which is a d^5 metal ion. Cyanide ion is a very strong field ligand, so this is a strong field complex and hence is a low spin complex, with $S = 1/2$ and one unpaired electron. The configuration is t_{2g}^5 and the LFSE is $-2.0\Delta_o + 2P$.

d) $[Cr(NH_3)_6]^{3+}$. The complex contains six neutral NH_3 ligands, so chromium is Cr^{3+}, a d^3 metal ion. The configuration is t_{2g}^3, and so there are three unpaired electrons and $S = 3/2$. Note that, for octahedral complexes, only d^4–d^7 metal ions have the possibility of being either high spin or low spin. For $[Cr(NH_3)_6]^{3+}$, the LFSE = $3(-0.4\Delta_o) = -1.2\Delta_o$. (For d^1–d^3, d^8, and d^9 metal ions in octahedral complexes, only one spin state is possible.)

e) $[W(CO)_6]$. Carbon monoxide (i.e., the carbonyl ligand) is neutral, so this is a complex of W(0). The W atom in this octahedral complex is d^6. Since CO is such a strong field ligand (it is even higher in the spectrochemical series than CN^-), $W(CO)_6$ is a strong field complex and hence is low spin, with no unpaired electrons (the configuration is t_{2g}^6). The LFSE = $6(-0.4\Delta_o) + 2P = -2.4\Delta_o + 2P$.

f) Tetrahedral $[FeCl_4]^{2-}$. The iron ion in this complex, which contains four negatively charged Cl^- ion ligands, is Fe^{2+}, which is a d^6 metal ion. All tetrahedral complexes are high spin because Δ_T is much smaller than Δ_o ($\Delta_T = (4/9)\Delta_o$ if the metal ion, the ligands, and the metal–ligand distances are kept constant), so for this complex $S = 2$ and there are four unpaired electrons. The configuration is $e^3t_2^3$. The LFSE is $3(-0.6\Delta_T) + 3(0.4\Delta_T) = -0.6\ \Delta_T$.

g) Tetrahedral $[Ni(CO)_4]$? The neutral CO ligands require that the metal centre in this complex is Ni^0, which is a d^{10} metal atom. Regardless of geometry, complexes of d^{10} metal atoms or ions will never have any unpaired electrons and will always have LFSE = 0, and this complex is no exception.

E20.2 It is clear that π-acidity cannot be a requirement for a high position in the spectrochemical series, since H^- is a very strong field ligand but is not a π-acid (it has no empty *low-energy* acceptor orbitals of local π-symmetry). However, ligands that are very strong σ-bases will increase the energy of the e_g orbitals in an octahedral complex relative to the t_{2g} orbitals. Thus, there are two ways for a complex to develop a large value of Δ_o, by possessing ligands that are π-acids *or* by possessing ligands that are strong σ-bases (of course some ligands, like CN^-, exhibit both π-acidity and moderately strong σ-basicity). A class of ligands that are also very high in the spectrochemical series are alkyl anions, R^- (e.g., CH_3^-). These are not π-acids but, like H^-, are very strong bases.

E20.3 The formula for the spin-only moment is $\mu/\mu_B = [(N)(N + 2)]^{1/2}$ where N is the number of unpaired electrons. Therefore, the spin-only contributions are:

Complex	N	$\mu/\mu_B = [(N)(N + 2)]^{1/2}$
$[Co(NH_3)_6]^{3+}$	0	0
$[Fe(OH_2)_6]^{2+}$	4	4.9
$[Fe(CN)_6]^{3-}$	1	1.7
$[Cr(NH_3)_6]^{3+}$	3	3.9
$[W(CO)_6]$	0	0
$[FeCl_4]^{2-}$	4	4.9
$[Ni(CO)_4]$	0	0

E20.4 We can approach this exercise in two ways: (i) calculate m for each of the given complexes or (ii) calculate the number of unpaired electrons corresponding to each given m value and then do the matching. Exercise 20.3 outlines the first procedure. Here well follow the second procedure for additional practice.

Using the spin-only formula, $\mu = [N(N+2)]^{1/2}$, for magnetic moment we can calculate the number of unpaired electrons that correspond to each given μ value. Solving this equation for N gives a quadratic equation

$$N^2 + 2N - \mu^2 = 0.$$

The solutions (roots) for $\mu = 3.8$:

$$N^2 + 2N - (3.8)^2 = 0$$

$$N^2 + 2N - 14.44 = 0$$

$$N_{1,2} = \frac{-2 \pm \sqrt{2^2 + 4 \times 14.44}}{2}$$

$N_1 = 3$, and $N_2 = -5$.

Clearly, the second root has no physical sense, so the number of unpaired electrons that corresponds to $\mu = 3.8$ is three.

Following the same procedure, we find the N for other magnetic moments:

$$\mu = 0, N = 0,$$

$$\mu = 1.8, N = 1 \text{ and}$$

$$\mu = 5.9, N = 5.$$

The first complex, $[Fe(CN)_6]^{3-}$, is a low spin complex—it contains a strong field ligand and 3+ charge on the metal. The electronic configuration of Fe^3 is d^5, and in low spin complex results in $t_{2g}{}^5e_g{}^0$ configuration with one unpaired electron in t_{2g} level. Therefore, $\mu = 1.8$ corresponds to this complex.

Similar analysis would show that $[Fe(H_2O)_6]^{3+}$ is associated with $5.9\mu_b$ moment, $[CrO_4]^{2-}$ with $0\mu_b$ and $[Cr(H_2O)_6]^{3+}$ with $3.8\mu_b$.

E20.5 The colours of metal complexes are frequently caused by ligand-field transitions involving electron promotion from one subset of d orbitals to another (e.g., from t_{2g} to e_g for octahedral complexes or from e to t_2 for tetrahedral complexes). Of the three complexes given, the lowest energy transition probably occurs for $[CoCl_4]^{2-}$ because it is tetrahedral ($\Delta_T = (4/9)(\Delta_o)$) and because Cl^- is a weak field ligand. This complex is blue because a solution of it will absorb low-energy red light and reflect blue—that is, the complement of red. Of the two complexes that are left, $[Co(NH_3)_6]^{2+}$ probably has a higher energy transition than $[Co(OH_2)_6]^{2+}$ because NH_3 is a stronger field ligand than H_2O (see Table 20.1). The complex $[Co(NH_3)_6]^{2+}$ is yellow because only a small amount of visible light, at the blue end of the spectrum, is absorbed by a solution of this complex. By default, you should conclude that $[Co(OH_2)_6]^{2+}$ is pink.

E20.6 **a) $[Cr(OH_2)_6]^{2+}$ or $[Mn(OH_2)_6]^{2+}$.** The chromium complex is expected to have a larger crystal field stabilization because of the $t_{2g}{}^3e_g{}^1$ configuration $[3 \times (-0.4\Delta_o) + (1 \times 0.6\Delta_o) = -0.6\Delta_o]$ compared to the manganese complex with an electronic configuration $t_{2g}{}^3e_g{}^2$ $[3 \times (-0.4\Delta_o) + (2 \times 0.6\Delta_o) = 0]$.

b) $[Mn(OH_2)_6]^{2+}$ or $[Fe(OH_2)_6]^{3+}$. Although Fe^{3+} and Mn^{2+} are iso-electronic, the higher charge on the Fe ion leads to larger LFSE.

c) $[Fe(OH_2)_6]^{3+}$ or $[Fe(CN)_6]^{3-}$. Since water is a weak field ligand compared to CN^-, the electronic configurations of these two complexes differ. For $[Fe(CN)_6]^{3-}$ the configuration will be $t_{2g}{}^5e_g{}^0$ $[5 \times (-0.4\Delta_o) = -2.0\Delta_o]$, whereas the corresponding configuration for the aquo complex will be $t_{2g}{}^3e_g{}^2$ $[3 \times (-0.4\Delta_o) + (2\times0.6\Delta_o) = 0]$. Hence $[Fe(CN)_6]^{3-}$ will have higher LFSE.

d) $[Fe(CN)_6]^{3-}$ or $[Ru(CN)_6]^{3-}$. The LFSE increases down the group and hence the ruthenium complex will have higher LFSE.

e) Tetrahedral $[FeCl_4]^{2-}$ or tetrahedral $[CoCl_4]^{2-}$. In general, tetrahedral complexes form high-spin complexes. Fe^{2+} with an electronic configuration of $e^3t_2^3$ will have smaller stabilization $[3 \times (-0.6\Delta_T) + (3 \times 0.4\Delta_T) = 0]$ compared to Co^{2+} with a configuration of $e^4\ t_2^3$ $[4 \times (-0.6\Delta_T) + (3 \times 0.4\Delta_T) = -1.2\ \Delta_T]$.

E20.7 As in the answer to Self-Test 20.3, there are two factors that lead to the values given in this question and plotted below: decreasing ionic radius from left to right across the d block, leading to a general increase in ΔH_{latt} from CaO to NiO, and LFSE, which varies in a more complicated way for high-spin metal ions in an octahedral environment, increasing from d^0 to d^3, then decreasing from d^3 to d^5, then increasing from d^5 to d^8, then decreasing again from d^8 to d^{10}. The straight line through the black squares is the trend expected for the first factor, the decrease in ionic radius (the last black square is not a data point, but simply the extrapolation of the line between ΔH_{latt} values for CaO and MnO, both of which have LFSE = 0). The deviations of ΔH_{latt} values for TiO, VO, FeO, CoO, and NiO from the straight line are a manifestation of the second factor, the non-zero values of LFSE for Ti^{2+}, V^{2+}, Fe^{2+}, Co^{2+}, and Ni^{2+}. TiO and VO have considerable metal–metal bonding, and this factor also contributes to their stability.

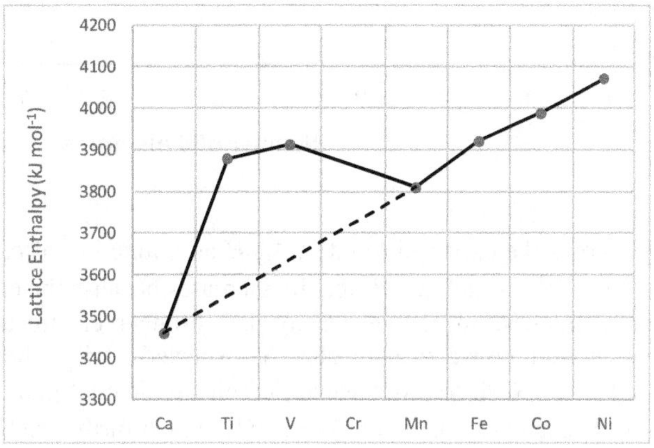

E20.8 **a)** The plot is shown below:

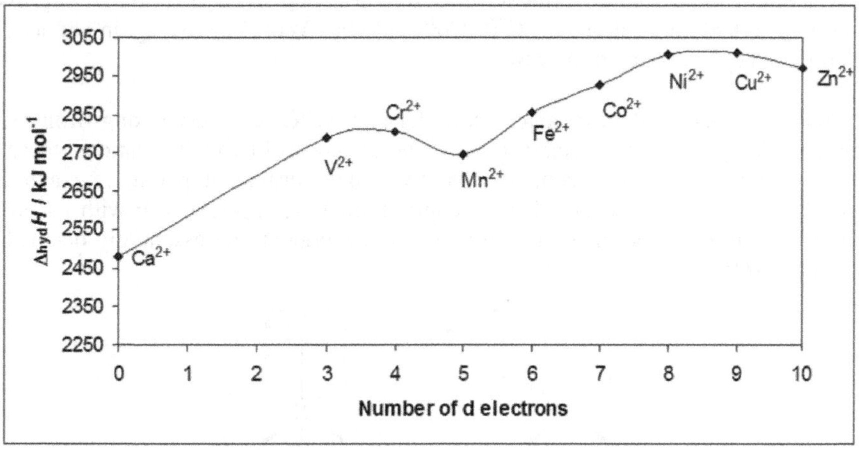

b) The following table lists LFSE in kJ mol^{-1}:

Ion	**Ca^{2+}**	**V^{2+}**	**Cr^{2+}**	**Mn^{2+}**	**Fe^{2+}**	**Co^{2+}**	**Ni^{2+}**	**Cu^{2+}**	**Zn^{2+}**
Δ_o (in cm^{-1})	0	12600	13900	7800	10400	9300	8300	12600	0
Δ_o (in kJ mol^{-1})	0	150.5	166.1	93.2	124.2	111.1	99.2	150.5	0
$t_{2g}\ e_g$ configuration	$t_{2g}^0 e_g^0$	$t_{2g}^3 e_g^0$	$t_{2g}^3 e_g^1$	$t_{2g}^3 e_g^2$	$t_{2g}^4 e_g^2$	$t_{2g}^5 e_g^2$	$t_{2g}^6 e_g^2$	$t_{2g}^6 e_g^3$	$t_{2g}^6 e_g^4$
LFSE (kJ mol^{-1})	0	–180.6	–99.6	0	–49.7	–88.9	–119	–90.3	0

c) After the correction for LFSE is applied (add LFSE to $\Delta_{hyd}H$) the following plot is obtained. The dashed line, which shows almost linear relationship between d electron count and hydration enthalpies, is for LFSE-corrected $\Delta_{hyd}H$ values:

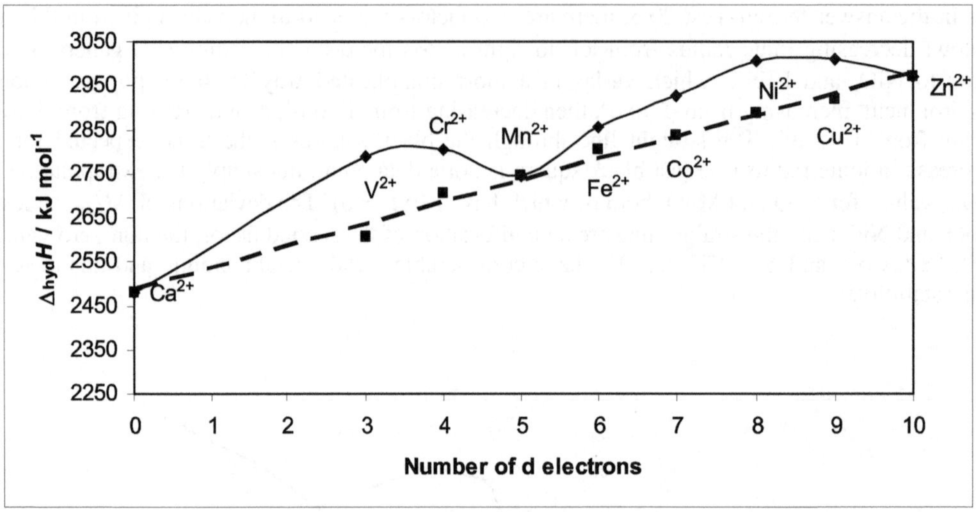

The straight line now shows the expected trend in $\Delta_{hyd}H$ as a function of cationic radius only without electronic effects. As we move from Ca^{2+} to Zn^{2+}, $\Delta_{hyd}H$ steadily increases because the cation radii decrease. A decrease in the ionic radii results in an increase in charge density and stronger electrostatic interaction between cations and partially negative oxygen atom on water molecules. As expected, Ca^{2+} without d electrons has no LFSE and there is no effect on its $\Delta_{hyd}H$ value. With V^{2+} we have a $t_g^3e_g^0$ electronic configuration and three electrons in low-energy t_{2g} level contribute $-0.4\Delta_o$ each ($-1.2\Delta_o$ in total) to LFSE, which further stabilize the hydrated cation over V^{2+}(g). Moving to Cr^{2+} we have to add one electron in high-energy e_g level, and LFSE is increased for $0.6\Delta_o$—as a result we see a decrease in corrected $\Delta_{hyd}H$. Mn^{2+} has a $t_g^3e_g^2$ configuration with both t_{2g} and e_g half-filled and thus no LFSE and no correction to $\Delta_{hyd}H$. As we move from Mn^{2+} we keep adding electrons in t_{2g} level. This results in an increase in stability of hydrated cations for Fe^{2+}, Co^{2+} and Ni^{2+}. With Cu^{2+} we again add an electron into e_g level. Zn^{2+} with filled t_{2g} and e_g levels has zero LFSE.

E20.9 Perchlorate, ClO_4^-, is a very weak basic anion (consider that $HClO_4$ is a very strong Brønsted acid). Therefore, in the compound containing Ni(II), the neutral macrocyclic ligand, and two ClO_4^- anions, there are probably a four-coordinate square planar Ni(II) (d^8) complex and two non-coordinated perchlorate anions. Square planar d^8 complexes are diamagnetic because they have a configuration $(xz, yz)^4(z^2)^2(xy)^2$ with all electrons paired. When SCN^- ligands are added, they coordinate to the nickel ion, producing an essentially octahedral complex that has two unpaired electrons (configuration $t_{2g}^6e_g^2$).

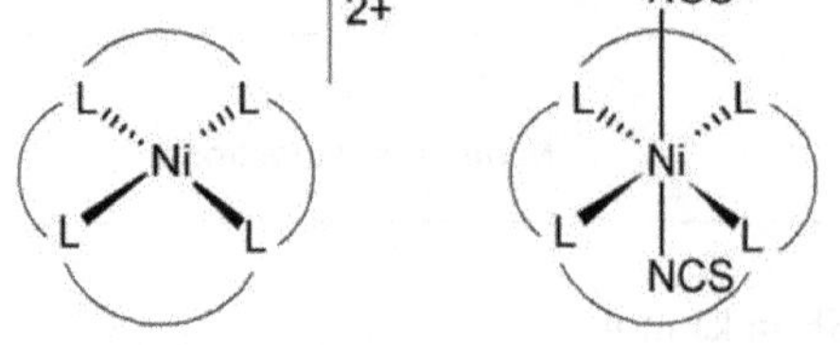

E20.10 The main consequence of the Jahn–Teller theorem is that a nonlinear molecule or ion with a degenerate ground state is not as stable as a distorted version of the molecule or ion if the distortion removes the degeneracy. The high-spin d^4 complex $[Cr(OH_2)_6]^{2+}$ has the configuration $t_{2g}^3e_g^1$, which is orbitally degenerate because the single e_g electron can be in either the d_{z^2} or the $d_{x^2-y^2}$ orbital. Therefore, by the Jahn–Teller theorem, the complex should not have O_h symmetry. A tetragonal distortion, whereby two *trans* metal–ligand bonds are elongated and the other four are shortened, removes the degeneracy. This is the most common distortion observed for octahedral complexes of high-spin d^4, low-spin d^7, and d^9 metal ions, all of which possess e_g degeneracies and exhibit measurable Jahn–Teller distortions. The predicted structure of the $[Cr(OH_2)_6]^{2+}$ ion, with the elongation of the two *trans* Cr–O bonds greatly exaggerated, is shown below.

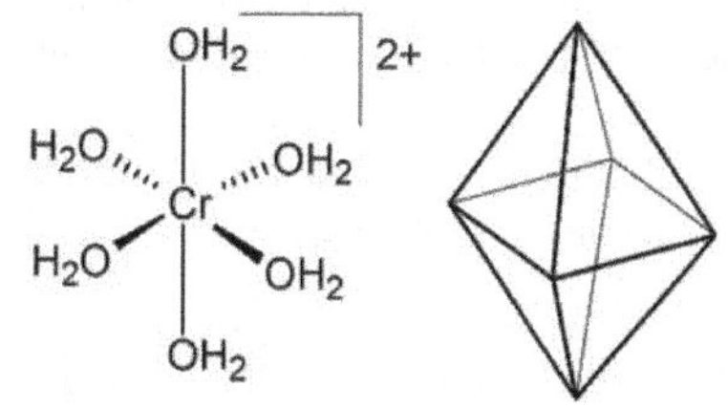

E20.11 The ground state of $[Ti(OH_2)_6]^{3+}$, which is a d^1 complex, is not one of the configurations that usually leads to an observable Jahn–Teller distortion (the three main cases are listed in the answer E20.8). However, the electronic excited state of $[Ti(OH_2)_6]^{3+}$ has the configuration $t_{2g}^0e_g^1$, and so the excited state of this complex possesses an e_g degeneracy. Therefore, the "single" electronic transition is really the superposition of two transitions, one from an O_h ground-state ion to an O_h excited-state ion, and a lower energy transition from an O_h ground-state ion to a lower energy distorted excited-state ion (probably D_{4h}). Since these two transitions have slightly different energies, the unresolved superimposed bands result in an asymmetric absorption peak.

E20.12 **a) $L = 0, S = 5/2$.** For $L = 0$, the term symbol is S. The multiplicity of state with $S = 5/2$ is $2S + 1 = 2 \times 5/2 + 1 = 6$. Thus, the term symbol is ^{6}S.

b) $L = 3, S = 3/2$. The term symbol is ^{4}F.

c) $L = 2, S = 1/2$. This set of quantum numbers is described by the term symbol ^{2}D.

d) $L = 1, S = 1$. This set of quantum numbers is described by the term symbol ^{3}P.

E20.13 **a) ^{3}F, ^{3}P, ^{1}P, ^{1}G?** Recall Hund's rules: (1) the term with the greatest multiplicity lies lowest in energy; (2) for a given multiplicity, the greater the value of L of a term, the lower the energy. Therefore, the ground term in this case will be a triplet, not a singlet (rule 1). Of the two triplet terms, ^{3}F lies lower in energy than ^{3}P: $L = 3$ for ^{3}F, $L = 1$ for ^{3}P (rule 2). Therefore, the ground term is ^{3}F.

b) ^{5}D, ^{3}H, ^{3}P, ^{1}G, ^{1}I? The ground term is ^{5}D because this term has the highest multiplicity.

c) ^{6}S, ^{4}G, ^{4}P, ^{2}I? The ground term is ^{6}S because this term has the highest multiplicity.

E20.14 **a) $4s^1$.** You can approach this exercise in the way described in Section 20.2 for the d^2 configuration (see Table 20.6). You first write down all possible microstates for the s^1 configuration, then write down the M_L and M_S values for each microstate, then infer the values of L and S to which the microstates belong. In this case, the procedure is not lengthy because there are only two possible microstates, (0^+) and (0^-). The only possible values of M_L and M_S are 0 and 1/2, respectively. If M_L can only be 0, then L must be 0, which gives an S term (remember that L can take on all values M_L, $(M_L - 1)$, …, 0, …, $-M_L$). Similarly, if M_S can only be +1/2 or –1/2, then S must be 1/2, which gives a multiplicity $2S + 1 = 2$. Therefore, the one and only term that arises from a $4s^1$ configuration is ^{2}S.

b) $3p^2$. This case is more complicated because there are 15 possible microstates:

	$M_S = -1$	$M_S = 0$	$M_S = 1$
$M_L = 2$		$(1^+, 1^-)$	
$M_L = 1$	$(1^-, 0^-)$	$(1^+, 0^-)$, $(1^-, 0^+)$	$(1^+, 0^+)$
$M_L = 0$	$(1^-, -1^-)$	$(1^+, -1^-)$, $(0^+, 0^-)$, $(-1^+, 1^-)$	$(1^+, -1^+)$
$M_L = -1$	$(-1^-, 0^-)$	$(-1^+, 0^-)$, $(-1^-, 0^+)$	$(-1^+, 0^+)$
$M_L = -2$		$(-1^+, -1^-)$	

A term that contains a microstate with $M_L = 2$ must be a D term ($L = 2$). This term can only be a singlet because if both electrons have $m_l = 1$ they must be spin-paired. Therefore, one of the terms of the $3p^2$ configuration is ^{1}D, which contains five microstates. To account for these, you can cross out $(1^+, 1^-)$, $(-1^+, -1^-)$, and one microstate from each of the other three rows under $M_S = 0$. That leaves ten microstates to be accounted for. The maximum value of M_L of the remaining microstates is 1, so you next consider a P term ($L = 1$). Since M_S can be –1, 0, or 1, this term will be ^3P, which contains nine of the ten remaining microstates. The last remaining microstate is one of the original three for which $M_L = 0$ and $M_S = 0$, which is the one and only microstate that belongs to a ^1S term ($L = 0$, $S = 0$). Of the three terms ^1D, ^3P, and ^1S that arise from the $3p^2$ configuration, the ground term is ^3P because it has a higher multiplicity than the other two terms (see Hund's rule 1).

E20.15 The diagram below outlines the relative energies of the ^3F, ^1D, and ^3P terms. It can be seen that the 10 642 cm^{-1} energy gap between the ^3F and ^1D terms is $5B + 2C$, while the 12 920 cm^{-1} energy gap between the ^3F and ^3P terms is $15B$. From the two equations

$$5B + 2C = 10\,642\ \text{cm}^{-1} \quad \text{and} \quad 15B = 12\,920\ \text{cm}^{-1}$$

you can determine that $B = (12\,920\ \text{cm}^{-1})/(15) = 861.33\ \text{cm}^{-1}$ and $C = 3\,167.7\ \text{cm}^{-1}$.

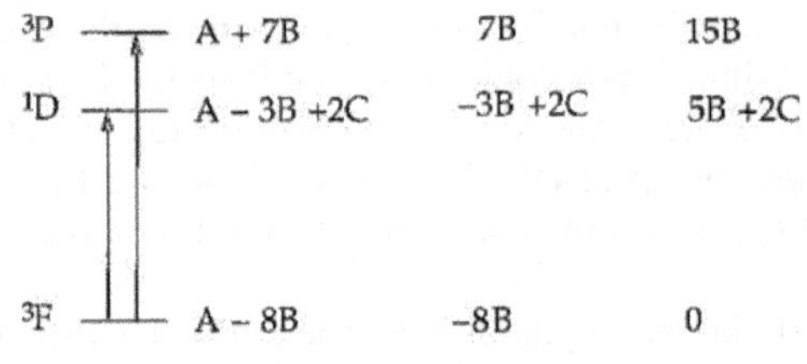

Relative Energies

E20.16 **a)** The oxidation number for Rh in $[Rh(NH_3)_6]^{3+}$ complex is +3, and the electronic configuration of the metal is d^6. According to the d^6 Tanabe–Sugano diagram (see Resource Section 6), the ground term for a low-spin t_{2g}^6 metal ion is $^1A_{1g}$.

b) The Ti^{3+} ion in the octahedral $[Ti(H_2O)_6]^{3+}$ ion has a d^1 configuration. A correlation diagram for d^1 metal ions is shown in Figure 20.3 (Resource Section 6 does not include the d^1 Tanabe–Sugano diagram). According to this diagram, the ground term for a t_{2g}^1 metal ion is $^2T_{2g}$.

c) The Fe^{3+} ion in the octahedral $[Fe(H_2O)_6]^{3+}$ ion has a d^5 configuration. According to the d^5 Tanabe–Sugano diagram, the ground term for a high-spin $t_{2g}^3e_g^2$ metal ion is $^6A_{1g}$.

E20.17 **a) $[Ni(H_2O)_6]^{2+}$ (absorptions at 8 500, 15 400, and 26 000 cm^{-1}).** According to the d^8 Tanabe–Sugano diagram (Resource Section 6), the absorptions at 8 500 cm^{-1}, 15 400 cm^{-1}, and 26 000 cm^{-1} correspond to the following spin-allowed transitions, respectively: $^3T_{2g} \leftarrow {^3A_{2g}}$, $^3T_{1g} \leftarrow {^3A_{2g}}$, and $^3T_{1g} \leftarrow {^3A_{2g}}$. The ratios 15 400/8 500 = 1.8 and 26 000/8 500 = 3.0 can be used to estimate $\Delta_o/B \approx 11$. Using this value of Δ_o/B and the fact that $E/B = \Delta_o/B$ for the lowest-energy transition, $\Delta_o = 8\,500$ cm^{-1} and $B \approx 770$ cm^{-1}. Note that B for a gas-phase Ni^{2+} ion is 1 080 cm^{-1}. The fact that B for the complex is only ~70% of the free ion value is an example of the nephalauxetic effect.

b) $[Ni(NH_3)_6]^{2+}$ (absorptions at 10 750, 17 500, and 28 200 cm^{-1}). The absorptions for this complex are at 10 750 cm^{-1}, 17 500 cm^{-1}, and 28 200 cm^{-1}. The ratios in this case are 17 500/10 750 = 1.6 and 28 200/10 750 = 2.6, and lead to $\Delta_o/B \approx 15$. Thus, $\Delta_o = 10\,750$ cm^{-1} and $B \approx 720$ cm^{-1}. It is sensible that B for $[Ni(NH_3)_6]^{2+}$ is smaller than B for $[Ni(H_2O)_6]^{2+}$ because NH_3 is higher in the nephalauxetic series than is H_2O.

E20.18 Table 20.9 in your textbook links the extinction coefficient values to the electronic band types. The tetrahedral $[MnO_4]^-$ anion contains Mn in oxidation state +7 with d^0 electronic configuration. Any colour this anion has (it is indeed a deep, deep purple) must be due to charge transfer bands because there are no d electrons to cause d-d transitions. Charge transfer, in this case more specifically ligand-to-metal charge transfer, is a completely allowed transition (it is both Laporte and spin allowed) and has the highest ε_{max} value—in this case > 10^4 dm^3 mol^{-1} cm^{-1}.

Fe^{3+} is a d^5 metal cation and in an acidic aqueous solution it exists as a high-spin $[Fe(H_2O)_6]^{3+}$ complex. Consulting the appropriate Tanabe–Sugano diagram, we find the ground term to be $^6A_{1g}$. The same diagram shows no other terms of the same multiplicity. Hence, any absorption in this case is due to both spin and Laporte forbidden transition. These transitions have very low probability of occurring and they have the lowest ε_{max} value—in this case <1 dm^3 mol^{-1} cm^{-1}.

Co^{2+} is also present as an aqua complex, $[Co(H_2O)_6]^{2+}$, in an acidic aqueous solution. This is also a high-spin complex with, for a d^7 metal centre, $t_{2g}^5e_g^2$ electronic configuration. Inspection of the d^7 Tanabe–Sugano diagram, high-spin region, reveals a $^4T_{1g}$ ground state and several spin-allowed d-d transitions. According to Table 20.9, these are weak transitions, but stronger than completely forbidden ones. Thus, this complex has ε_{max} of 10 dm^3 mol^{-1} cm^{-1}.

This leaves us with ε_{max} = 100 dm^3 mol^{-1} cm^{-1} for $[CoCl_4]^{2-}$. This tetrahedral complex has $e_g^4t_2^3$ electronic configuration, and higher ε_{max} can be explained by the non-centrosymmetric structure of this anion.

(Why are all species in this exercise, except MnO_4^-, in acidic solution?)

E20.19 If $[Co(NH_3)_6]^{3+}$, a d^6 complex, were high spin, the only spin-allowed transition possible would be $^5E_g \leftarrow {}^5T_{2g}$ (refer to the d^6 Tanabe–Sugano diagram). On the other hand, if it were low spin, several spin-allowed transitions are possible, including $^1T_{1g} \leftarrow {}^1A_{1g}$, $^1T_{2g} \leftarrow {}^1A_{1g}$, $^1E_g \leftarrow {}^1A_{1g}$, etc. The presence of *two* moderate-intensity bands in the visible/near-UV spectrum of $[Co(NH_3)_6]^{3+}$ suggests that it is low spin. The first two transitions listed above correspond to these two bands. The very weak band in the red corresponds to a spin-forbidden transition such as $^3T_{2g} \leftarrow {}^1A_{1g}$.

E20.20 The d^5 Fe^{3+} ion in the octahedral hexafluorido complex must be high spin (F^- is a weak field ligand). According to the d^5 Tanabe–Sugano diagram, a high-spin complex has no higher energy terms of the same multiplicity as the $^6A_{1g}$ ground term. Therefore, since no spin-allowed transitions are possible, the complex is expected to be colourless (i.e., only very weak spin-forbidden transitions are possible). If this Fe(III) complex were low spin, spin-allowed transitions such as $^2T_{1g} \leftarrow {}^2T_{2g}$, $^2A_{2g} \leftarrow {}^2T_{2g}$, etc. would render the complex coloured. The d^6 Co^{3+} ion in $[CoF_6]^{3-}$ is also high spin, but in this case a single spin-allowed transition, $^5E_g \leftarrow {}^5T_{2g}$, makes the complex coloured and gives it a one-band spectrum.

E20.21 NH_3 and CN^- ligands are quite different with respect to the types of bonds they form with metal ions. Ammonia and cyanide ion are both σ-bases, but cyanide is also a π-acid. This difference means that NH_3 can form molecular orbitals only with the metal e_g orbitals, while CN^- can form molecular orbitals with the metal e_g and t_{2g} orbitals. The formation of molecular orbitals is the way that ligands "expand the clouds" of the metal d orbitals—the bonding between Co^{3+} and the ligand has more covalent character in $[Co(CN)_6]^{3-}$ than in $[Co(NH_3)_6]^{3+}$.

E20.22 The intense band, with $\varepsilon_{max} = 2 \times 10^4$ M^{-1} cm^{-1}, at relatively high energy is undoubtedly both a spin- and Laporte allowed charge-transfer transition because it is too intense to be a ligand field, Laporte-forbidden (d–d) transition (refer to Table 20.9 in the textbook). Furthermore, it is probably an LMCT transition, not an MLCT transition because the ligands do not have empty orbitals necessary for an MLCT transition. The two bands with ε_{max} = 60 and 80 M^{-1} cm^{-1} are probably spin-allowed but Laporte-forbidden ligand field (d–d) transitions. Even though the complex is not strictly octahedral, the ligand field bands are still not very intense. The *very* weak band with ε_{max} = 2 M^{-1} cm^{-1} is most likely a spin-forbidden ligand field (d–d) transition.

E20.23 The Fe^{3+} ions in question are d^5 metal ions. If they were low spin, several spin-allowed ligand field transitions would give the glass a colour even when viewed through the wall of the bottle (see the Tanabe–Sugano diagram for d^5 metal ions). Therefore, the Fe^{3+} ions are high spin, and as such have no spin-allowed transitions (the ground state of an octahedral high-spin d^5 metal ion is $^6A_{1g}$, and there are no sextet excited states). The faint green colour, which is only observed when looking through a *long* pathlength of bottle glass, is caused by spin-forbidden ligand field transitions (recall the Lambert–Beer's law, $A = \varepsilon l c$, where l is the optical pathlength).

E20.24 The blue-green colour of the Cr^{3+} ions in $[Cr(H_2O)_6]^{3+}$ is caused by spin-allowed but Laporte-forbidden ligand field transitions. The relatively low-molar-absorption coefficient, ε, which is a manifestation of the Laporte-forbidden nature of the transitions, is the reason that the intensity of the colour is weak. The oxidation state of chromium in tetrahedral chromate dianion is Cr(VI), which is d^0. Therefore, no ligand field transitions are possible. The intense yellow colour is due to LMCT transitions (i.e., electron transfer from the oxide ion ligands to the Cr(VI) metal centre). Charge transfer transitions are intense because they are both spin-allowed and Laporte-allowed.

E20.25 The d_{z^2} orbital in $[CrCl(NH_3)_5]^{2+}$ complex is left unchanged by each of the symmetry operations of the C_{4v} point group. It therefore has A_1 symmetry. The Cl^- lone electron pairs can form π molecular orbitals with d_{xz} and d_{yz}. These metal atomic orbitals are π-antibonding MOs in $[CoCl(NH_3)_5]^{2+}$ (they are nonbonding in $[Co(NH_3)_6]^{3+}$), and so they will be raised in energy relative to their position in $[Co(NH_3)_6]^{3+}$, in which they were degenerate with d_{xy}.

Since Cl^- ion is not as strong a σ-base as NH_3 is, the d_{z^2} orbital in $[CoCl(NH_3)_5]^{2+}$ will be at lower energy than in $[Co(NH_3)_6]^{3+}$, in which it was degenerate with $d_{x^2-y^2}$. A qualitative d-orbital splitting diagram for both complexes is shown below (L = NH_3).

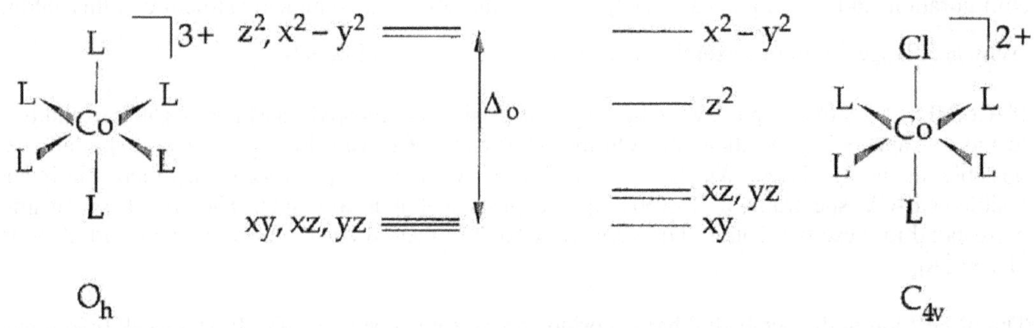

E20.26 As discussed in Section 20.5 *Charge-transfer bands*, ligand field transitions can occur for d-block metal ions with one or more, but fewer than ten, electrons in the metal e and t_2 orbitals. However, the oxidation state of manganese in permanganate anion is Mn(VII), which is d^0. Therefore, no ligand field transitions are possible. The metal e orbitals are the LUMOs, and can act as acceptor orbitals for LMCT transitions. These fully allowed transitions give permanganate its characteristic *intense* purple colour. You can see that two LMCT transitions are possible, E ← A_1 and T_2 ← A_1. These two transitions give rise to the two absorption bands observed at 18 500 cm^{-1} and 32 200 cm^{-1}. The difference in energy between the two transitions, $E(T_2) - E(E)$ = 13700 cm^{-1}, is just equal to Δ_T, as shown in the diagram below:

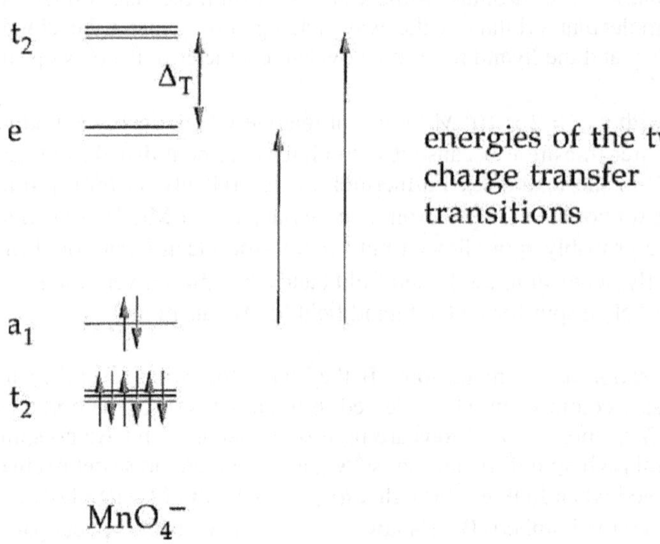

Guidelines for Selected Tutorial Problems

T20.3 See Problem 20.4 for guidelines.

T20.4 From Table RS5.1 in Resource Section 5 we can see how metal orbitals transform in D_{4h} point group. Thus, d_{z^2} and s orbitals transform as A_{1g}, d_{yz} and d_{xz} as E_g, d_{xy} as B_{2g}, $d_{x^2-y^2}$ as B_{1g}, p_z as A_{2u}, and finally p_x and p_y as E_u. Looking in D_{4h} column we can also find ligand orbitals: singly degenerate A_{1g} and B_{1g} and doubly degenerate E_u. These are the only possible ligand symmetry adapted linear combinations that are relevant for purely σ bonding. The molecular orbital diagram is shown on the next page. Note that the metallic d_{xy}, d_{xz}, d_{yz}, and p_z atomic orbitals have no symmetry match on ligand side and they remain non-bonding in a pure σ bonding scenario. These orbitals do have a match if we include π bonding. Analysis of D_{4h} column for ligand SALC reveals a singly degenerate A_{2u} set (a match for p_z metallic orbital) and a doubly degenerate E_g set (a match for d_{xz} and d_{yz}). With inclusion of p bonding, only d_{xy} (of B_{2g} class) remains non-bonding.

Note that this diagram can serve as a basis for tutorial Problem 20.3 as well. *Trans*-[ML_4X_2] complex also has a D_{4h} symmetry, so the metal orbital classes remain the same. It can be derived from square planar ML_4 (for which the diagram applies) by adding two X ligands along *z*-axis. Thus, what we have to add on the ligand side of the molecular orbital diagram is a singly degenerate A_1 set that is a good match for both d_{z^2} and s metallic orbitals. The new ligand orbitals should be *higher* in energy than the ones shown in the figure below because X is lower than L in the electrochemical series. Consult also the solution for Exercise 20.23.

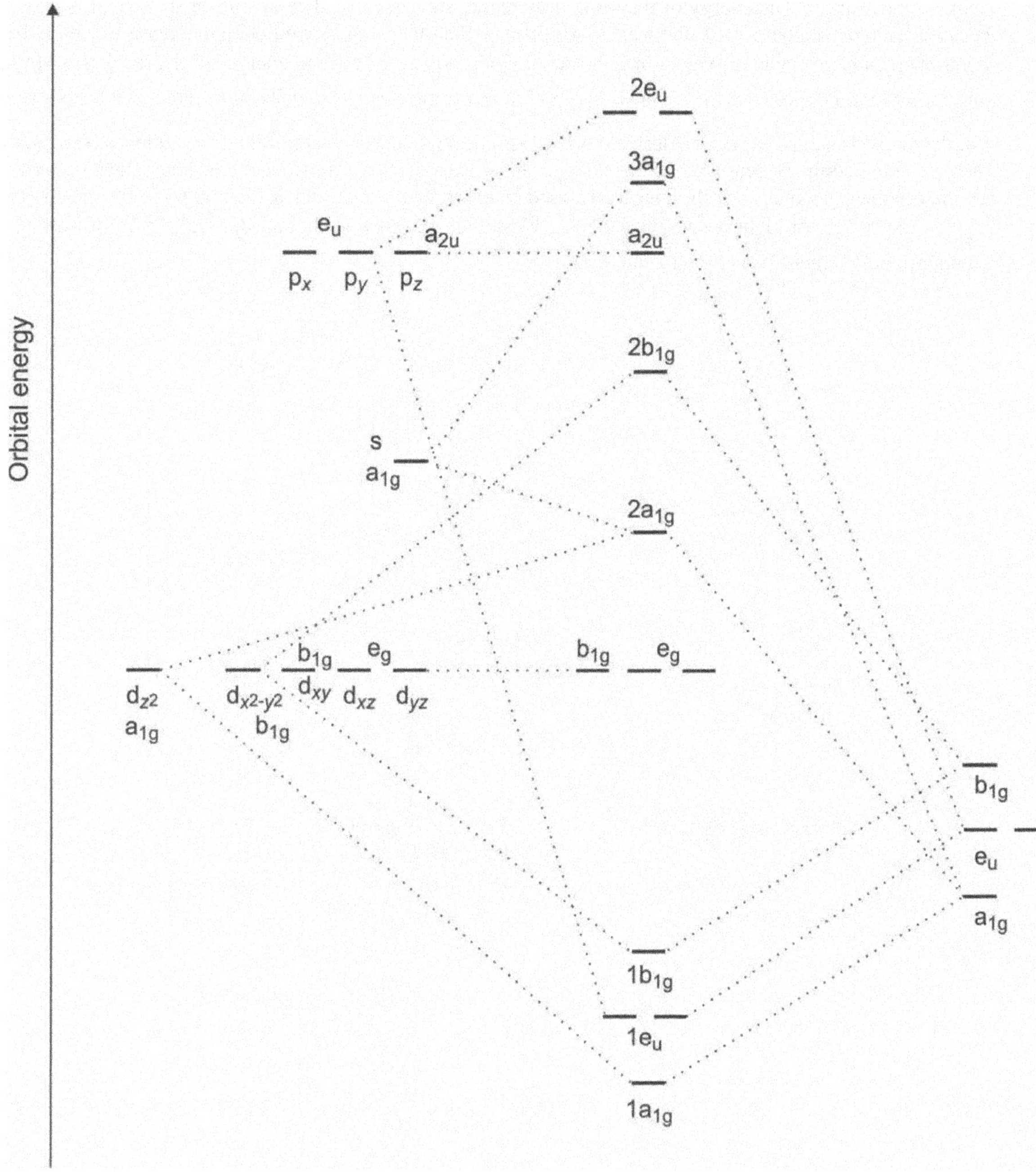

T20.5 The figure below shows the rough orbital diagrams indicating changes in energy as we go from an octahedral complex (O_h) to a tetragonally distorted (D_{4h}) to finally a square planar one when the ligands along *z*-axis have been completely removed. If we elongate the bonds along *z*-axis, the repulsion between ligands (as point charges) and electrons in all d orbitals that have *z* component (d_{z^2}, d_{xz}, and d_{zy}) is going to decrease. Therefore, these d metallic orbitals are going to drop in energy in a tetragonally distorted complex and will break the degeneracy of t_{2g} and e_g sets (the Jahn–Teller theorem). The ligands along *x*- and *y*-axes now can approach metal a bit closer (due to less steric crowding), will repel the electrons in $d_{x^2-y^2}$ and d_{xy} more than in O_h symmetry, and will consequently increase in energy in comparison to starting octahedral complex. Square planar geometry results when we completely remove the ligands along *z* axis. In this case the ligands are approaching the metal centre along *x* and *y* axes and strongly repel with the electrons in $d_{x^2-y^2}$ orbital (remember that this orbital has its lobes located exactly along *x* and *y* axes). The energy of this orbital increases significantly. The energy of d_{xy} orbital is also increasing because its lobes are between *x* and *y* axes. All orbitals that have *z*-axis component decrease in energy because the point charges have been removed along this axis. Note, however, that the energy of d_{z^2} orbital is not the lowest one, as we might expect. This is because d_{z^2} orbital has some electron density in *xy* plane in a torus shape.

Similar analysis can be applied to derive d orbital splitting for linear two coordinate complexes if we, starting from the octahedral complex, remove ligands along *x*- and *y*-axes. In this case we can consider a squished octahedron as an intermediate geometry (with axial M–L bonds shorter than equatorial M–L bonds). Your result should have d_{z^2} as the orbital of highest energy, followed by doubly degenerate d_{xz} and d_{yz}, and finally lowest in energy is another doubly degenerate set of d_{xy} and $d_{x^2-y^2}$.

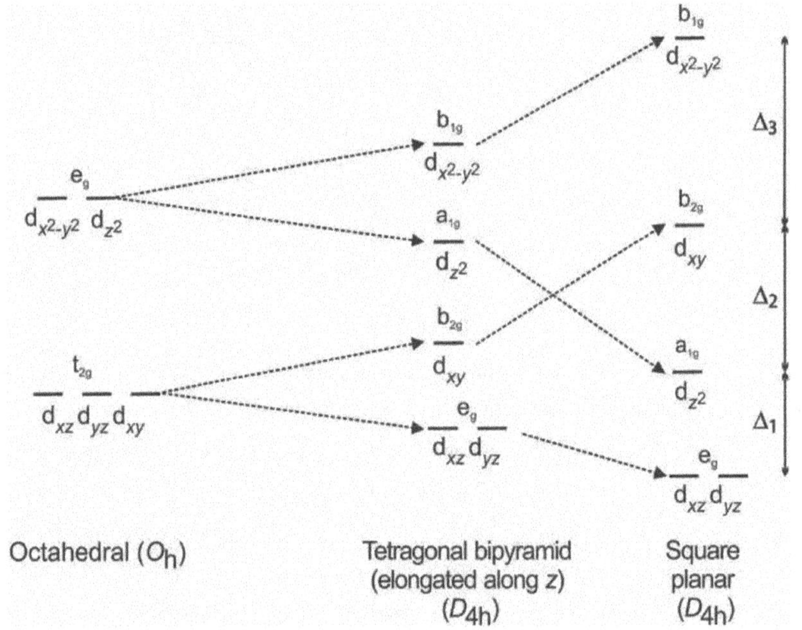

Chapter 21 Coordination Chemistry: Reactions of Complexes

Self-Tests

S21.1 We can use Equation 21.6 to calculate the rate constant:

$$\log k_2(NO_2^-) = Sn_{Pt}(NO_2^-) + C$$

where S is the nucleophilic discrimination factor for this complex, $n_{Pt}(NO_2^-)$ is the nucleophilicity parameter of nitrite ion, and C is the logarithm of the second order rate constant for the substitution for Cl^- in this complex by MeOH. S was determined to be 0.41 in the example, and $n_{Pt}(NO_2^-)$ is given as 3.22. C is a constant for a given complex and is –0.61 in this case, as determined in the example also. Therefore, $k_2(NO_2^-)$ can be calculated as follows:

$$\log k_2(NO_2^-) = (0.41) \times (3.22) - 0.61 = 0.71$$

$$k_2(NO_2^-) = 10^{0.71} = 5.1 \text{ dm}^3 \text{ mol}^{-1} \text{ s}^{-1}.$$

S21.2 For the three ligands in question, Cl^-, NH_3, and PPh_3, the *trans* effect series is $NH_3 < Cl^- < PPh_3$. This means that a ligand *trans* to Cl^- will be substituted at a faster rate than a ligand *trans* to NH_3, and a ligand *trans* to PPh_3 will be substituted at a faster rate than a ligand *trans* to Cl^-. Since our starting material is $[PtCl_4]^{2-}$, two steps involving substitution of Cl^- by NH_3 or PPh_3 must be used.

If you first add NH_3 to $[PtCl_4]^{2-}$, you will produce $[PtCl_3(NH_3)]^-$. Now if you add PPh_3, one of the mutually *trans* Cl^- ligands will be substituted faster than the Cl^- ligand *trans* to NH_3, and the *cis* isomer will be the result. If you first add PPh_3 to $[PtCl_4]^{2-}$, you will produce $[PtCl_3(PPh_3)]^-$. Now if you add NH_3, the Cl^- ligand *trans* to PPh_3 will be substituted faster than one of the mutually *trans* Cl^- ligands, and the *trans* isomer will be the result.

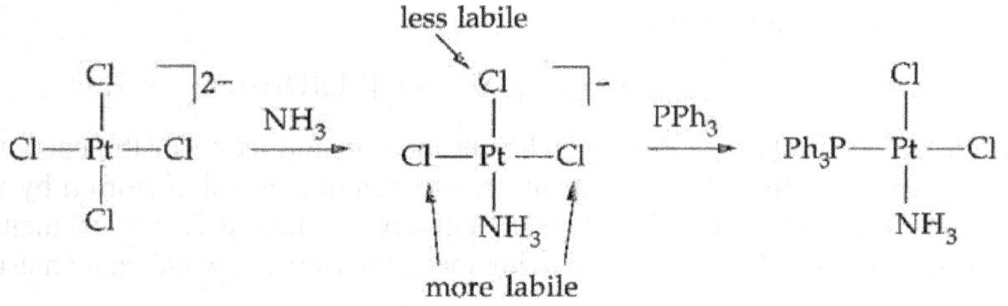

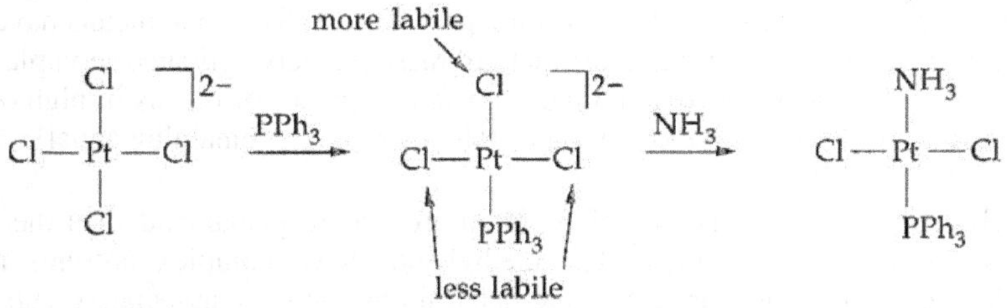

S21.3 As discussed in Section 21.6, the Eigen–Wilkins mechanism for substitution in octahedral complexes suggests the formation of an encounter complex $\{[V(H_2O)_6]^{2+}, Cl^-\}$, in the first, fast step:

$$[V(H_2O)_6]^{2+} + Cl^- \rightarrow \{[V(H_2O)_6]^{2+}, Cl^-\}.$$

The second, slow step of the mechanism is producing the product:

$$\{[V(H_2O)_6]^{2+}, Cl^-\} \rightarrow [VCl(H_2O)_5]^+ + H_2O.$$

The observed rate constant, k_{obs}, is given by:

$$k_{obs} = kK_E$$

and in the case of the substitution of H_2O by Cl^- in $[V(H_2O)_6]^{2+}$ $k_{obs} = 1.2 \times 10^2\ dm^3\ mol^{-1}\ s^{-1}$. You will be able to calculate k if you can estimate a proper value for K_E. Inspection of Table 21.6 shows that $K_E = 1\ dm^3\ mol^{-1}$ for the encounter complex formed by F^- or SCN^- and $[Ni(H_2O)_6]^{2+}$. This value can be used for the reaction in question because (i) the charge and size of Cl^- are similar to those of F^- and SCN^-, and (ii) the charge and size of $[Ni(H_2O)_6]^{2+}$ are similar to those of $[V(H_2O)_6]^{2+}$. Therefore, $k = k_{obs}/K_E = (1.2 \times 10^2\ dm^3\ mol^{-1}\ s^{-1})/(1\ dm^3\ mol^{-1}) = 1.2 \times 10^2\ s^{-1}$.

Exercises

E21.1 If the mechanism of substitution were to be associative, the nature of the incoming ligand should affect the rate of the reaction. This is because the rate-limiting step would require the formation of M–X bonds (X = incoming ligand). In the present case, however, the rate of the reaction does not vary much with the nature or the size of the incoming ligand. Therefore, the more likely mechanism would be *dissociative*.

E21.2 The rate of an associative process depends on the identity of the entering ligand and, therefore, it is not an inherent property of $[M(OH_2)_6]^{n+}$.

E21.3 Since the rate of substitution is the same for a variety of different entering ligands L (i.e., phosphines and or phosphites), the activated complex in each case must not include any significant bond making to the entering ligand. Thus, the reaction must be *d*. If the rate-determining step included any Ni–L bond making, the rate of substitution would change as the electronic and steric properties of L were changed.

E21.4 The rate law for this substitution reaction is:

$$\text{rate} = (kK_E[Mn(OH_2)_6^{2+}][X^-])/(1 + K_E[X^-])$$

where K_E is the equilibrium constant for the formation of the encounter complex $\{[Mn(OH_2)_6]^{2+}, X^-\}$, and k is the first-order rate constant for the reaction:

$$\{[Mn(OH_2)_6^{2+}], X^-\} \rightarrow [Mn(OH_2)_5X]^+ + H_2O.$$

The rate law will be the same regardless of whether the transformation of the encounter complex into products is dissociatively or associatively activated. However, you can distinguish *d* from *a* by varying the nature of X^-. If k varies as X^- varies, then the reaction is *a*. If k is relatively constant as X^- varies, then the reaction is *d*. Note that k cannot be measured directly. It can be found using the expression $k_{obs} = kK_E$ and an estimate of K_E.

E21.5 If ligand substitution takes place by a *d* mechanism, the strength of the metal-leaving group bond is directly related to the substitution rate. Metal centres with high oxidation numbers will have stronger bonds to ligands than metal centres with low oxidation numbers. Furthermore, period 5 and 6 d-block metals have stronger metal ligand bonds (see Section 21.1). Therefore, for reactions that are dissociatively activated, complexes of period 5 and 6 metals are less labile than complexes of period 3 metals, and complexes of metals in high oxidation states are less labile than complexes of metals in low oxidation states (all other factors remaining equal).

E21.6 The two Pt(II) complexes are shown below. Both are square planar and thus the ligand substitution reactions would follow associative mechanism. The methyl-substituted complex presents more steric hindrance to an incoming Cl^- ion nucleophile. Since the rate-determining step for associative substitution of X^- by Cl^- is the formation of a Pt–Cl bond, the more hindered complex will react more slowly.

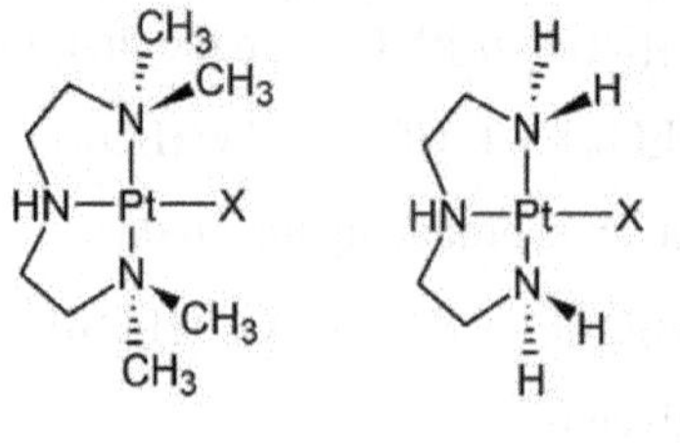

E21.7 Since the rate of loss of chlorobenzene, PhCl, from the tungsten complex becomes faster as the cone angle of L increases, this is a case of dissociative activation (i.e., steric crowding in the transition state accelerates the rate).

E21.8 Since the volume of activation is positive (+11.3 $cm^3\ mol^{-1}$), the activated complex takes up more volume in solution than the reactants, as shown in the drawing below. Therefore, the mechanism of substitution must be dissociative.

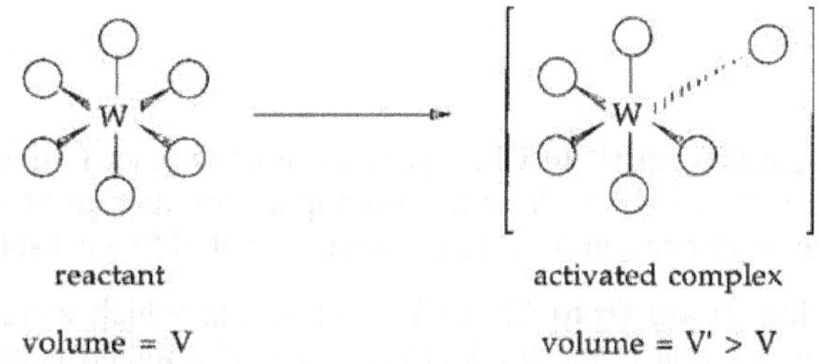

E21.9 The fact that the five-coordinate complex $[Ni(CN)_5]^{3-}$ can be detected does indeed explain why substitution reactions of the four-coordinate complex $[Ni(CN)_4]^{2-}$ are fast. The reason is that, for a detectable amount of $[Ni(CN)_5]^{3-}$ to build up in solution, the forward rate constant k_f must be numerically close to or greater than the reverse rate constant k_r:

$$[Ni(CN)_4]^{2-} + CN^- \underset{k_r}{\overset{k_f}{\rightleftharpoons}} [Ni(CN)_5]^{3-}$$

If k_f were much smaller than k_r, the equilibrium constant $K = k_f/k_r$ would be small and the concentration of $[Ni(CN)_5]^{3-}$ would be too small to detect. Therefore, because k_f is relatively large, you can infer that rate constants for the association of other nucleophiles are also large, with the result that substitution reactions of $[Ni(CN)_4]^{2-}$ are very fast.

E21.10 The entropy of activation is positive, ruling out an associative mechanism (associative mechanisms have uniformly negative entropies of activation). A possible mechanism is loss of one dimethyl sulfide ligand, followed by the coordination of 1,10-phenanthroline, giving an unstable, intermediate complex that can be either five-coordinate (both phen donor atoms coordinated) or four-coordinate (only one phen donor atom coordinated). In the final step, the second dimethyl sulfide ligand leaves and rearranges to form a square planar complex.

E21.11 Starting with $[PtCl_4]^{2-}$, you need to perform two separate ligand substitution reactions. In one, NH_3 will replace Cl^- ion. In the other, NO_2^- ion will replace Cl^- ion. The question is which substitution to perform first. According to the *trans* effect series shown in Section 21.4, the strength of the *trans* effect for the three ligands in question is $NH_3 < Cl^- < NO_2^-$. This means that a Cl^- ion *trans* to another Cl^- will be substituted faster than a Cl^- ion *trans* to NH_3, while a Cl^- ion *trans* to NO_2^- will be substituted faster than a Cl^- ion *trans* to another Cl^- ion.

If you first add NH_3 to $[PtCl_4]^{2-}$, you will produce $[PtCl_3(NH_3)]^-$. Now if you add NO_2^-, one of the mutually *trans* Cl^- ligands will be substituted faster than the Cl^- ligand *trans* to NH_3, and the *cis* isomer will be the result.
If you first add NO_2^- to $[PtCl_4]^{2-}$, you will produce $[PtCl_3(NO_2)]^{2-}$. Now if you add NH_3, the Cl^- ligand *trans* to NO_2^- will be substituted faster than one of the mutually *trans* Cl^- ligands, and the *trans* isomer will be the result. These two-step syntheses are shown below:

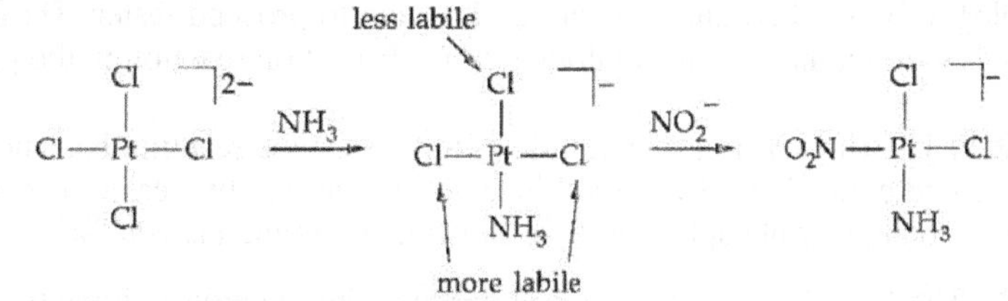

cis isomer

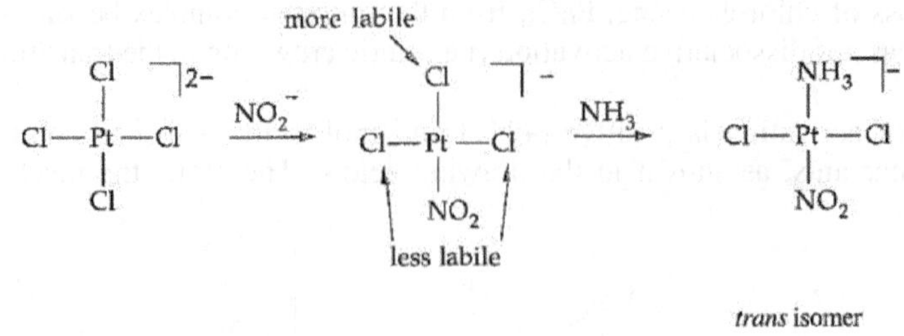

E21.12 **a) Changing a *trans* ligand from H^- to Cl^-.** Hydride ion lies higher in the *trans* effect series than does chloride ion. Thus, if the ligand *trans* to H^- or Cl^- is the leaving group, its rate of substitution will be decreased if H^- is changed to Cl^-. The change in rate can be as large as a factor of 10^4 (see Table 21.4).

b) Changing the leaving group from Cl^- to I^-. The rate at which a ligand is substituted in a square planar complex is related to its position in the *trans* effect series. If a ligand is high in the series, it is a *good* entering group (i.e., a good nucleophile) and a *poor* leaving group. Since iodide ion is higher in the *trans* effect series than chloride ion, it is a poorer leaving group than chloride ion. Therefore, the *iodido* complex will undergo I^- substitution more slowly than the *chlorido* complex will undergo Cl^- substitution.

c) Adding a bulky substituent to a *cis* ligand. This change will hinder the approach of the entering group and will slow the formation of the five-coordinate activated complex. The rate of substitution of a square planar complex with bulky ligands will be slower than that of a comparable complex with sterically smaller ligands.

d) Increasing the positive charge on the complex. If all other factors are kept equal, increasing the positive charge on a square planar complex will increase the rate of substitution. This is because the entering ligands are either anions or have the negative ends of their dipoles pointing at the metal ion. If the charge density of the complex decreases in the activated complex, as would happen when an anionic ligand adds to a cationic complex, the solvent molecules will be *less* ordered around the complex (the opposite of the process called electrostriction). The increased disorder of the solvent makes ΔS less negative (compare the values of ΔS for the Pt(II) and Au(III) complexes in Table 21.5).

E21.13 The general trend is easy to explain: the octahedral $[Co(H_2O)_6]^{3+}$ complex undergoes dissociatively activated ligand substitution. The rate of substitution depends on the nature of the bond between the metal and the leaving group because this bond is partially broken in the activated complex. The rate is independent of the nature of the bond to the entering group because this bond is formed in a step after the rate-determining step. The anomalously high rate of substitution by OH^- signals an alternate path, that of base hydrolysis, as shown below (see Section 21.7):

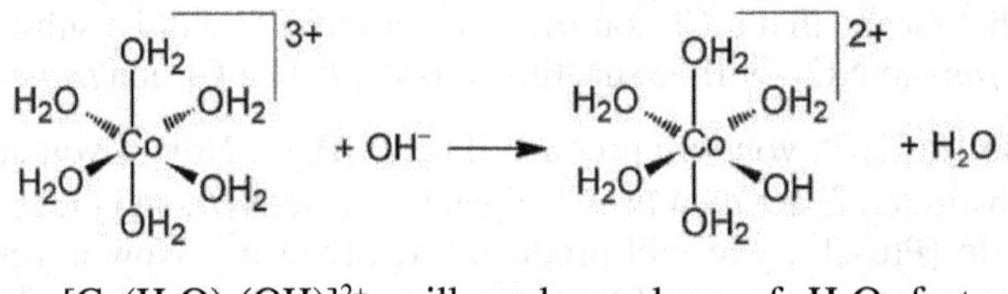

The deprotonated complex $[Co(H_2O)_5(OH)]^{2+}$ will undergo loss of H_2O faster than the starting complex $[Co(H_2O)_6]^{3+}$, because the anionic OH^- ligand is both a better σ and π donor than H_2O. Therefore, the bond *trans* to Co–OH bond is going to be weaker, and substitution is going to proceed faster. The implication is that a complex without protic ligands will not undergo anomalously fast substitution reaction in the presence of OH^-.

E21.14 **a) $[Pt(PR_3)_4]^{2+} + 2Cl^-$.** The first Cl^- ion substitution will produce the reaction intermediate $[Pt(PR_3)_3Cl]^+$. This will be attacked by the second equivalent of Cl^- ion. Since phosphines are higher in the *trans* effect series than chloride ion, a phosphine *trans* to another phosphine will be substituted, giving *cis*-$[PtCl_2(PR_3)_2]$.

b) $[PtCl_4]^{2-} + 2PR_3$. The first PR_3 substitution will produce the reaction intermediate $[PtCl_3(PR_3)]^-$. This will be attacked by the second equivalent of PR_3. Since phosphines are higher in the *trans* effect series than chloride ion, the Cl^- ion *trans* to PR_3 will be substituted, giving *trans*-$[PtCl_2(PR_3)_2]$.

c) *cis*-$[Pt(NH_3)_2(py)_2]^{2+} + 2Cl^-$. You should assume that NH_3 and pyridine are about equal in the *trans* effect series. The first Cl^- ion substitution will produce one of two reaction intermediates, *cis*-$[PtCl(NH_3)(py)_2]^+$ or *cis*-

$[PtCl(NH_3)_2(py)]^+$. Either of these will produce the same product when they are attacked by the second equivalent of Cl^- ion. This is because Cl^- is higher in the *trans* effect series than NH_3 or pyridine. Therefore, the ligand *trans* to the first Cl^- ligand will be substituted faster, and the product will be *trans*-$[PtCl_2(NH_3)(py)]$.

E21.15 The three amine complexes will undergo substitution more slowly than the two aqua complexes. This is because of the higher overall complex charge and their low-spin d^6 configurations. You should refer to Table 21.8, which lists ligand field activation energies (LFAE) for various d^n configurations. While low-spin d^6 is not included, note the similarity between d^3 (t_{2g}^3) and low-spin d^6 (t_{2g}^6). A low-spin octahedral d^6 complex has LFAE = $0.4\Delta_{oct}$, and consequently is inert to substitution. Of the three amine complexes listed above, the iridium complex is the most inert, followed by the rhodium complex, which is more inert than the cobalt complex. This is because Δ_{oct} increases on descending a group in the d-block. Of the two aqua complexes, the Ni complex, with LFAE = $0.2\Delta_{oct}$, undergoes substitution more slowly than the Mn complex, with LFAE = 0. Thus, the order of increasing rate is $[Ir(NH_3)_6]^{3+} < [Rh(NH_3)_6]^{3+} < [Co(NH_3)_6]^{3+} < [Ni(OH_2)_6]^{2+} < [Mn(OH_2)_6]^{2+}$.

E21.16 **a) An increase in the positive charge on the complex.** Since the leaving group (X) is invariably negatively charged or the negative end of a dipole, increasing the positive charge on the complex will retard the rate of M–X bond cleavage. For a dissociatively activated reaction, this change will result in a decreased rate.

b) Changing the leaving group from NO_3^- to Cl^-. Change of the leaving group from nitrate ion to chloride ion results in a decreased rate. The explanation offered in Section 21.6 is that the Co–Cl bond is stronger than the Co–ONO_2 bond. For a dissociatively activated reaction, a stronger bond to the leaving group will result in a decreased rate.

c) Changing the entering group from Cl^- to I^-. This change will have little or no effect on the rate. For a dissociatively activated reaction, the bond between the entering group and the metal is formed subsequent to the rate-determining step.

d) Changing the *cis* ligands from NH_3 to H_2O. These two ligands differ in their σ basicity. The less basic ligand, H_2O, will decrease the electron density at the metal and will destabilize the coordinatively unsaturated activated complex. Therefore, this change from NH_3 to the less basic ligand H_2O will result in a decreased rate (see Section 21.6(b)).

E21.17 The inner-sphere pathway is shown below (solvent = H_2O):

$$[Co(N_3)(NH_3)_5]^{2+} + [V(OH_2)_6]^{2+} \rightarrow \{[Co(N_3)(NH_3)_5]^{2+}, [V(OH_2)_6]^{2+}\}$$

$$\{[Co(N_3)(NH_3)_5]^{2+}, [V(OH_2)_6]^{2+}\} \rightarrow \{[Co(N_3)(NH_3)_5]^{2+}, [V(OH_2)_5]^{2+}, H_2O\}$$

$$\{[Co(N_3)(NH_3)_5]^{2+}, [V(OH_2)_5]^{2+}, H_2O\} \rightarrow [\underset{Co(III)}{(NH_3)_5Co}\text{–N=N=N–}\underset{V(II)}{V(OH_2)_5}]^{4+} + H_2O$$

$$[\underset{Co(III)}{(NH_3)_5Co}\text{–N=N=N–}\underset{V(II)}{V(OH_2)_5}]^{4+} \rightarrow [\underset{Co(II)}{(NH_3)_5Co}\text{–N=N=N–}\underset{V(III)}{V(OH_2)_5}]^{4+}$$

$$H_2O + [\underset{Co(II)}{(NH_3)_5Co}\text{–N=N=N–}\underset{V(III)}{V(OH_2)_5}]^{4+} \rightarrow [Co(NH_3)_5(OH_2)]^{2+} + [V(N_3)(OH_2)_5]^{2+}$$

The pathway for outer-sphere electron transfer is shown below:

$$[Co(N_3)(NH_3)_5]^{2+} + [V(OH_2)_6]^{2+} \rightarrow \{[Co(N_3)(NH_3)_5]^{2+}, [V(OH_2)_6]^{2+}\}$$

$$\{[Co(N_3)(NH_3)_5]^{2+},[V(OH_2)_6]^{2+}\} \rightarrow \{[Co(N_3)(NH_3)_5]^{+}, [V(OH_2)_6]^{3+}\}$$

$$\{[Co(N_3)(NH_3)_5]^{+},[V(OH_2)_6]^{3+}\} + H_2O \rightarrow [Co(NH_3)_5(OH_2)]^{2+} + [V(OH_2)_6]^{3+} + N_3^-$$

In both cases, the cobalt-containing product is the aqua complex because H_2O is present in abundance, and high-spin d^7 complexes of Co(II) are substitution labile. However, something that distinguishes the two pathways is the composition of the vanadium-containing product. If $[V(N_3)(OH_2)_5]^{2+}$ is the product, then the reaction has proceeded via an inner-sphere pathway. If $[V(OH_2)_6]^{3+}$ is the product, then the electron-transfer reaction is outer-sphere. The complex $[V(N_3)(OH_2)_5]^{2+}$ is inert enough to be experimentally observed before the water molecule displaces the azide anion to give $[V(OH_2)_6]^{3+}$.

E21.18 The direct transfer of a ligand from the coordination sphere of one redox partner (in this case the oxidizing agent, $[Co(NCS)(NH_3)_5]^{2+}$) to the coordination sphere of the other (in this case the reducing agent, $[Fe(OH_2)_6]^{2+}$) signals an inner-sphere electron-transfer reaction. Even if the first formed product $[Fe(NCS)(OH_2)_5]^{2+}$ is short lived and undergoes hydrolysis to $[Fe(OH_2)_6]^{3+}$, its fleeting existence demands that the electron was transferred across a Co–(NCS)–Fe bridge.

E21.19 In the case of reactions with Co(II), the difference in the rate of reduction can be explained by different mechanisms for the two reactions. The reduction of $[Co(NH_3)_5(OH)]^{2+}$ proceeds via inner-sphere mechanism: Co(II) complexes are labile (see Fig. 21.1) whereas $[Co(NH_3)_5(OH)]^{2+}$ has OH^- ligand that can serve as a bridging ligand. The $[Co(NH_3)_5(H_2O)]^{3+}$ has no bridging ligands and the only mechanistic pathway for the electron-transfer is outer sphere.

On the other hand, both reactions with $[Ru(NH_3)_6]^{2+}$ as a reducing agent likely proceed via the same mechanism, the outer-sphere mechanism because Ru(II) complex, being a complex of the second row transition element with high LFSE, is more inert than Co(II) complex.

E21.20 Using the Marcus cross-relation (Equation 21.16), we can calculate the rate constants. In this equation $[k_{12} = [k_{11} \cdot k_{22} \cdot K_{12} \cdot f_{12}]^{1/2}$, the values of k_{11} and k_{22} can be obtained from Table 21.13. We can assume f_{12} to be unity. The redox potential data allows us to calculate K_{12} because $E° = [RT/\nu F]\ln K$. The value of $E°$ can be calculated by subtracting the anodic reduction potential (the V^{3+}/V^{2+} couple serves as the anode) from the cathodic one.

a) $[Ru(NH_3)_6]^{3+}$ $k = 4.53 \times 10^3$ dm^3 mol^{-1} s^{-1}

b) $[Co(NH_3)_6]^{3+}$ $k = 1.41 \times 10^{-2}$ dm^3 mol^{-1} s^{-1}

Relative sizes? The reduction of the Ru complex is more thermodynamically favoured than the reaction of the Co complex, and it is faster as evident from its larger k value.

E21.21 Using the Marcus cross-relation (Equation 21.16), we can calculate the rate constants. In this equation $[k_{12} = [k_{11} \cdot k_{22} \cdot K_{12} \cdot f_{12}]^{1/2}$, the values of k_{11} and k_{22} can be obtained from Table 21.13. We can assume f_{12} to be unity. The redox potential data allows us to calculate K_{12} because $E° = [RT/\nu F]\ln K$. The value of $E°$ can be calculated by subtracting the anodic reduction potential (the Cr^{3+}/Cr^{2+} couple serves as the anode) from the cathodic one.

a) $k_{11}(Cr^{3+}/Cr^{2+}) = 1 \times 10^{-5}$ dm^3 mol^{-1} s^{-1}; k_{22} (Ru^{3+}/Ru^{2+} for the hexamine complex) $= 6.6 \times 10^3$ dm^3 mol^{-1} s^{-1}; $f_{12} = 1$; $K_{12} = e^{[nF\varepsilon^o/RT]}$ where $\varepsilon^o = 0.07$ V – (–0.41V) = 0.48 V; $n = 1$; $F = 96485$ C; $R = 8.31$ J mol^{-1} K^{-1} and $T = 298$ K. Using these values, we get $K_{12} = 1.32 \times 10^8$. Substitution of these values in the Marcus cross relationship gives $k_{12} = 2.95 \times 10^3$ dm^3 mol^{-1} s^{-1}.

b) $k_{11}(Cr^{3+}/Cr^{2+}) = 1\times10^{-5}$ dm^3 mol^{-1} s^{-1}; $k_{22}(Fe^{3+}/Fe^{2+}$, aqua complex) = 1.1 dm^3 mol^{-1} s^{-1}; $f_{12} = 1$; $K_{12} = e^{[\nu F E°/RT]}$ where $E° = 0.77$ V – (–0.41V) = +1.18 V; $\nu = 1$; $F = 96485$ C mol^{-1}; $R = 8.31$ J mol^{-1} K^{-1} and $T = 298$ K. Using these values, we get $K_{12} = 9.26 \times 10^{19}$. Substitution of these values in the Marcus cross relation gives $k_{12} = 3.19 \times 10^7$ dm^3 mol^{-1} s^{-1}.

c) k_{11} $(Cr^{3+}/Cr^{2+}) = 1 \times 10^{-5}$ dm^3 mol^{-1} s^{-1}; $k_{22}(Ru^{3+}/Ru^{2+}$ for the bpy complex) $= 4 \times 10^8$ dm^3 mol^{-1} s^{-1}; $f_{12} = 1$; $K_{12} = e^{[\nu F E°/RT]}$ where $E° = 1.26$ V – (–0.41V) = 1.67 V; $\nu = 1$; $F = 96485$ C mol^{-1}; $R = 8.31$ J mol^{-1} K^{-1} and $T = 298$ K. Using these values, we get $K_{12} = 1.81 \times 10^{28}$. Substitution of these values in the Marcus cross-relation gives $k_{12} = 8.51 \times 10^{15}$ dm^3 mol^{-1} s^{-1}.

Note: The assumption that $f_{12} = 1$ is used in all cases; however, a more precise value would be needed for cases (b) and (c) because of higher values of the equilibrium constant.

To compare the obtained value, we first note that k_{11} is the same for all three reactions because the reducing agent is the same and that we approximated f_{12} to be 1. Thus, the differences lie in the reduction potential differences (essentially K_{12} with all reactions being one electron processes) and k_{22} values. Although the iron aqua complexes have a slower self-exchange rate than Ru hexamine ones, the $E_{red} - E_{ox}$ difference (and hence larger K_{12} value) for the reduction of $[Fe(H_2O)_6]^{3+}$ is large enough (and hence K_{12}) to significantly influence the reaction rate with Fe complex. In the case of Ru bpy complexes both the reduction potential difference and the self-exchange rate are large and this reaction has the highest k_{12}.

E21.22 Since the intermediate is believed to be $[W(CO)_5]$, the properties of the entering group (triethylamine versus triphenylphosphine) should not affect the quantum yield of the reaction, which is a measure of the rate of formation of $[W(CO)_5]$ from the excited state of $[W(CO)_5(py)]$. The product of the photolysis of $[W(CO)_5(py)]$ in the presence of excess triethylamine will be $[W(CO)_5(NEt_3)]$, and the quantum yield will be 0.4. This photosubstitution is initiated from a ligand field excited state, not an MLCT excited state. A metal–ligand charge transfer increases the oxidation state of the metal, which would strengthen, not weaken, the bond between the metal and a σ-base–like pyridine.

E21.23 The intense band at ~250 nm is an LMCT transition (specifically a Cl^--to-Cr^{3+} charge transfer). Irradiation at this wavelength should produce a population of $[CrCl(NH_3)_5]^{2+}$ ions that contain Cr^{2+} ions and Cl atoms instead of Cr^{3+} ions and Cl^- ions. Irradiation of the complex at wavelengths between 350 and 600 nm will not lead to photoreduction. The bands that are observed between these two wavelengths are ligand field transitions: the electrons on the metal are rearranged, and the electrons on the Cl^- ion are not involved.

Guidelines for Selected Tutorial Problems

T21.2 As PEt_3 is added, an exchange between coordinated and free PEt_3 would affect the shape of the hydride signal. Once a large excess of PEt_3 is added it is likely that a pentacoordinate complex $[Pt(H)(PEt_3)_4]^+$ is formed. If this complex is fluxtional, then rapid Berry pseudorotation can make any coupling between hydride and phosphorus atom unobservable.

T21.3 The data in the table is related to the Figure 21.1 and Table 21.1 in the main text. All values should be discussed in the context of this section except for $[Cr(H_2O)_5(OH)]^{2+}$ for which more details are given in Exercise 21.13.

T21.8 See Exercises 21.20 and 21.21 for instructions.

Chapter 22 d-Metal Organometallic Chemistry

Self-Tests

S22.1 A Mo atom (group 6) has six valence electrons, and each CO ligand is a two-electron donor. Therefore, the total number of valence electrons on the Mo atom in this compound would be 6 + 7(2) = 20. Since organometallic compounds with more than 18 valence electrons on the central metal are unstable, $[Mo(CO)_7]$ is *not* likely to exist. On the other hand, the compound $[Mo(CO)_6]$, with exactly 18 valence electrons, is very stable.

S22.2 For the donor-pair method, treat Cl as a chloride anion, Cl^-, and $CH_2{=}CH_2$ as a neutral, two-electron donor as stated in the exercise (and Table 22.2). Considering that the anion of Zeise's salt has –1 overall charge, the charge on Pt centre must be +2. Since Pt has 10 valence electrons, Pt(II) must have eight and electronic configuration d^8. Then, the electron count is 8 (from Pt(II)) + 3 × 2e⁻ (from 3 Cl^-) + 1 × 2e⁻ (from $CH_2{=}CH_2$) = 16 electrons. This is a common electron count for d^8 square planar complexes like this one.

S22.3 The formal name of $[Ir(Br)_2(CH_3)(CO)(PPh_3)_2]$ is dibromidocarbonylmethylbis(triphenylphosphine)iridium(III).

S22.4 $Fe(CO)_5$ will have the higher CO stretching frequency and the longer bond. As noted in Example 22.4, PEt_3 is a good σ donor and causes increased M→CO backbonding, a lower CO stretching frequency, and a shorter Fe–C bond.

S22.5 **a) $[(\eta^6\text{-}C_7H_8)Mo(CO)_3]$ (54).** The η^6-C_7H_8 provides six electrons, as does a formally neutral Mo atom. Each carbonyl provides another two electrons, giving a total of 18.

b) $[(\eta^7\text{-}C_7H_7)Mo(CO)_3]^+$ (56)? The η^7-$C_7H_7^+$ group provides six electrons, leaving a formally neutral Mo atom and the three carbonyl groups each providing a further six, giving a total of 18.

S22.6 Consider the reactions of carbonyl complexes discussed in Section 22.18. The Mn–Mn bond in the starting material, $[Mn_2(CO)_{10}]$, can be reductively cleaved with sodium, forming $Na[Mn(CO)_5]$:

$$Mn_2(CO)_{10} + 2Na \rightarrow 2Na[Mn(CO)_5].$$

The anionic carbonyl complex, $[Mn(CO)_5]^-$, is a relatively good nucleophile. When treated with CH_3I, it displaces I^- to form $[Mn(CH_3)(CO)_5]$:

$$Na[Mn(CO)_5] + CH_3I \rightarrow [Mn(CH_3)(CO)_5] + NaI.$$

Many alkyl-substituted metal carbonyls undergo a migratory insertion reaction when treated with basic ligands. The alkyl group (methyl in this case) migrates from the Mn atom to an adjacent C atom of a CO ligand, leaving an open coordination site for the entering group (PPh_3 in this case) to attack:

$$[Mn(CH_3)(CO)_5] + PPh_3 \rightarrow [Mn(COCH_3)(CO)_4(PPh_3)].$$

S22.7 The CO stretching bands at 1857 cm^{-1} and 1897 cm^{-1} are both lower in frequency than typical terminal CO ligands (recall that for terminal CO ligands, $\nu(CO) > 1900\ cm^{-1}$). Therefore, it seems likely that $[Ni_2(\eta^5\text{–}Cp)_2(CO)_2]$ only contains bridging CO ligands. The presence of two bands suggests that the bridging CO ligands are probably *not* colinear because only one band would be observed if they were.

S22.8 The neutral ferrocene contains 18 valence electrons (six from Fe^{2+} and 12 from the two Cp^- ligands). The MO diagram for ferrocene is shown in Figure 22.13. The 18 electrons will fill the available molecular orbitals up to the second a_1' orbital. Since this orbital is the highest occupied molecular orbital (HOMO), oxidation of ferrocene will result in removal of an electron from it, leaving the 17-electron ferricinium cation, $[Fe(\eta^5\text{–}Cp)_2]^+$. If this orbital were strongly bonding, removal of an electron would result in weaker Fe–C bonds. If this orbital were strongly antibonding, removal of an electron would result in stronger Fe–C bonds. However, this orbital is essentially nonbonding (see Figure 22.13). Therefore, oxidation of $[Fe(\eta^5\text{–}Cp)_2]$ to $[Fe(\eta^5\text{–}Cp)_2]^+$ will not produce a substantial change in the Fe–C bond order or the Fe–C bond length.

S22.9 There are four relevant pieces of information provided. First, the fact that $[Fe_4(Cp)_4(CO)_4]$ is a highly coloured compound suggests that it contains metal–metal bonds. Second, the composition can be used to determine the cluster valence electron count, which can be used to predict which polyhedral structure is the likely one:

$(Fe^+)_4$	$4(7e^-)$	= 28 valence e^-
$(Cp^-)_4$	$4(6e^-)$	= 24 valence e^-
$(CO)_4$	$4(2e^-)$	= 8 valence e^-
Total		= 60 valence e^- (a tetrahedral cluster)

Third, the presence of only one line in the ^1H NMR spectrum suggests that the Cp ligands are all equivalent and are η^5. Finally, the single CO stretch at 1640 cm^{-1} suggests that the CO ligands are triply bridging. Also, the facts that there is only one line in ^1H NMR and only one CO stretch suggest a structure of high symmetry (otherwise both NMR and IR spectra would show more lines). A likely structure for $[Fe_4Cp_4(CO)_4]$ is shown below.

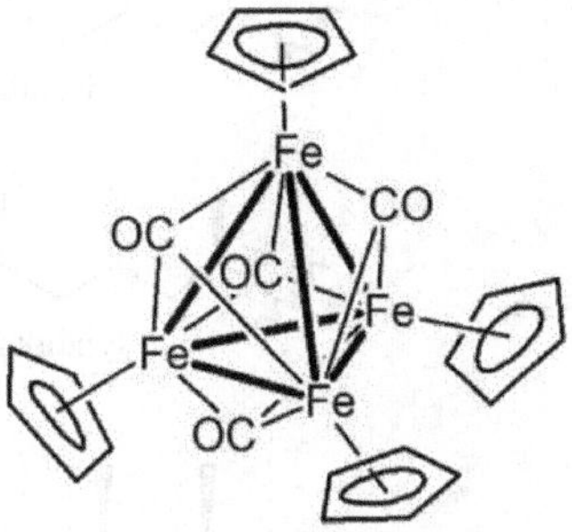

The tetrahedral Fe_4 core of this cluster exhibits six equivalent Fe–Fe bond distances and corresponds to the CVE count from above. Each Fe atom has a Cp ligand with η^5 bonding mode, each able to freely rotate around Fe–Cp bond even at low temperatures (hence one line in ^1H NMR), and each of the four triangular faces of the Fe_4 tetrahedron are capped by a triply bridging CO ligand—the CO ligands are all symmetrical and give one IR stretch.

S22.10 Since $[Mo(CO)_3L_3]$ is a highly substituted complex, the effects of steric crowding must be considered. The steric crowding is particularly important in this case because six-coordinate tricarbonyl complexes adopt a *fac* geometry —all three L ligands would be "squeezed" forming one side of the coordination octahedron. In this case, the two ligands should be very similar electronically but very different sterically. The cone angle for $P(CH_3)_3$, given in Table 22.11, is 118°. The cone angle for $P(t\text{-}Bu)_3$, also given in the table, is 182°. Therefore, because of its smaller size, PMe_3 would be preferred.

S22.11 Fluorenyl compounds are more reactive than indenyl because fluorenyl compounds have two aromatic resonance forms when they bond in the η^3 mode.

S22.12 The starting, six-coordinate palladium compound is an 18-electron Pd(IV) species. The four-coordinate product is a 16-electron Pd(II) species. The decrease in both coordination number and oxidation number by 2 identifies the reaction as a reductive elimination.

S22.13 The ethyl group in $[Pt(PEt_3)_2(Et)(Cl)]$ is prone to β-hydride elimination giving a hydride complex with the loss of ethene, whereas the methyl group in $[Pt(PEt_3)_2(Me)(Cl)]$ is not.

S22.14 In the case of cyclohexene, only one hydroformylation product is possible: cyclohexanecarboxaldehyde. Unlike the unsaturated compounds mentioned in the text and in Example 25.1, cyclohexene, due to its cyclic structure, cannot produce linear and branched isomers.

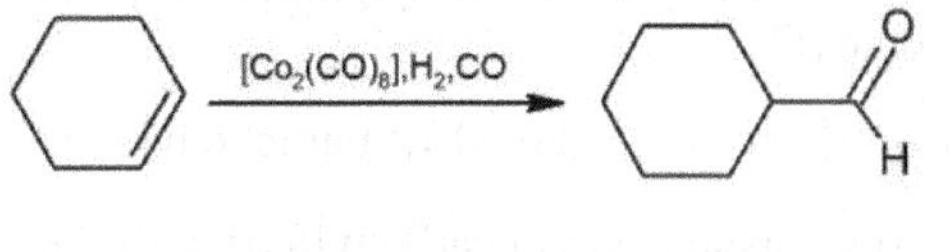

S22.15 The complex $[RhH(CO)(PPh_3)_3]$ is an 18-electron species that must lose a phosphine ligand before it can enter the catalytic cycle:

$$\underset{18\ e^-}{[RhH(CO)(PPh_3)_3]} \rightleftharpoons \underset{16e^-}{[RhH(CO)(PPh_3)_2]} + PPh_3$$

The coordinatively and electronically unsaturated complex $[RhH(CO)(PPh_3)_2]$ can add the alkene that is to be hydroformylated. Added phosphine will shift the above equilibrium to the left, resulting in a lower concentration of the catalytically active 16-electron complex. Thus, you can predict that the rate of hydroformylation will be decreased by added phosphine.

S22.16 The polymerization of mono-substituted alkenes introduces stereogenic centres along the carbon chain at every other position. Without R groups attached to the Zr centre there is no preference for specific binding of new alkenes during polymerization and thus the repeat is random or atactic (shown below).

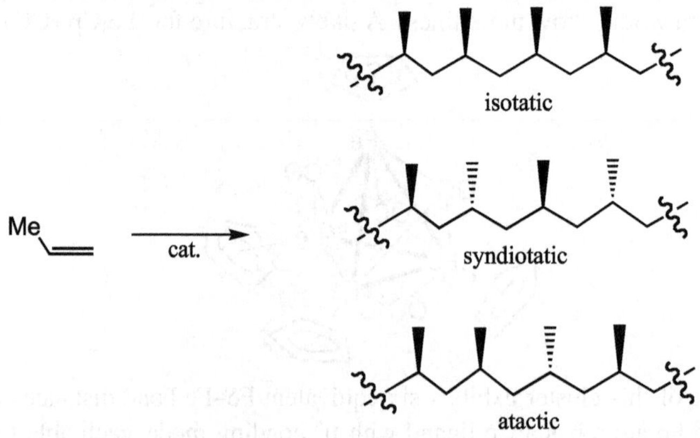

As you can see, the atactic is the most random polymer, meaning that nothing is driving the stereochemistry of the polymer. $[Cp_2ZrCl_2]$ has simple cyclopentadienes as ancillary ligands, which generally do not have enough steric bulk to drive the formation of a specific isomer; therefore, primarily atactic polypropene is produced. Placing bulky alkyl groups on the Cp^- rings offers one way to obtain a specific geometry and thus mediate the properties of polypropene.

Exercises

E22.1 a) **$[Fe(CO)_5]$.** Pentacarbonyliron(0), $18e^-$.

b) **$[Mn_2(CO)_{10}]$.** Decacarbonyldimanganese(0), $18e^-$.

c) **$[V(CO)_6]$.** Hexacarbonylvanadium(0), $17e^-$; can be easily reduced to $[V(CO)_6]^-$.

d) **$[Fe(CO)_4]^{2-}$.** Tetracarbonylferrate(II), $18e^-$.

e) **$[La(\eta^5\text{-}Cp^*)_3]$.** Tris(pentamethylcyclopentadienyl)lanthanum(III), $18e^-$.

f) **$[Fe(\eta^3\text{-allyl})(CO)_3Cl]$.** η^3-allyltricarbonylchloridoiron(II), $18e^-$.

g) **$[Fe(CO)_4(PEt_3)]$.** Tetracarbonyltriethylphosphineiron(0) $18e^-$.

h) **$[Rh(CO)_2(Me)(PPh_3)]$.** Dicarbonylmethyltriphenylphosphinerhodium(I),$16e^-$; a square planar complex, undergoes oxidative addition easily.

i) **$[Pd(Cl)(Me)(PPh_3)_2]$.** Chloridomethylbis(triphenylphosphine)palladium(II), $16e^-$; a square planar complex, undergoes oxidative addition easily.

j) **$[Co(\eta^5\text{-}C_5H_5)(\eta^4\text{-}C_4Ph_4)]$.** η^5-cyclopentadienyl-η^4-tetraphenylcyclobutadienecobalt(I), $18e^-$.

k) **$[Fe(\eta^5\text{-}C_5H_5)(CO)_2]^-$.** η^5-cyclopentadienyldicarbonylferrate(0), $18e^-$.

l) [Cr(η^6-C_6H_6)(η^6-C_7H_8)]. η^6-benzene-η^6-cycloheptatrienechromium(0), 18e$^-$.

m) [Ta(η^5-C_5H_5)$_2$$Cl_3$]. Trichloridobis($\eta^5$-cyclopentadineyl)tantalum(V), 18e$^-$.

n) [Ni(η^5-C_5H_5)NO]. η^5-cyclopentadineylnitrosylnickel(II),16e$^-$; a rather reactive complex.

E22.2

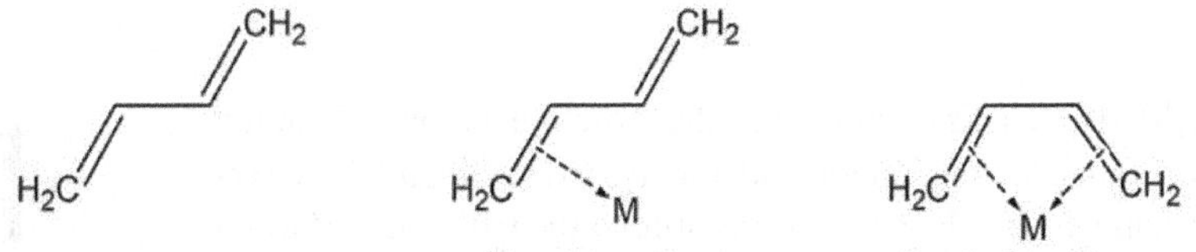

1,3-butadiene η^2 – 1,3-butadiene η^4 – 1,3-butadiene

E22.3

a) C_2H_4? Ethylene coordinates to d-block metals in only one way, using its π-electrons to form a metal–ethylene σ-bond (there may also be a significant amount of back donation if ethylene is substituted with electron-withdrawing groups; see Section 22.9). Therefore, C_2H_4 is always η^2.

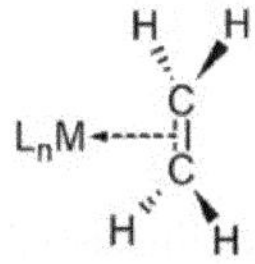

b) Cyclopentadienyl? This is a very versatile ligand that can be η^5 (a six-electron donor in the donor-pair method for electron counting), η^3 (a four-electron donor similar to simple allyl ligands), or η^1 (a two-electron donor similar to simple alkyl and aryl ligands). These three bonding modes are shown below.

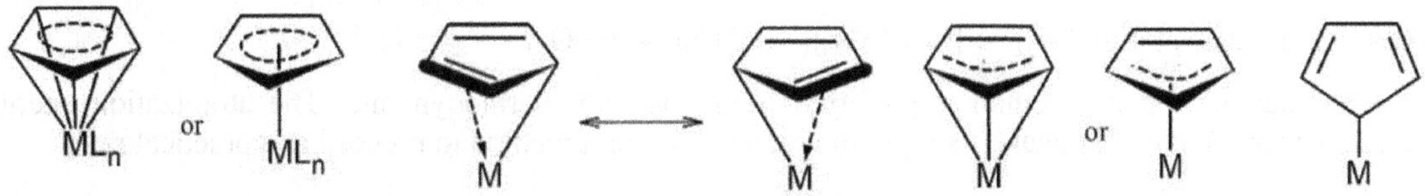

η^5–C_5H_5 η^3–C_5H_5 η^1-C_5H_5

c) C_6H_6? This is also a versatile ligand, which can form η^6, η^4, and η^2 complexes. In such complexes, the ligands are, respectively, six-, four-, and two-electron donors. The structures are very similar to the cyclopentadienyl ones above–there is one extra CH group in the ring.

d) Cyclooctadiene? This ligand can form η^2 (two-electron donor) and η^4 (four-electron donor, shown below) complexes.

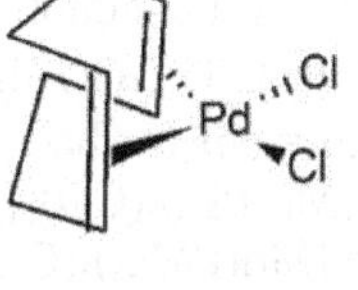

e) Cyclooctatetraene? This ligand contains four C=C double bonds, any combination of which can coordinate to a d-block (or f-block) metal. Thus *cyclo*-C_8H_8 can be η^8 (an eight-electron donor), η^6 (a six-electron donor), η^4 (a four-electron donor, very similar to the one above for cyclooctadiene), and η^2 (a two-electron donor, in which it would resemble an η^2-ethylene complex, except that three of the four C=C double bonds would remain uncoordinated).

E22.4

a) $[Ni(\eta^3\text{-}C_3H_5)_2]$. Bis(allyl)nickel has the structure shown. Each allyl ligand, which is planar, is a four-electron donor in donor-pair method, so the number of valance electrons around the Ni^{2+} cation (Group 10) is 8 + 2 × (4) = 16. Sixteen-electron complexes are very common for Group 9 and Group 10 elements, especially for Rh^+, Ir^+, Ni^{2+}, Pd^{2+}, and Pt^{2+} (all of which are d^8).

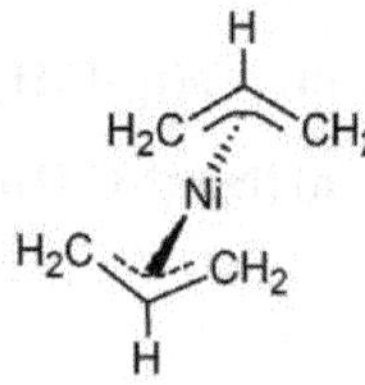

b) $[Co(\eta^4\text{-}C_4H_4)(\eta^5\text{-}C_5H_5)]$. This complex has the structure shown. Since the $\eta^4\text{-}C_4H_4$ ligand is a four-electron donor and the $\eta^5\text{–}C_5H_5^-$ ligand is a six-electron donor, the number of valence electrons around the Co^+ cation (Group 9) is 8 + 4 + 6 = 18.

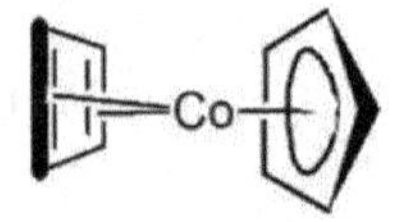

c) $[Co(\eta^3\text{-}C_3H_5)(CO)_3]$. This complex has the structure shown. The electron count for the Co^+ cation is eight, giving a total of 8 + 4 + 3 × 2 = 18 electrons.

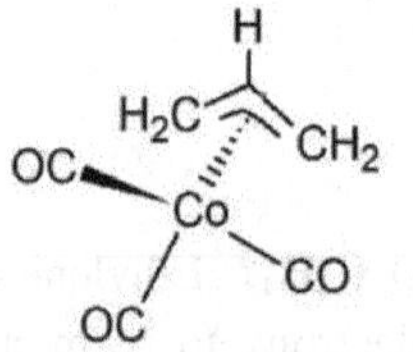

E22.5 As discussed in Section 22.18, the two principal methods for the preparation of simple metal carbonyls are (1) direct combination of CO with a finely divided metal and (2) reduction of a metal salt under CO pressure with a presence of a reducing agent. Two examples are shown below, the preparation of hexacarbonylmolybdenum(0) and octacarbonyldicobalt(0). Other examples are given in the text.

(1) $Mo(s) + 6CO(g) \rightarrow [Mo(CO)_6](s)$ (high temperature and pressure required; the product sublimes easily)

(2) $2CoCO_3(s) + 2H_2(g) + 8CO(g) \rightarrow [Co_2(CO)_8](s) + 2CO_2 + 2H_2O$

The reason that the second method is preferred is kinetic, not thermodynamic. The atomization energy (i.e., sublimation energy) of most metals is simply too high for the first method to proceed at a practical rate.

E22.6 In general, the lower the symmetry of an $[M(CO)_n]$ fragment, the greater the number of CO stretching bands in the IR spectrum. Therefore, given complexes with $[M(CO)_3]$ fragments that have C_{3v}, D_{3h}, or C_s symmetry, the complex that has C_s symmetry will have the greatest number of bands. When you consult Table 22.7, you will see that an $[M(CO)_3]$ fragment with D_{3h} symmetry will give rise to one band, a fragment with C_{3v} symmetry will give rise to two bands, and finally a fragment with C_s symmetry will give rise to three bands.

E22.7 **a) $[Mo(PF_3)_3(CO)_3]$ 2040, 1991 cm^{-1} vs. $[Mo(PMe_3)_3(CO)_3]$ 1945, 1851 cm^{-1}.** The two CO bands of the trimethylphosphine complex are 100 cm^{-1} or more lower in frequency than the corresponding bands of the trifluorophosphine complex because PMe_3 is a σ-donor ligand, whereas PF_3 is a strong π-acid ligand (PF_3 is the ligand that most resembles CO in π acidity). The CO ligands in $[Mo(PF_3)_3(CO)_3]$ have to compete with the PF_3 ligands for electron density from the Mo atom for back-donation. Therefore, less electron density is transferred from the Mo atom to the CO ligands in $[Mo(PF_3)_3(CO)_3]$ than in $[Mo(PMe_3)_3(CO)_3]$. This makes the Mo–C bonds in $[Mo(PF_3)_3(CO)_3]$ weaker than those in $[Mo(PMe_3)_3(CO)_3]$, but it also makes the C–O bonds in $[Mo(PF_3)_3(CO)_3]$ stronger than those in $[Mo(PMe_3)_3(CO)_3]$, and stronger C–O bonds will have higher CO stretching frequencies.

b) $[MnCp(CO)_3]$ 2023, 1939 cm^{-1} vs. $[MnCp^*(CO)_3]$ 2017, 1928 cm^{-1}? The two CO bands of the Cp* complex are slightly lower in frequency than the corresponding bands of the Cp complex. Recall that the Cp* ligand is η^5-C_5Me_5 (pentamethylcyclopentadienyl). Due to the presence of five electron donating CH_3 groups, this ligand is a stronger electron donor than unsubstituted Cp (the inductive effect of the five methyl groups). Therefore, the Mn atom in the Cp* complex has a greater electron density than in the Cp complex, and hence there is more back-donation to the CO ligands in the Cp* complex than in the Cp complex. As explained in part **(a)** above, more back-donation leads to lower stretching frequencies.

E22.8 **a)** The structure below is consistent with the IR spectroscopic data. This D_{3h} complex has all three η^5-C_5H_5 ligands in identical environments. Furthermore, it has two CO ligands bridging three Ni centres, which agrees with the observed single CO stretching band at a relatively low CO stretching frequency.

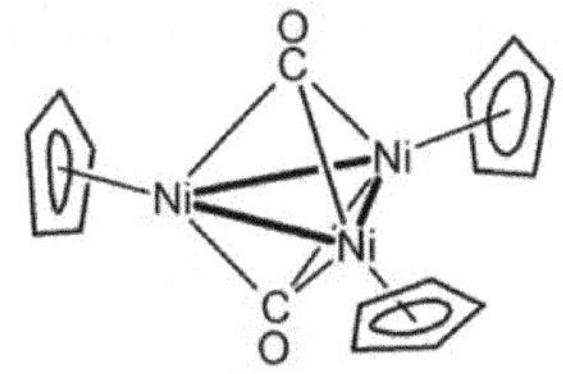

(b) The three Ni atoms are in identical environments, so you only have to determine the number of valence electrons for one of them:

Ni^+	9 valence e^-
η^5–$C_5H_5^-$	6
2 Ni–Ni	2
1/3 (2 CO)	4/3
Total	18 1/3

Therefore, the Ni atoms in this trinuclear complex do not obey the 18-electron rule. Deviations from the rule are common for cyclopentadienyl complexes to the right of the d-block.

E22.9 The 18-electron $[W(CO)_6]$ complex undergoes ligand substitution by a dissociative mechanism. The rate-determining step involves cleavage of a relatively strong W–CO bond. In contrast, the 16-electron $[Ir(CO)Cl(PPh_3)_2]$ complex undergoes ligand substitution by an associative mechanism, which does not involve Ir–CO bond cleavage in the activated complex. Accordingly, $[IrCl(CO)(PPh_3)_2]$ undergoes faster exchange with ^{13}CO than does $[W(CO)_6]$.

E22.10 **a)** The dianionic complex $[Fe(CO)_4]^{2-}$ should be the more basic. The trend involved is the greater affinity for a cation toward a species with a higher negative charge, all other factors being equal. In this case the "other factors" are (1) same set of ligands, (2) same structure (tetrahedral), and (3) same electron configuration (d^{10}) at the metal.

b) The rhenium complex is the more basic. The trend involved is the greater M–H bond enthalpy for a period 6 metal ion relative to a period 4 metal ion in the same group, all other factors being equal. In this case the "other factors" are (1) same set of ligands, (2) same structure (trigonal bipyramidal), and (3) same metal oxidation number (–1). Remember that in the d-block, bond enthalpies such as M–M, M–H, and M–R *increase* down a group. This behaviour is opposite to that exhibited by the p-block elements.

E22.11 **a)** In $[W(\eta^6$–$C_6H_6)(CO)_n]$ $n = 3$ (W contributes six electrons and the benzene ring gives six electrons; three carbonyls at two electrons each gives us 18 electrons for the complex);

b) In $[Rh(\eta^5$–$C_5H_5)(CO)_n]$ $n = 2$ (Rh^+ contributes eight electrons and the Cp^- ring gives six electrons; two COs at two electrons each yield an 18-electron complex;

c) In $[Ru_3(CO)_n]$ $n = 12$ (This is a cluster compound with three Ru atoms in the centre. Table 22.9 predicts a CVE count of 48 electrons. Each Ru contributes eight electrons, giving 24 electrons; to reach 48 we need 12 COs that each contribute two electrons.)

E22.12 i) Reduce $[Mn_2(CO)_{10}]$ with Na to give $[Mn(CO)_5]^-$; react this anion with MeI to give $[MnMe(CO)_5]$:

$$[Mn_2(CO)_{10}] + 2Na \rightarrow 2Na[Mn(CO)_5]$$

$$Na[Mn(CO)_5] + MeI \rightarrow [MnMe(CO)_5] + NaI$$

ii) Oxidize $[Mn_2(CO)_{10}]$ with Br_2 to give $[MnBr(CO)_5]$; displace the bromide with MeLi to give $[MnMe(CO)_5]$:

$$[Mn_2(CO)_{10}] + Br_2 \rightarrow [MnBr(CO)_5]$$

$$[MnBr(CO)_5] + MeLi \rightarrow [MnMe(CO)_5] + LiBr$$

E22.13 The product of the first reaction contains a –C(=O)Ph ligand formed by the nucleophilic attack of Ph^- on one of the carbonyl C atoms:

$$[Mo(CO)_6] + PhLi \rightarrow Li[Mo(CO)_5(C(O)Ph)].$$

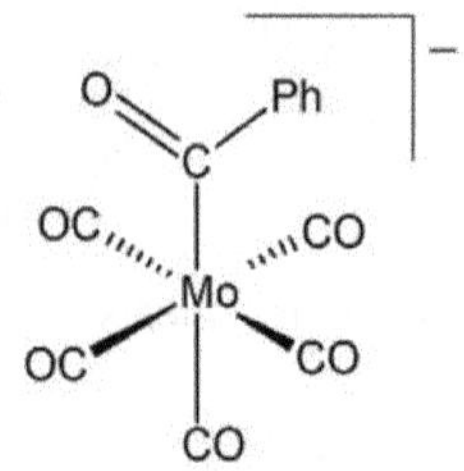

The most basic site on this anion is the acyl oxygen atom, and it is the site of attack by the methylating agent $CH_3OSO_2CF_3$:

$$Li[Mo(CO)_5(COPh)] + CH_3OSO_2CF_3 \rightarrow [Mo(CO)_5(C(OCH_3)Ph)] + LiSO_3CF_3.$$

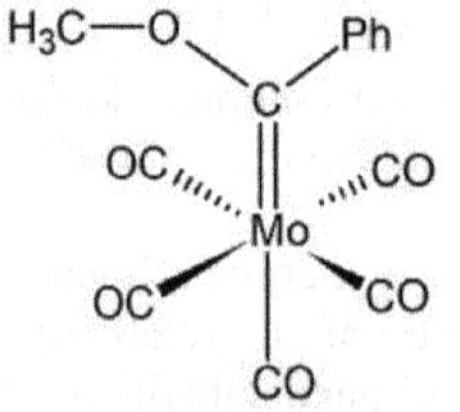

The final product is a carbene (alkylidene) complex. Since the carbene C atom has an oxygen-containing substituent, it is an example of a Fischer carbene (see Section 22.15). The structures of two reaction products are shown above.

E22.14 The compound **A** is prepared from the starting $Na[W(\eta^5\text{–}C_5H_5)(CO)_3]$ complex and 3-chloroprop-1-ene and has the molecular formula $[W(C_3H_5)(C_5H_5)(CO)_3]$. We can write the following chemical equation for this reaction:

$$Na[W(\eta^5\text{–}C_5H_5)(CO)_3] + Cl\text{–}CH_2\text{–}CH{=}CH_2 \rightarrow [W(C_3H_5)(C_5H_5)(CO)_3] + NaCl.$$

Thus, from the reaction, C_3H_5 is an anionic ligand. Recall that both anionic ligands in this complex, C_3H_5 (allyl) and C_5H_5 (cyclopentadienyl), can have multiple hapticities. To decide on the most likely structure of the complex **A** and hence hapticities of the ligands, we should look at the electron count for several possible hapticities. At maximum hapticity (i.e., η^3) $C_3H_5^-$ contributes four electrons, at maximum hapticity (η^5) $C_5H_5^-$ contributes six electrons, three CO ligands contribute six electrons, and W(II) provides four electrons. This would give a total of 20 electrons, that is, an unstable species. This means that either C_3H_5 or C_5H_5 have lower hapticity than originally assumed. Considering that we are just adding a ligand to the complex that already contains $\eta^5\text{–}C_5H_5$ it is reasonable to expect that in **A** $C_5H_5^-$ is going to remain η^5 but that $C_3H_5^-$ is going to be η^1 not η^3. Lowering the hapticity would lower the overall electron count from 20 to 18 resulting in a stable species. Thus compound **A** is likely to be $[W(\eta^1\text{-}C_3H_5)(\eta^5\text{-}C_5H_5)(CO)_3]$ with the structure shown below:

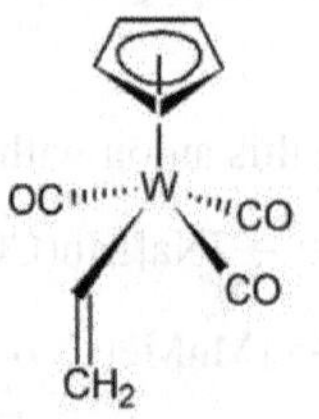

Compound **A**: $[W(\eta^1\text{-}C_3H_5)(\eta^5\text{-}C_5H_5)(CO)_3]$

Going from compound **A** to compound **B** involves the loss of one CO ligand. If we were to just remove one CO from the structure of **A**, without changing the hapticities of either $C_3H_5^-$ or $C_5H_5^-$ ligands, the electron count for the compound **B** would be 16—an unstable compound. Thus, it is very likely that in **B**, the hapticity of $C_3H_5^-$ increases from η^1 to η^3 and thus compensates for the two electrons lost with CO dissociation. Compound **B** is therefore $[W(\eta^3\text{-}C_3H_5)(\eta^5\text{-}C_5H_5)(CO)_2]$ with the likely structure shown below.

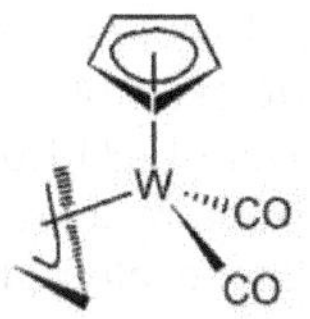

Compound **B**:
$[W(\eta^3\text{-}C_3H_5)(\eta^5\text{-}C_5H_5)(CO)_2]$

Compound **C**, with the molecular formula $[W(C_3H_6)(C_5H_5)(CO)_3][PF_6]$, is formed when **A** is treated with HCl followed by the addition of KPF_6. Note that in **C** $C_3H_5^-$ ligand has been protonated to give neutral, two electron donor $H_3C\text{-}CH{=}CH_2$ ligand. The only simple possibility in this case is η^2 bonding mode. The likely structure of compound **C** is shown below.

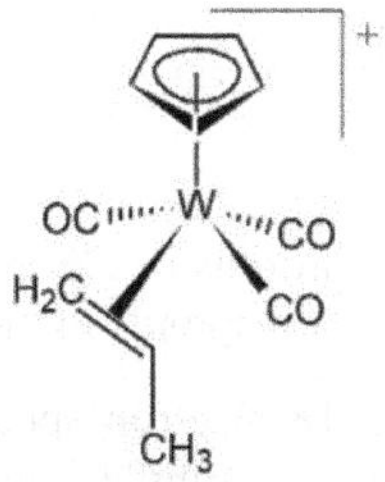

Compound **C**:
$[W(H_2C{=}CH\text{-}CH_3)(\eta^5\text{-}C_5H_5)(CO)_3]$

E22.15 Suggest syntheses of (a) and (b)?

a) $[Mo(\eta^7\text{-}C_7H_7)(CO)_3]BF_4$ from $[Mo(CO)_6]$. Reflux $[Mo(CO)_6]$ with cycloheptatriene to give $[Mo(\eta^6\text{-}C_7H_8)(CO)_3]$; treat this product with the trityl tetrafluoroborate to abstract a hydride from coordinated cycloheptatriene and give the final $[Mo(\eta^7\text{-}C_7H_7)(CO)_3]BF_4$.

$$[Mo(CO)_6] + C_7H_8 \rightarrow [Mo(\eta^6\text{-}C_7H_8)(CO)_3] + 3CO$$

$$[Mo(\eta^6\text{-}C_7H_8)(CO)_3] + [CPh_3][BF_4] \rightarrow [Mo(\eta^7\text{-}C_7H_7)(CO)_3][BF_4] + HCPh_3.$$

b) $[IrCl_2(COMe)(CO)(PPh_3)_2]$ from $[IrCl(CO)(PPh_3)_2]$. React $[IrCl(CO)(PPh_3)_2]$ with MeCl to give $[IrMeCl_2(CO)(PPh_3)_2]$ (oxidative addition) and then expose to CO atmosphere to induce the $-CH_3$ group migration and produce $[IrCl_2(COMe)(CO)(PPh_3)_2]$:

$$[IrCl(CO)(PPh_3)_2] + MeCl \rightarrow [IrMeCl_2(CO)(PPh_3)_2]$$

$$[IrMeCl_2(CO)(PPh_3)_2] + CO \rightarrow [IrCl_2(COMe)(CO)(PPh_3)_2].$$

E22.16 The reaction between $[Fe(CO)_5]$ and cyclopentadiene (C_5H_6) gives compound **A** about which we know the following: it has empirical formula $C_8H_6O_3Fe$, is an 18-electron species, and has a complicated ^1H NMR spectrum. From the empirical formula we can see that **A** must still have three CO ligands—the only way we can explain three O atoms in the empirical formula because cyclopentadiene does not contain oxygen. Also, the only source of hydrogen can be cyclopentadiene, C_5H_6. Both facts tell us that cyclopentadiene is not deprotonated during the reaction, rather it coordinates through either one or both localized C=C bonds in its structure; that is, it can be either η^2 or η^4. Of these, hapticity four gives an 18-electron species: Fe provides eight electrons, three CO ligands six electrons, and η^2 C_5H_6 provides four electrons. Thus, the most likely structure for compound **A** is shown below:

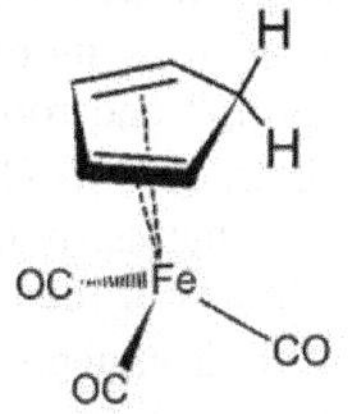

Compound **A**:
$[Fe(\eta^4\text{-}C_5H_6)(CO)_3]$

A complicated ^1H NMR spectrum arises because there are four different hydrogen environments in the structure of **A**—note that the two hydrogens of CH_2 group are not equivalent. This results in four chemical shifts that are not so different in ppm values but have extensive couplings. In the case of **A**, the empirical formula matches the molecular formula.

About compound **B** we know the following: it is formed from **A** and no other reagents/reactants, it has one CO ligand less than **A,** it is an 18-electron species and has two different proton environments—the less shielded one contains five more protons than the shielded one. Simple removal of CO from **A** results in a 16-electron species, thus it does not produce **B**. However, the CO loss could be followed by a proton migration from CH_2 group on the coordinated C_5H_6 to the Fe centre. This pathway produces cyclopentadienyl and hydride ligands. The cyclopentadienyl must be η^5: six electrons from η^5-$C_5H_5^-$, two electrons from H^-, four electrons from the remaining two CO ligands, and six electrons from the Fe^{2+} centre give a total of 18 electrons. The most likely structure is shown below:

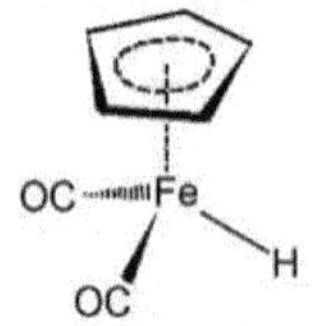

Compound **B**:
[Fe(η^5-C_5H_5)$(CO)_2$(H)]

The above structure also explains the observed ^{1}H NMR. The protons of Cp^- ring will appear as a singlet at 5 ppm due to the rapid rotation of the ring; the deshielded resonance corresponds to the hydride ligand.

Compound **C**, with empirical formula $C_7H_5O_2Fe$, is also an 18-electron species but has only one singlet in ^{1}H NMR and is formed when compound **B** loses H_2. In this case, a simple removal of the coordinated H^- from **B** would result in a 16-electron species. Another possibility is to combine two 17-electron [Fe(η^5-C_5H_5)$(CO)_2$] fragments, obtained after removal of H· radical (two of which produce H_2), and obtain a dimer [Fe(η^5-C_5H_5)$(CO)_2]_2$ that has Fe–Fe bond and an 18-electron count:

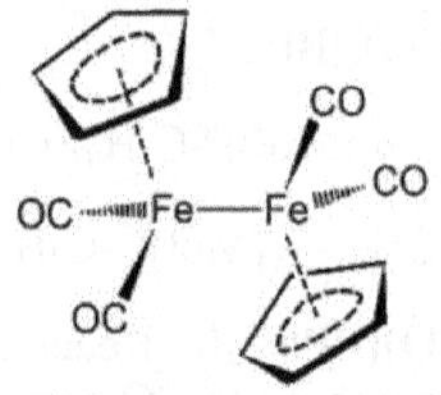

Compound **C**:
[Fe(η^5-C_5H_5)$(CO)_2]_2$

Note that the above structure of compound **C** is a very fluxional one—not only can the Cp rings freely rotate, but the dimer can rotate around Fe–Fe bond, and the CO ligands can change from terminal to bridging mode, resulting only in one signal in the ^{1}H NMR spectrum. Actually, compound **C** probably contains two bridging CO and two terminal CO ligands. However, to really support this bridging structure we would need additional IR data (see Figure 22.12 in your textbook). While compound **A** had molecular and empirical formulas the same, the molecular formula of compound **C** is twice its empirical formula.

E22.17 In either case compounds formed are tetraalkyl titanium(IV) complexes, TiR_4. Notice that neither the methyl nor trimethylsilyl groups have β hydrogens. Thus, unlike [$TiEt_4$], [$TiMe_4$] and [Ti$(CH_2SiMe_3)_4$], cannot undergo a low energy β-hydride elimination decomposition reaction.

E22.18 Recall that π-bound aromatic cyclic ligands (such as Cp^- and C_6H_6) are not only free to rotate around ring–metal axis, but can also “slip” from higher to lower hapticity modes. It is reasonable to expect tetracyclopentadienyl-titanium(IV) to be the product in the reaction between $TiCl_4$ and NaCp. However, due to small radius of Ti^{4+}, it is very unlikely that all four Cp^- rings are η^5; this coordination mode for all four would place 20 CH groups around small Ti^{4+} and result in significant steric crowding. Thus it is more likely that some Cp^- rings have lower hapticities (i.e., 3 or 1). Considering that Cp^- can both “slip” and rotate, at room temperature we see only one resonance as an effect of averaging the proton environments. At –40°C we observe two different signals of equal intensity. This tells us that there are two different environments, both of which contain two Cp^- rings. If we continue cooling, *one* of these signals is split into three signals of intensity 1 : 2 : 2. The ratio corresponds to the η^1-Cp coordination mode that has three distinct proton environments. Based on the above analysis, the actual product is most likely [Ti(η^1-C_5H_5)$_2$(η^5-C_5H_5)$_2$] with the low-temperature structure shown below:

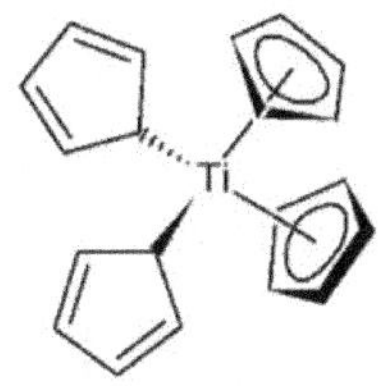

E22.19 Give the equation for a workable reaction for the conversion of

a) [Fe(η^5-C_5H_5)$_2$] to [Fe(η^5-C_5H_5)(η^5-$C_5H_4COCH_3$)]. The Cp^- rings in cyclopentadienyl complexes behave like simple aromatic compounds such as benzene, and so are subject to typical reactions of aromatic compounds such as Friedel–Crafts alkylation and acylation. If you treat ferrocene with acetyl chloride and some aluminum(III) chloride as a catalyst, you will obtain the desired compound:

$$[Fe(\eta^5\text{-}C_5H_5)_2] + CH_3COCl \xrightarrow{AlCl_3} [Fe(\eta^5\text{-}C_5H_5)(\eta^5\text{-}C_5H_4COCH_3)] + HCl$$

b) [Fe(η^5-C_5H_5)$_2$] to [Fe(η^5-C_5H_5)(η^5-$C_5H_4CO_2H$)]. The following sequence of reactions is the best route:

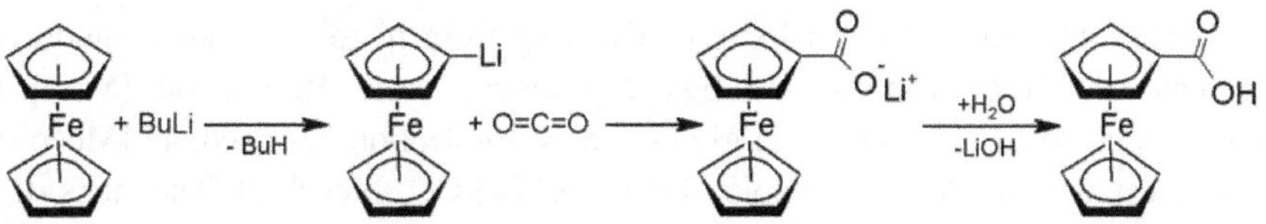

E22.20 The symmetry-adapted orbitals of the two eclipsed C_5H_5 rings in a metallocene are shown in Resource Section 5. On the metal the d_{z^2} and s orbitals have an a_1' symmetry and thus have correct symmetry to form the MOs with the a_1' symmetry-adapted orbital on Cp rings. Other atomic orbitals on the metal do not have the correct symmetry. Therefore, three a_1' MOs will be formed. Figure 22.13 shows the energies of the three a_1' MOs: one is the most stable orbital of a metallocene involving metal–ligand bonding, one is relatively nonbonding (it is the HOMO in [$FeCp_2$]), and one is a relatively high-lying antibonding orbital.

E22.21 Protonation of [$FeCp_2$] at iron does not change its number of valence electrons: both [$FeCp_2$] and [$FeCp_2H$]$^+$ are 18-electron species:

[$FeCp_2$]		**[$FeCp_2H$]$^+$**	
Fe^{2+}	6 electrons	Fe^{3+}	5 electrons
2 Cp^- (6 electrons each)	12 electrons	2 Cp^- (6 electrons each)	12 electrons
		H^-	2 electrons
		Positive charge	–1 electron
Total	**18 electrons**	**Total**	**18 electrons**

Since [$NiCp_2$] is a 20-electron complex, the hypothetical metal-protonated species [$NiCp_2H$]$^+$ would also be a 20-electron complex. On the other hand, protonation of [$NiCp_2$] at a Cp carbon atom produces the 18-electron complex [NiCp(η^4-C_5H_6)]$^+$. Therefore, the reason that the Ni complex is protonated at a carbon atom is that a more stable (i.e., 18-electron) product is formed.

E22.22 a) [Mn(CO)$_5$(CF_2)]$^+$ + H_2O → [Mn(CO)$_6$]$^+$ + 2HF. The =CF_2 carbene ligand is similar electronically to a Fischer carbene (see Section 22.15). The electronegative F atoms render the C atom subject to nucleophilic attack, in this case by a water molecule. The initially formed [(CO)$_5$Mn(CF_2(OH_2))]$^+$ complex (as a result of nucleophilic attack of H_2O) can then eliminate two equivalents of HF, as shown in the following mechanism:

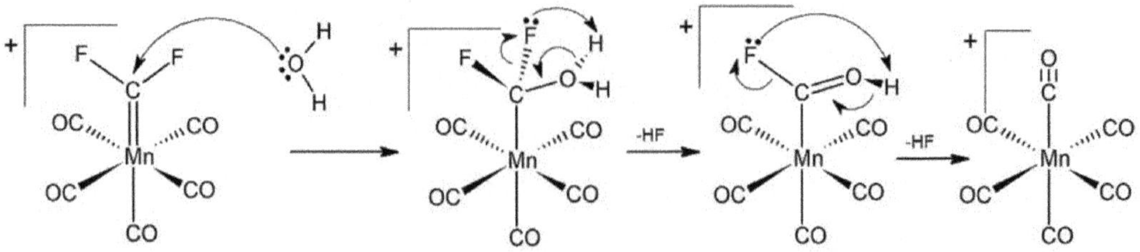

b) $[Rh(C_2H_5)(CO)(PR_3)_2] \rightarrow [RhH(CO)(PR_3)_2] + C_2H_4$. This is an example of a β-hydrogen elimination reaction, discussed at the beginning of Section 22.26. This reaction is believed to proceed through a cyclic intermediate involving a (3c,2e) M–H–C interaction, as shown in the mechanism below:

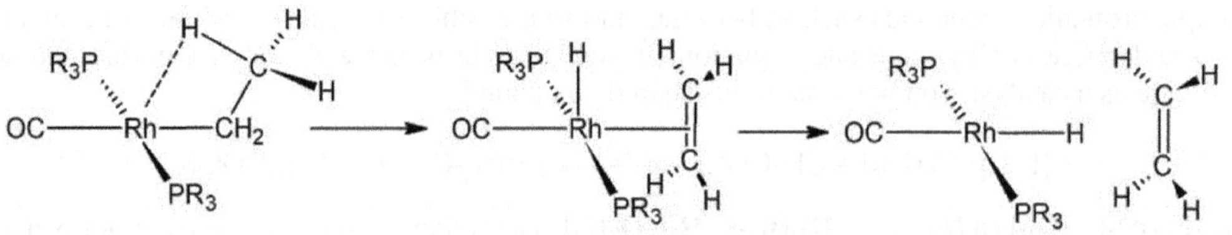

E22.23 One of the possible routes involves initial slip of Cp^- ring from η^5 to η^3 bonding mode creating a 16-electron complex that can coordinate a phosphine ligand producing new 18-electron $[Mo(\eta^3\text{-}C_5H_5)(CO)_3Me(PR_3)]$ complex. This complex loses one CO ligand giving a new 16-electron intermediate $[Mo(\eta^3\text{-}C_5H_5)(CO)_2Me(PR_3)]$ that converts to the stable 18-electron complex $[Mo(\eta^5\text{-}C_5H_5)(CO)_2Me(PR_3)]$. The steps of this route are shown below:

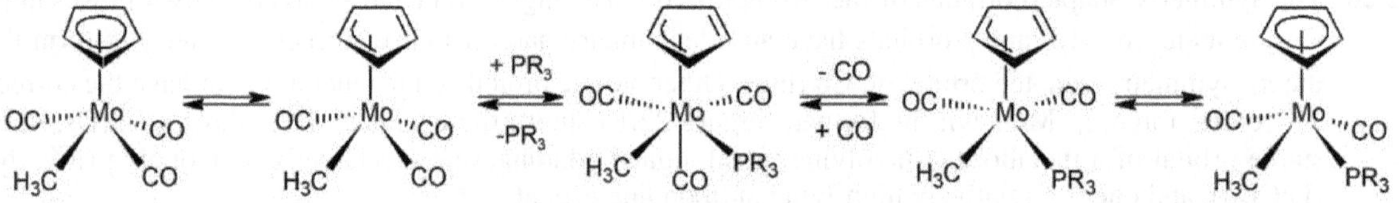

The second route starts with CO insertion (or methyl group migration) to produce a 16-electron acyl complex. This complex can coordinate one PR_3 ligand to give an 18-electron species, which if heated can lose one CO ligand and undergo reverse alkyl group migration. The steps for this route are outlined below:

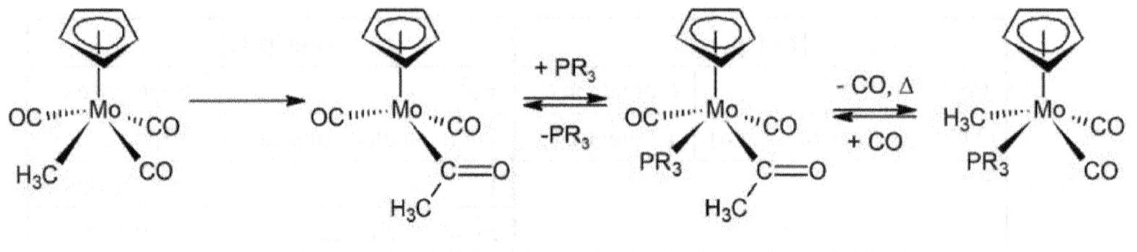

E22.24 **a) CVE count characteristic of octahedral and trigonal prismatic complexes.** According to Table 22.9, an octahedral M_6 will have 86 cluster valence electrons and a trigonal prismatic M_6 cluster will have 90.

b) Can these CVE values be derived from the 18-electron rule? In the case of the trigonal prismatic arrangement, the CVE value can be derived from the 18-electron rule: (6 M centres × 18 electrons at each) – (9 M–M bonds in the cluster × 2 electrons in each M–M bond) = 90 electrons. However, for the octahedral arrangement, the CVE value cannot be derived from the 18-electron rule.

c) Determine the probable geometry of $[Fe_6(C)(CO)_{16}]^{2-}$ and $[Co_6(C)(CO)_{15}]^{2-}$. The iron complex has 86 CVEs, whereas the cobalt complex has 90 (see the calculations table). Therefore, the iron complex probably contains an octahedral Fe_6 array, whereas the cobalt complex probably contains a trigonal-prismatic Co_6 array.

$[Fe(C)(CO)_{16}]^{2-}$		$[Co(C)(CO)_{15}]^{2-}$	
6 Fe (8 electrons each)	48 electrons	6 Co (9 electrons each)	54 electrons
C (4 electrons)	4 electrons	C (4 electrons)	4 electrons
16 CO (2 electrons each)	32 electrons	15 CO (2 electrons each)	30 electrons
2– charge	2 electrons	2– charge	2 electrons
Total	86 electrons	**Total**	90 electrons

E22.25 **a) $[Co_3(CO)_9CH]$ → OCH_3, $N(CH_3)_2$, or $SiCH_3$.** Isolobal groups have the same number and shape of valence orbitals *and* the same number of electrons in those orbitals. The CH group has three sp^3 hybrid orbitals that contain a single electron each. The $SiCH_3$ group has three similar orbitals, similarly occupied, so it is isolobal with CH and would probably replace CH in $[Co_3(CO)_9CH]$ to form $[Co_3(CO)_9SiCH_3]$. In contrast, the OCH_3 and $N(CH_3)_2$ groups are not isolobal with CH. Instead, they have, respectively, three sp^3 orbitals that contain a pair of electrons each and two sp^3 orbitals that contain a pair of electrons each.

b) $[(OC)_5MnMn(CO)_5]$ → I, CH_2, or CCH_3? The $Mn(CO)_5$ group has a single σ-type orbital that contains a single electron. An iodine atom is isolobal with it because I atom also has a singly occupied s orbital. Therefore, you can expect the compound $[Mn(CO)_5I]$ to be reasonably stable. In contrast, the CH_2 and CCH_3 are not isolobal with $Mn(CO)_5$. The CH_2 group has either a doubly occupied σ orbital and an empty p orbital or a singly occupied σ orbital and a singly occupied p orbital. The CCH_3 group has three singly occupied σ orbitals (note that it is isolobal with $SiCH_3$).

E22.26 If the rate-determining step in the substitution is cleavage of one of the metal–metal bonds in the cluster, the cobalt complex will exhibit faster exchange. This is because metal–metal bond strengths *increase* down a group in the d-block. Therefore, Co–Co bonds are weaker than Ir–Ir bonds if all other factors (such as geometry and ligand types) are kept the same.

E22.27 The catalytic cycle for homogeneous hydrogenation of alkenes by Wilkinson's catalyst, $[RhCl(PPh_3)_3]$, is shown in Figure 22.18. For the olefin substrate to coordinate to the Rh centre in complex **C** (a 16-electron species), complex **B** (an 18-electron species) must first lose one PPh_3 ligand. Although this is not usually shown when depicting catalytic cycles, we have to keep in mind that all steps are (generally) reversible, and complex **C** and free PPh_3 are in equilibrium with complex **B**. Increasing the concentration of PPh_3 shifts the equilibrium toward complex **B**, which is not able to bind the olefin and the hydrogenation rate will drop.

E22.28 The data show that hydrogenation is faster for hexene than for *cis*-4-methyl-2-pentene (a factor of 2910/990 ≈ 3). It is also faster for cyclohexene than for 1-methylcyclohexene (a factor of 3160/60 ≈ 53). In both cases, the alkene that is hydrogenated slowly has a greater degree of substitution and so is sterically more demanding—hexene is a monosubstituted alkene, whereas *cis*-4-methyl-2-pentene is a disubstituted alkene; cyclohexene is a disubstituted alkene, whereas 1-methylcyclohexene is a trisubstituted alkene. According to the catalytic cycle for hydrogenation shown in Figure 22.18, different alkenes could affect the equilibrium (B) → (C) or the reaction (C) → (D). A sterically more demanding alkene could result in (i) a smaller equilibrium concentration of (D) $[RhClH_2L_2(alkene)]$, or (ii) a slower rate of conversion of (C) to (D). Since the reaction (C) → (D) is the rate-determining step for hydrogenation by $RhCl(PPh_3)_3$, either effect would lower the rate of hydrogenation.

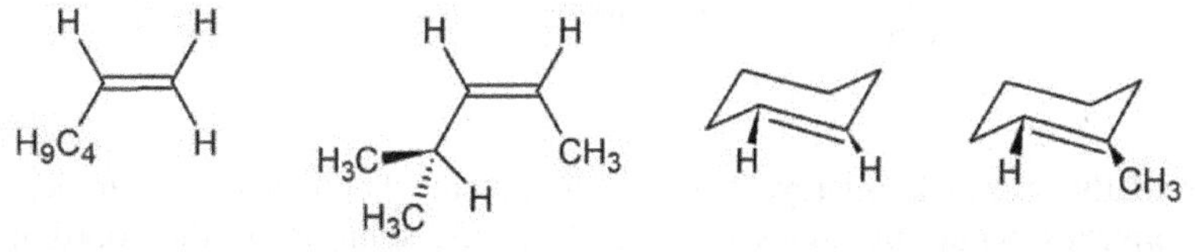

hexene: 2910 L mol^{-1} s^{-1} *cis*-4-methyl-2-pentene: 990 L mol^{-1} s^{-1} cyclohexene: 3160 L mol^{-1} s^{-1} 1-methylcyclohexene: 60 L mol^{-1} s^{-1}

E22.29 The catalytic cycle for the formation of both *n* (linear) and *iso* (branched) products is shown below. The starting point of the cycle is the active catalyst [$HCo(CO)_4$] formed from hydrogenolysis of [$Co_2(CO)_8$]. The linear product (*n* isomer) is butanal, whereas the branched product (*iso* isomer) is 2-methylpropanal. The selectivity step is the hydride migration onto the coordinated prop-1-ene in [$HCo(CO)_4(H_2C{=}CHCH_3)$] complex. Note that migration can produce either primary (outer path) or secondary (inner path) alkyl coordinated to the Co centre.

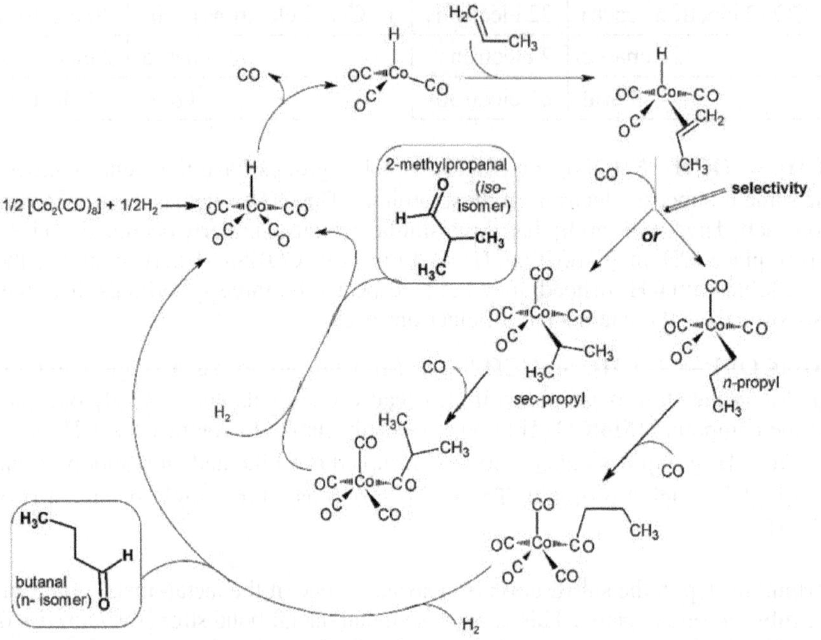

E22.30 Since compound (E) in Figure 22.20 is observed under catalytic conditions, its formation must be faster than its transformation into products. If this were not true, a spectroscopically observable amount of it would not build up. Therefore, the transformation of (E) into [$CoH(CO)_4$] and product must be the rate-determining step in the absence of added PBu_3. In the presence of added PBu_3, neither compound (E) nor its phosphine-substituted equivalent, that is, [$Co(C(=O)C_4H_9)(CO)_3(PBu_3)$], is observed, requiring that their formation must be slower than their transformation into products. Thus, in the presence of added PBu_3, the formation of either (A) or (E), or their phosphine-substituted equivalents, is the rate-determining step.

E22.31 **a)** Attack by dissolved hydroxide.

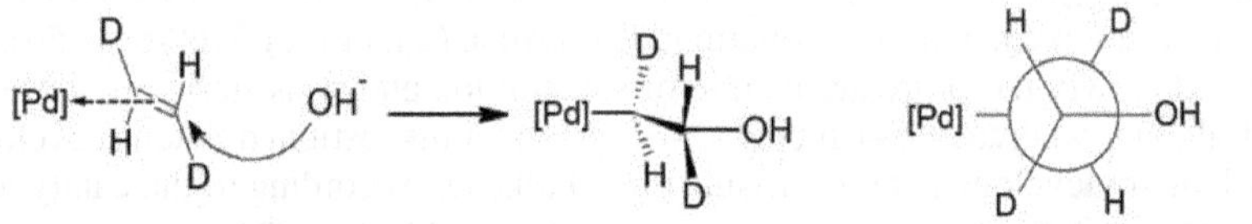

b) Attack by coordinated hydroxide.

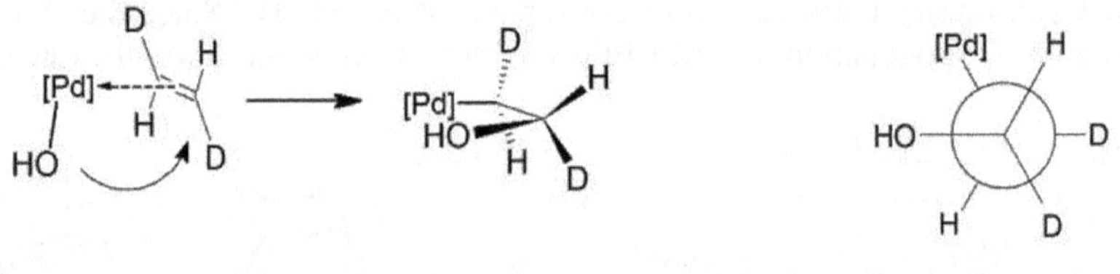

c) Differentiate of the stereochemistry. Yes. Detailed stereochemical studies indicate that the hydration of the alkene-Pd(II) complexes occurs by attack of water from the solution on the coordinated ethene rather than the insertion of coordinated OH.

E22.32 The catalytic cycle for the Ziegler–Natta polymerization of propene is shown below. In this case the catalyst is prepared from $TiCl_4$, but $TiCl_3$ also produces the active catalyst in a similar fashion. Either are commonly bound to $MgCl_2$ as a support (note that $TiCl_4$ is a very moisture-sensitive liquid, hence easier to handle when supported).

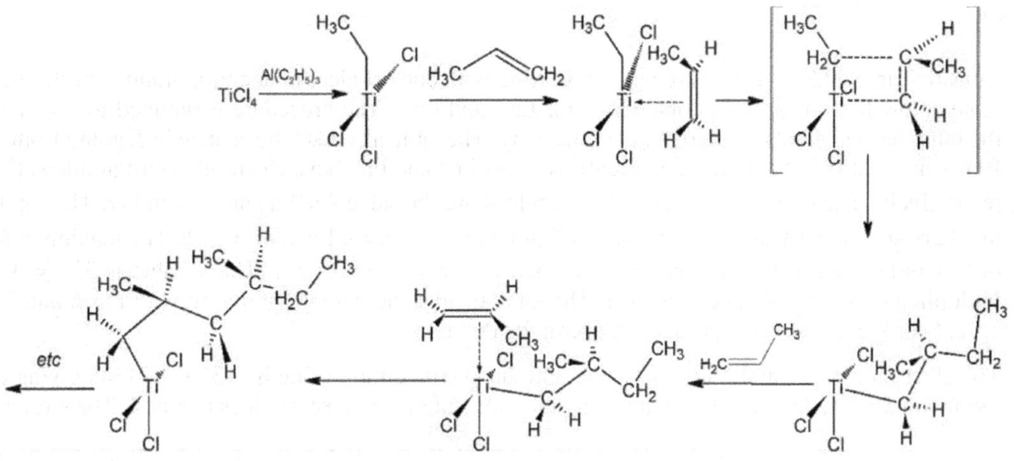

The mechanism starts with alkylation of a Ti atom on the catalyst surface. The alkylating agent is usually $AlEt_3$. The Ti on the surface of the catalyst is coordinatively unsaturated and an alkene molecule (in this case propene) coordinates. The ethyl group bonded to titanium migrates, probably via a cyclic intermediate shown in square brackets, to the coordinated propene. This not only initiates the propagation of polymer chain but also opens a new coordination site on Ti. Another propene molecule coordinates to this position and undergoes the migratory insertion; consequently, a new coordination spot opens, and the chain lengthens and the process continues. The resulting polymer is going to be isotactic: it is going to have a high melting point and will be semicrystalline.

Chapter 23 The f-Block Metals

Self-Tests

S23.1 Consult Table 23.3 to make sure you are starting with correct electronic configurations of the elements listed. The table below lists the spectroscopic terms for Ln^{2+} and Ln^{3+}. The procedure is outlined for Pr^{2+} and Pr^{3+} only, since the other term symbols are derived the same way. The table also lists the change in L going from Ln^{2+} to Ln^{3+}.
Pr has an electronic configuration [Xe]$4f^3 6s^2$; thus Pr^{2+} and Pr^{3+} have electronic configurations [Xe]$4f^3$ and [Xe]$4f^2$ respectively. The three f electrons in Pr^{2+} would share the same $\ell = 3$ quantum number. Their individual m_l and m_s numbers are: $m_\ell = +3$ and $m_s = +½$, $m_\ell = +2$ and $m_s = +½$; $m_\ell = +1$ and $m_s = +½$. The maximum M_L value is (+3) + (+2) + (+1) = +6 that must come from a state with $L = 6$, an I term. The S value is 3/2 (= +½ + ½ + ½) with multiplicity $2S + 1 = 4$. The term is ^4I. The total angular momentum of a term with $L = 6$ and $S = 3/2$ will be J = 15/2, 6, or 9/2. Finally, the ground state term for Pr^{2+} is $^4I_{9/2}$.

The electronic configuration of Pr^{3+} is $4f^2$ and the maximum M_L value is (+3) + (+2) = 5 giving us $L = 5$ and an H term with $S = 1$, a triplet: ^3H. The total angular momentum can take J values 6, 5 or 4. The term is hence $^3H_4$.

Ln	**Pr**	**Nd**	**Pm**	**Dy**	**Ho**	**Er**
El. Config./[Xe]	$4f^3 6s^2$	$4f^4 6s^2$	$4f^5 6s^2$	$4f^{10} 6s^2$	$4f^{11} 6s^2$	$4f^{12} 6s^2$
Ln^{2+}	**Pr^{2+}**	**Nd^{2+}**	**Pm^{2+}**	**Dy^{2+}**	**Ho^{2+}**	**Er^{2+}**
El. Config./[Xe]	$4f^3$	$4f^4$	$4f^5$	$4f^{10}$	$4f^{11}$	$4f^{12}$
Term	$^4I_{9/2}$	$^5I_4$	$^6H_{5/2}$	$^5I_8$	$^4I_{15/2}$	$^3H_6$
Ln^{3+}	**Pr^{3}**	**Nd^{3+}**	**Pm^{3+}**	**Dy^{3+}**	**Ho^{3+}**	**Er^{3+}**
El. Config./[Xe]	$4f^2$	$4f^3$	$4f^4$	$4f^9$	$4f^{10}$	$4f^{11}$
Term	$^3H_4$	$^4I_{9/2}$	$^5I_4$	$^6H_{15/2}$	$^5I_8$	$^4I_{15/2}$
Change in L	Decrease	The same	Increase	Decrease	The same	Increase

From the table, we see that the L does not change for ionizations $Nd^{2+}(g) \rightarrow Nd^{3+}(g) + e^-(g)$ and $Ho^{2+}(g) \rightarrow Ho^{3+}(g) + e^-(g)$, a quarter- and three quarter-shell effects—positions where we find small irregularities in I_3.

S23.2 GdO should be a slat-like compound—looking at Figure 23.8 we can see that Gd^{2+} 4f orbitals are significantly lower in energy than its 5d orbital. This means that the occupancy of higher energy 5d orbitals is not favourable making this oxide an insulating one. Considering that gadolinium's second and third ionization energies are close, this oxide should be considered relatively unstable.

S23.3 A bridge is going to be formed if there is a ligand capable of bridging and if metal centres do not have complete coordination. Lanthanoid cations have high coordination numbers and hydride ligands can make bridges, so the fact that the product is a hydride-bridged dimer is not surprising. To prevent bridge formation, and keep hydride ligands we have to find a way to increase the coordination number around Ln^{3+} cation. Considering that these cations are hard cations, a choice of a hard-donor ligand is the best. For example, THF—with oxygen donor atom—could be a good choice. Many Ln^{3+} complexes with coordinated THF are known.

S23.4 From the Frost diagram in Figure 23.17, the most stable oxidation state of uranium in aqueous acid is U^{4+} (i.e., it has the most negative free energy of formation, the quantity plotted on the y-axis). However, the reduction potentials for the UO_2^{2+}/UO_2^+ and UO_2^+/U^{4+} couples are quite small, +0.170 and +0.38 V, respectively (hence, the UO_2^+/U^{4+} potential is +0.275 V). Therefore, because the O_2/H_2O reduction potential is 1.229 V, the most stable uranium ion is UO_2^{2+} if sufficient oxygen is present:

$$2U^{4+} + O_2 + 2H_2O \rightarrow 2UO_2^{2+} + 4H^+ \; E^\circ = +1.229\ V - (+0.275\ V) = +0.0954\ V.$$

S23.4 The e_{2g} interactions are sketched below. Note that this overlap involves a 6d atomic orbital on the actionoid atom and that the overlap is δ type.

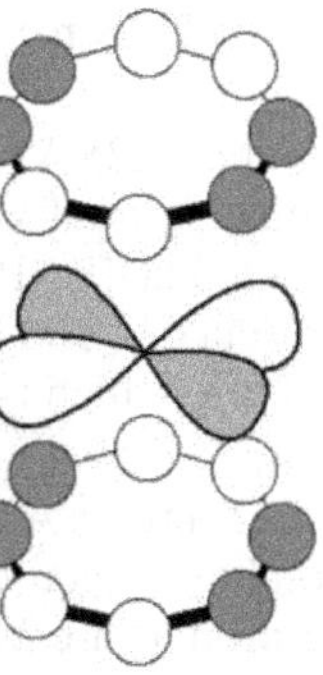

Exercises

E23.1 **a) Balanced equation.** All lanthanoids are very electropositive and reduce water (or protons) to H_2 while being oxidized to the trivalent state. If Ln is used as the symbol for a generic lanthanoid element, the balanced equation is:

$$2Ln(s) + 6H_3O^+(aq) \rightarrow 2Ln^{3+}(aq) + 3H_2(g) + 6H_2O(l).$$

b) Reduction potentials. The reduction potentials for the Ln^{3+}/Ln^0 couples in acid solution range from –1.99 V for europium to –2.38 V for lanthanum, a remarkably self-consistent set of values spanning 15 elements. Since the potential for the H_3O^+/H_2 reduction is 0 V, the $E°$ values for the equation shown in part (**a**) for all lanthanides range from 1.99 to 2.38 V, a very large driving force.

c) Two unusual lanthanides. The usual oxidation state for the lanthanide elements in aqueous acid is +3. There are two lanthanides that deviate slightly from this trend. The first is Ce^{4+}, which, while being a strong oxidizing agent, is kinetically stable in aqueous acid. Since Ce^{3+} is $4f^1$ and Ce^{4+} is $4f^0$, you can see that the special stability of tetravalent cerium is because of its stable $4f^0$ electron configuration. The second deviation is Eu^{2+}, which is a strong reducing agent but exists in a growing number of compounds. Since Eu^{2+} is $4f^7$ and Eu^{3+} is $4f^6$, you can see that the special stability of divalent europium is because of its stable $4f^7$ electron configuration (half-filled f subshell).

E23.2 The ionic radius of Lu^{3+} is significantly smaller than La^{3+} because of the incomplete shielding of the 4f electrons of the former.

E23.3 The unusual oxidation states displayed by two of the lanthanoids, +4 for Ce and +2 for Eu (see Exercise 23.1) were exploited in separation procedures because their charge/radius ratios are very different than those of the typical Ln^{3+} ions. Tetravalent cerium can be precipitated as $Ce(IO_3)_4$, leaving all other Ln^{3+} ions in solution because their iodates are soluble. Divalent europium, which resembles Ca^{2+}, can be precipitated as $EuSO_4$ leaving all other Ln^{3+} ions in solution because their sulfates are soluble.

E23.4 Ln^{3+} cyclopentadienyl complexes exist in η^5–η^1 equilibrium. The η^1-Cp^- bonding mode can easily be substituted by another ligand, such as Cl^-. Larger U(III) triscyclopentadienyl complex does not have this equilibrium—all Cp^- rings are η^5 bonded and thus less labile. Not only that the Cp^- rings do not slip and are not substituted, we could also expect ability to add another ligand to $[U(C_5H_5)_3]$.

E23.5 The chemistry of the transition metals is dominated by their ability to access several oxidation states. For example, recall that Mn can have oxidation states from +2 to +7 and that osmium can reach +8. Another characteristic feature of their chemistry is extensive use of d orbitals in bonding. This is similar to the actinoid chemistry—early actinoids also have several oxidation states accessible (consider for example U) and their 6d orbitals are more diffuse than lanthanoid's 5d. Lanthanoids have limited range of oxidation states, mostly 3+, and 5d orbitals are core-like.

E23.6 UF_6 is a molecular solid—its solid-state structure contains isolated, octahedral UF_6 molecules held together via relatively weak intermolecular forces. The intermolecular forces are very easy to break and this compound sublimes easily. On the other hand, UF_3 and UF_4 form network solids in which each U atom is surrounded by nine and eight fluorines respectively. Therefore, UF_3 and UF_4 have high melting points.

E23.7 The reduction potentials for Pu^{4+}/Pu, Pu^{3+}/Pu are −1.25 V and −2.00 V respectively, indicating that Pu metal is going to readily (at least from the thermodynamic point of view) dissolve in HCl. However, the composition of the solution can be complicated and could contain Pu(III), Pu(IV), Pu(V), and even Pu(VI) species because the potentials for the couples Pu(VI)/Pu(V), Pu(V)/Pu(IV), and Pu(IV)/Pu(III) are very similar (ca. +1 V). Addition of F^- to the solution is likely going to precipitate PuF_3, which is insoluble in water, like UF_3.

E23.8 Refer to Figure 23.18 for a molecular orbital diagram for AnO_2^{2+} cations. A short analysis reveals that 6p electrons from the An atom do not contribute to the bonding orbitals, that is, they remain core electrons largely located on the An atom. Thus, the important An valence orbitals are 5f and 6d. We can also note that all bonding molecular orbitals are filled. The 12 electrons shown in these orbitals all come from the two oxygen atoms bonded to the An atom that has an oxidation state +6. This is the electronic configuration found in UO_2^{2+} because U(VI) has an electronic configuration $5f^0 6d^0$. Therefore, the overall bond order would be ½ [12 electrons in bonding orbitals] = 6. If we further divide this with two U-O bonds we get a bond order of 3 *per* U-O bond. Moving to NpO_2^{2+} we have to add one electron to the diagram because Np(VI) has electronic configuration $5f^1 6d^0$. This extra electron would occupy a ϕ_u molecular orbital. However, this orbital is nonbonding with respect to the NpO_2^{2+} unit. Hence, the bond order should not change and Np-O bonds should have a bond order equal to three. Similar analysis for PuO_2^{2+} (Pu(VI) $5f^2\ 6d^0$) and AmO_2^{2+} (Am(VI) $5f^3 6d^0$) shows that in all these species the An-O bond order is three.

All these fragments are linear because this geometry maximizes the strong σ-overlap between 2p orbitals on the oxygen atoms and $6d_{z^2}$ and $5f_{z^3}$ orbitals on the An atom.

E23.9 Refer to Resource Section 2 for the electronic configuration of neutral atoms.

Tb^{3+} is a $4f^8$ system. Therefore $M_L = +3$ and $S = 3$. The term will be ^7F. The expected term symbol is $^7F_6$.

Nd^{3+} is a $4f^3$ system. Therefore $M_L = +6$ and $S = 3/2$. The term will be ^4I. The term symbol is $^4I_{9/2}$.

Ho^{3+} is a $4f^{10}$ system. Therefore $M_L = +6$ and $S = 2$. The term will be ^5I. The term symbol is $^5I_8$.

Er^{3+} is a $4f^{11}$ system. Therefore $M_L = +6$ and $S = 3/2$. The term will be ^4I. The term symbol is $^4I_{15/2}$.

Lu^{3+} is a $4f^{14}$ system. Therefore $M_L = 0$ and $S = 0$. The term will be ^1S. The term symbol is $^1S_0$.

Compare your answer with Table 23.5. See also Self-Test 23.1.

E23.10 Statement **(a)** is correct because the half-filled shell contributes zero angular momentum, that is, $J = S$.

Statement **(b)** is incorrect—Ln^{3+} cations exchange water molecules very rapidly, on the nanosecond timescale. This is about the same timescale as the water substitution rates for Zn^{2+}, but much faster than for the transition metal ions of the same charge (i.e., Fe^{3+} or Ti^{3+}).

Statement **(c)** is correct—Ln^{3+} cations do form stable complexes only with multidentate ligands, particularly if they are negatively charged.

E23.11 Apart from the fact that the discovery of a Ln(V) species would open a whole new avenue of chemistry, it would also shake the theoretical foundation of the claim that the lanthanide 4f subshell is rather inert. A good candidate could be Pr. This element is early in the lanthanoid series, meaning that its 4f orbitals are still comparatively high in energy. Also, removal of an electron from Pr(IV), a somewhat stable species, would produce a $4f^0$ electronic configuration for Pr(V), the same $4f^0$ configuration that stabilizes Ce(IV).

E23.12 Eu is similar to Mn because both have 2+ as a stable oxidation state. In both cases the stability of +2 oxidation state can be attributed to a half-filled subshell—in the case of Eu^{2+} that is $4f^7$, whereas in the case of Mn^{2+} that is $3d^5$. Consequently, both elements have very high third ionization energies. A similar parallel exists between Gd and Fe. The third ionization energies for these two elements are relatively low because in each case the removal of the third electron leads to the stable half-filled subshell: Gd^{3+} has $4f^7$, whereas Fe^{3+} has the $3d^5$ configuration.

E23.13 Stable carbonyl compounds need backbonding from metal orbitals of the appropriate symmetry into the antibonding π MO on CO. Thus, a metal centre has to be electron rich to stabilize M–CO bond. With the lanthanoids, the 5d orbitals are empty, and the 4f orbitals are deeply buried in the inert [Xe] core and cannot participate in bonding.

E23.14 Neptunocene ($NpCp_4$) can be prepared from $NpCl_4$ and an alkali metal cyclopentadienyl organometallic compound (e.g., NaCp) in an ether solvent (commonly THF) with cooling:

$$NpCl_4 + 4NaCp \xrightarrow[\text{THF, Low temp.}]{} NpCp_4 + 4NaCl.$$

E23.15 There is a considerable difference between the interactions of the 4f orbitals of the lanthanide ions with ligand orbitals and the 5f orbitals of the actinide ions with ligand orbitals. In the case of the 4f orbitals, interactions with ligand orbitals are negligible. Therefore, splitting of the 4f subshell by the ligands is also negligible and does not vary as the ligands vary. Since the colours of lanthanide ions arise due to 4f–4f electronic transitions, the colours of Eu^{3+} complexes are invariant as a function of ligand. In contrast, the 5f orbitals of the actinide ions interact more strongly with ligand orbitals, and the splitting of the 5f subshell, as well as the colour of the complex, varies as a function of ligand.

E23.16 The radius ratio is $\gamma = 96/146 = 0.657$. Using Table 4.6, the rock-salt structure is predicted as this ratio falls between 0.414 and 0.732.

E23.17 Similarity in bond lengths between U–O and Os–O bonds implies that U–O bond is of higher order than Os–O bond—if they were of the same order, then U–O should be longer due to the larger U(IV) radius. In Exercise 23.8 we concluded, using MO diagram in Figure 23.18, that U–O bond is 3. Thus, we would expect the Os–O bond order to be lower, about 2. The bond order is lowered because Os(VI) has still two valence electrons (electronic configuration $5d^2$) which would occupy antibonding MOs of OsO_2^{2+} fragment and thus lower the bond order.

Guidelines for Selected Tutorial Problems

T23.6 It would be a good idea (and practice as well) not only to compare the Latimer diagrams provided in Resource Section 3, but also to construct one Frost diagram with data for both Np and Re to compare the two elements directly. Also note which oxidation states are missing for Re and Np and the chemical composition of the species for each oxidation state of the elements—this gives you a more complete idea of the differences between the two elements.

T23.8 Consult Greenwood, N.N., Earnshaw, A. (1997). *Chemistry of the Elements*, 2nd ed. Elsevier (pp. 1260–1262), and references therein. Also, a good source is Morss, L. R., Edelstein, N. M., Fuger, J. (eds.). (2006). *Chemistry of the Acitinide and Transactinide Elements*, Vol. 2. Dordrecht, NLD: Springer.

Chapter 24 Materials Chemistry and Nanochemistry

Self-Tests

S24.1 **a)** $SrTiO_3$ can be prepared from $SrCO_3$ and TiO_2 in 1 : 1 ratio. The two are ground together and the well-mixed material is heated to the temperature above the temperature of $SrCO_3$ thermal decomposition.

b) $Sr_3Ti_2O_7$ can be prepared in similar manner to $SrTiO_3$ but starting from 3 : 2 stoichiometry ($SrCO_3$: TiO_2); alternatively, it can be prepared from $SrTiO_3$ and SrO in 2 : 1 stoichiometry.

c) $AlPO_4$ is aluminium salt of phosphoric acid (H_3PO_4). Thus, reasonable starting reagents could be $Al(OH)_3$ and H_3PO_4. Considering that aluminium hydroxide is a very weak base and cannot completely deprotonate the acid, just mixing the reagents would not produce the desired product—hydrothermal synthesis should do the trick. To ensure the product is porous, we should add a structure-directing agent, such as tetraalkylammonium hydroxide, to the reaction mixture.

S24.2 Increasing the pressure on a crystal compresses it, reducing the spacing between ions (e.g., subjecting NaCl to a pressure of 24 000 atm reduces the Na–Cl distance from 2.82 Å to 2.75 Å). In a rigid lattice such as β-alumina, larger ions migrate more slowly than smaller ions with the same charge (as discussed in the example). At higher pressures, with smaller conduction plane spacings, *all* ions will migrate more slowly than they do at atmospheric pressure. However, larger ions will be impeded to a greater extent by smaller spacings than will smaller ions. This is because the ratio (radius of migrating ion)/(conduction plane spacing) changes more, per unit change in conduction plane spacing, for a large ion than for a small ion. Therefore, increased pressure reduces the conductivity of K^+ more than that of Na^+ because K^+ is larger than Na^+.

S24.3 In the normal AB_2O_4 spinel structure, the A^{2+} ions (Fe^{2+} in this example) occupy tetrahedral sites and the B^{3+} ions (Cr^{3+} in this example) occupy octahedral sites. The fact that $FeCr_2O_4$ exhibits the normal spinel structure can be understood by comparing the ligand-field stabilization energy of high-spin octahedral Fe^{2+} and octahedral Cr^{3+}. With six d electrons, LFSE = $-0.4\Delta_o$ for high-spin Fe^{2+}. With only three d electrons, LFSE = $-1.2\Delta_o$ for Cr^{3+}. Since the Cr^{3+} ions experience more stabilization in octahedral sites than do the Fe^{2+} ions, the normal spinel structure is more stable than the inverse spinel structure, which would exchange the positions of the Fe^{2+} ions with half of the Cr^{3+} ions.

S24.4 A pure silica analogue of ZSM-5 would contain only weakly Brønsted acidic Si–OH groups and no strongly acidic Al–OH_2 groups found in aluminosilicates. Only strong Brønsted acids can protonate an alkene to form the carbocations that are necessary intermediates in benzene alkylation. Therefore, a pure silica analogue of ZSM-5 would not be an active catalyst for benzene alkylation.

S24.5 MCM-41 is a better choice as it has a range of pore sizes from 2 to 10 nm. These can better accommodate QDs with diameters ranging from 2 to 8 nm than typical ZSM-5 with pore sizes of less than 1 nm.

Exercises

E24.1 **a)** $MgCr_2O_4$ can be prepared by heating a 1 : 1 mixture of $MgCO_3$ (or $Mg(NO_3)_2$) and $(NH_4)_2Cr_2O_7$ at high temperature. Note that thermal decomposition of $(NH_4)_2Cr_2O_7$ produces Cr_2O_3 in a finely divided and reactive form:

$$(NH_4)_2Cr_2O_7(s) \rightarrow N_2(g) + 4H_2O(g) + Cr_2O_3(s).$$

Then

$$Cr_2O_3(s) + MgCO_3(s) \xrightarrow{1400^\circ C} MgCr_2O_4(s).$$

b) $LaFeO_3$ can be prepared by coprecipitating La^{3+} and Fe^{3+} hydroxides and then, after filtration, heat the precipitate and eliminate water:

$$La^{3+}(aq) + Fe^{3+}(aq) + 6OH^{-}(aq) \rightarrow La(OH)_3 \cdot Fe(OH)_3(s) \xrightarrow{450^\circ C} LaFeO_3(s) + 3H_2O(g).$$

Note the low final thermal decomposition temperature as a result of the fine atomic mixing achieved by the initial solution stage.

c) Ta_3N_5 can be prepared by heating Ta_2O_5 under a flow of NH_3 at high temperature:

$$3Ta_2O_5(s) + 10NH_3(g) \xrightarrow{600^\circ C} 2Ta_3N_5(s) + 15H_2O(g).$$

Note that the equilibrium is pushed to the right by the removal of steam in the gas stream.

d) $LiMgH_3$ can be prepared by heating a molar mixture of LiH and MgH_2 (under hydrogen or an inert gas) in a direct reaction, or by heating a mixture of the elements under hydrogen.

e) This complex fluoride can be prepared from solution of $CuBr_2$ and KF as a precipitate:

$$CuBr_2(aq) + 3\ KF(aq) \rightarrow KCuF_3(s) + 2KBr(aq).$$

f) The gallium analogue of zeolite A can be prepared starting from sodium gallate, $NaGaO_2$ (instead of sodium aluminate) and sodium silicate. The reaction would be undertaken hydrothermally in a sealed reaction vessel:

$$12NaGaO_2(s) + 12Na_2SiO_3(aq) \xrightarrow{150^\circ C} \rightarrow Na_{12}[Si_{12}Ga_{12}O_{48}] \cdot nH_2O(s) + 24NaOH(aq).$$

E24.2 **(a)** $LiCoO_2$, **(b)** Sr_2WMnO_6 (= $Sr(W_{0.5}Mn_{0.5})O_3$, a perovskite).

E24.3 The electronic conductivity of the solid increases owing to formation of $(Ni_{1-x}Li_x)O$ containing Ni(III) and a free electron that can hop through the structure from Ni(II) to Ni(III), thus promoting increased conductivity in the structure.

E24.4 Fe_xO contains vacancies. Since Fe(II) is easily oxidized to Fe(III), a substitution of three Fe^{2+} ions with two Fe^{3+} ions and a vacancy takes place.

E24.5 A solid solution would contain a random collection of defects, whereas a series of crystallographic shear plane structures would contain ordered arrays of crystallographic shear planes. Because of the lack of long-range order, the solid solution would give rise to an electron micrograph showing a random distribution of shear planes. In addition, the solid solution would not give rise to new X-ray diffraction peaks. In contrast, the ordered phases of the solid on the right would be detectable by electron microscopy and by the presence of a series of new peaks in the X-ray diffraction pattern, arising from the evenly spaced shear planes.

E24.6 Generally, to increase anion conductivity we must increase anion mobility by creating vacancies in anion sties. This can be achieved via cation substitution: doping the compound with a cation of similar size but lower charge will create anion vacancies.

a) To increase the anion conductivity in PbF_2 we must replace Pb^{2+} with an M^+ cation of a similar radius. Since PbF_2 has the fluorite structure, Pb^{2+} is eight coordinate and its radius is 129 pm. Thus, a good candidate for doping might be Na^+ with eight-coordinate radius of 132 pm.

b) The radius of a six-coordinate Bi^{3+} is 103 pm. A good candidate for a dopant in this case could be Ca^{2+} (six-coordinate radius is 100 pm) or again Na^+ (six-coordinate radius is 102 pm). Hg^{2+} is too poisonous, but with respect to radius, it is a good candidate as well (102 pm, six coordinate).

E24.7 The unit cell for ReO_3, which is shown in Figure 24.16(a), is reproduced in the figure on the left (the O^{2-} ions are the open spheres). As you can see, the structure is very open, with a very large hole in the centre of the cell, and it does appear to be sufficiently open to undergo Na^+ ion intercalation. We can calculate if this is indeed possible if we note that the length of the ReO_3 unit cell, a, can be calculated as $a = 2r(O^{2-}) + 2r(Re^{6+})$. Re^{6+} is six coordinate and, from Resource Section 1, its radius is 55 pm. There is no O^{2-} radius for the coordination two, but we can approximate it as about 140 pm. Then $a = 2 \times 55$ pm $+ 2 \times 140$ pm $= 390$ pm. The face diagonal of the cube, d, can be calculated as $d = 2^{1/2}a$ (we are looking at the face diagonal because any cation that would fill the hole would be in contact with larger O^{2-} anions; see figure on the right). Hence d = 1.414 × 390 pm = 551.5 pm. To really calculate the maximum radius of a metal that can fit in the central hole, we must subtract $2r(O^{2-})$ from d (because that is the part of the diagonal "occupied" by neighbouring O^{2-} ions) and then divide the difference by 2: $r(M^+) = (551.5\ \text{pm} - 2 \times 140\ \text{pm})/2 = 135.7\ \text{pm}$. Since the radius of Na^+ (eight coordinate, the maximum provided in Resource Section 1) is 132 pm, we can say that Na^+ can indeed fit inside the hole. An ReO_3 unit cell containing an Na^+ ion (large, heavily shaded sphere) at its centre is shown on the right below. For the 1 : 1 Na : Re ratio the structure type can be described as that of perovskite ($CaTiO_3$).

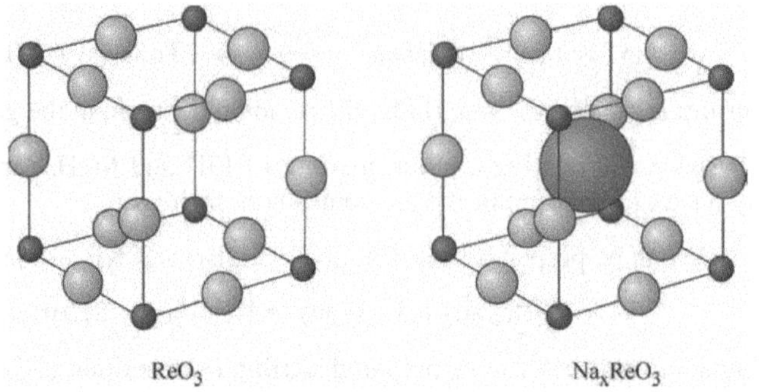

E24.8 The cation charges sum to 3+ for A = Rb^+ and A′ = Sr^{2+}, thus to keep electroneutrality of the compound the sum of charges for B and B′ should come to 9+. The possibilities are Fe^{3+} and Mo(VI) for B and B′ respectively. Both prefer coordination number six and are small enough to fit inside the perovskite structure. Another possible B candidate is Cr^{3+} while for B′ is W(VI).

E24.9 Sn^{2+} is smaller than Pb^{2+} so the $ASnI_3$ unit cell is expected to be smaller than $APbI_3$ unit cell. Thus, possible candidates for A^+ cation are Li^+ and Na^+—the smallest unipositive cations.

E24.10 The Néel temperature for CrO should be below 122 K (i.e., below the Néel temperature for MnO).

E24.11 A sulfide with the spinel structure could be $ZnCr_2S_4$ and a fluoride $NiLi_2F_4$.

E24.12 Copper-containing high-temperature superconductors contain mixed oxidation states of Cu. Thus, once the Cu oxidation states are worked out for each complex oxide given, all except $Gd_2Ba_2Ti_2Cu_2O_{11}$ (compound c), which contains only Cu^{2+}, are superconductors. The average oxidation state of copper in compound (a) is +2.25, in (b) +2.2, and in (d) is +2.24.

E24.13 White Ta_2O_5 reacts with ammonia at high temperature to produce red Ta_3N_5:

$$3Ta_2O_5(s) + 10NH_3(g) \rightarrow 2Ta_3N_5(s) + 15H_2O(g).$$

E24.14 According to Zachariasen, the formation of a glass is favoured for the substances that have polyhedral, corner-sharing oxygen atoms (rather than face- or edge-shared O atoms) and that can retain the coordination sphere around a glass-forming element while having variable bond angles around O atom. Compounds having strong covalent bonds are likely to conform to these rules because the bonds are not easy to break. CaF_2 is an example of a compound that does not have strong covalent bonds—it is an ionic compound. On cooling, mobile, and more-or-less free Ca^{2+} and F^- ions, simply pack into the fluorite structure. This is not the case with SiO_2—this oxide has a framework structure composed of SiO_4 tetrahedra connected through O atoms shared at corners. The Si–O bond has significant covalent character and is very strong. This network is not easily broken (and reformed) with temperature.

E24.15 BeO, B_2O_3, and to some extent GeO_2 because they involve metalloid and non-metal oxides. Transition metal and rare earth oxides are typically non-glass-forming oxides, many of which have crystalline phases.

E24.16 Potential glass-forming fluorides can be tin fluorides. Both tin fluorides are solids that contain corner-sharing fluoride anions and a significant degree of covalent bonding meaning that bond angles around F^- are flexible. Another possibility is SbF_3 (but not SbF_5 which is a liquid).

E24.17 The two methods for the preparation of the intercalation compound Li_xTiS_2 ($0 \leq x \leq 1$) are:

(a) direct combination of lithium and TiS_2 at elevated temperatures:

$$Li(g) + TiS_2(s) \rightarrow Li_xTiS_2(s).$$

(b) electrointercalation, with the following process occurring on electrodes:

anodic process: $Li(s) \rightarrow Li^+(aq) + e^-$

cathodic process: $TiS_2(s) + xLi^+(aq) + xe^- \rightarrow Li_xTiS_2$.

E24.18 The metallocene compounds undergo oxidation when intercalated. Both $[Co(\eta^5\text{-}C_5H_5)_2]$ and $[Co(\eta^5\text{-}C_5Me_5)_2]$ are 19-electron species ($C_5Me_5^-$ is pentamethylcyclopentadienyl anion) and can be easily oxidized to 18-electron species, $[Co(\eta^5\text{-}Cp)_2]^+$ and $[Co(\eta^5\text{-}C_5Me_5)_2]^+$ respectively. On the other hand, ferrocene $[Fe(\eta^5\text{-}C_5H_5)_2]$ is an 18-electron compound and cannot be easily oxidized. Since the metallocenes are "squeezed" between the sheets of the original structure, the unit cell parameters must increase—from 583 pm to ~1160 pm.

E24.19 The structure of Mo_6S_8 unit can be described as a cube of S atoms with an octahedron of Mo atoms inside so that each cube face has an Mo atom in the centre. The Mo_6S_8 units are further bonded via Mo–S bridges. The S atoms in one cube donate an electron pair from a filled p orbital to the empty $4d_{z^2}$ orbital on Mo located on a neighbouring cube.

E24.20 Be, Ga, Zn, and P all form stable tetrahedral units with oxygen. Mg would prefer an octahedral environment (i.e., the structure of brucite, $Mg(OH)_2$), whereas Cl would likely be an anion.

E24.21 The following zeotypes are possible: $(AlP)O_4$, $(BP)O_4$, $(ZnP_2)O_6$.

E24.22 **a)** When Al^{3+} replaces Si^{4+} on lattice site, the charge is balanced by H_3O^+, increasing the acidity of the solid catalyst.

b) Other 3+ that have similar radius and can adopt tetrahedral geometry, for example Ga^{3+}, Co^{3+}, or Fe^{3+}, would have a similar effect to Al^{3+}.

E24.23 Zeolites are porous aluminosilicates. The Al/Si framework defines channels and pores—the dimensions of these pores and channels dictate the molecular selectivity: only molecules of certain shape and size can enter the zeolite framework. This is the basis of their use as molecular sieves. The exchange of Si^{4+} for Al^{3+} also requires additional cations to be brought inside the framework to balance the charges. If the cation is H^+ (or better H_3O^+) the resulting zeolite behaves as a very strong acid (even more acidic than H_2SO_4). The acidity can be fine-tuned by replacement of some H^+ with other cations (for example lanthanoids). The Brønsted acidity can be converted into Lewis acidity by dehydration. Al^{3+} sites are responsible for this type of acidity. Thus, zeolites are heterogeneous catalysts in which the size selectivity and type of acidity can be designed.

The zeolite structure controls both reactant and product selectivity. Only the reactants of appropriate size and shape can enter the zeolite structure. Similarly, only products of certain shape and size can *exit* the structure. Another way in which zeolite structure can control the outcome of the reaction is through transition-state selectivity: specific orientation (dictated by the pore/channel shape and size) of reactive intermediates determines the products of reaction.

These points have been illustrated in the text (Section 24.13) using zeolite-catalysed isomerization of xylenes.

E24.24 MOFs (or metal-organic frameworks) require ambidentate ligands that are capable of binding two or more different metal centres simultaneously. An example of such ligand is CN^- which can bind one metal through carbon and another through a nitrogen atom. Most commonly, however, the ligands used to construct MOFs are based on organic compounds commonly with pyridine rings or carboxyl groups. 4,4′-bipyridine and 1,3,5-benzenetricarboxylate are the examples (for more examples, consult Figure 24.57):

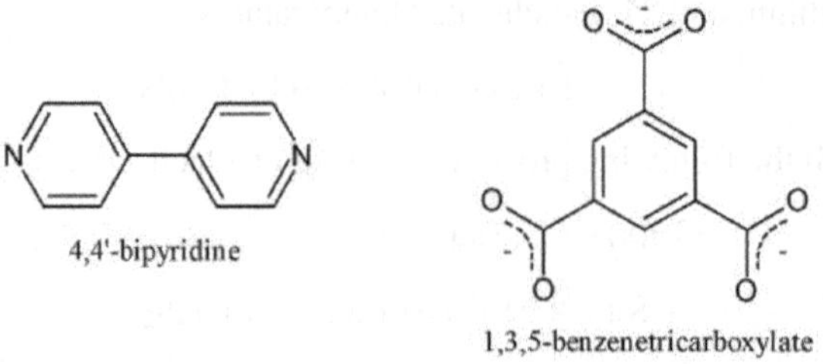

E24.25 The mass percent of hydrogen in $NaBH_4$ is 10.6%. This is above the range for technical targets (6–9%) and $NaBH_4$ could be a good candidate to consider. The problem, however, is that elevated temperatures for decomposition are required to liberate the hydrogen gas (typically above 500°C).

E24.26 The formula of Li- and Al-doped MgH_2 could be written as $Li_xMg_{1-x/2-2y/3}Al_yH_{2\pm z}$. The doped material can be prepared from the elements because all metals involved directly combine with H_2: heating an alloy with desired Li : Mg : Al ratio (which can be prepared by mulling) under H_2 atmosphere.

E24.27 With 18.1% per weight of hydrogen, BeH_2 would be an excellent candidate for hydrogen storage. Unfortunately, this compound is thermodynamically unstable ($\Delta_f H \sim 0$ kJ mol^{-1}) and very easily decomposes to give Be(s) and H_2(g). Consequently, the synthesis of BeH_2 is not straightforward like the synthesis of, for example, MgH_2.

E24.28 The Cu site in Egyptian blue is square planar and thus has a centre of symmetry (see Figure 24.66). The *d–d* transitions that give rise to the colour are, consequently, symmetry-forbidden (Laporte rule) and less intense. In copper aluminate spinel blue, the site is tetrahedral with no centre of symmetry and the transitions are not symmetry-forbidden. This leads to the increased intensity of blue colour in the spinel.

E24.29 The blue colour of the solution is due to the presence of anionic radical S_3^-, identical to that found as S_3^- in lapis lazuli described in Section 24.16. The cause of colour is the excitation of the unpaired electron in the electronic structure. In polysulfide solution, radical anions are very sensitive to O_2 from the atmosphere and upon oxidation the colour is lost.

E24.30 An ideal photocatalyst for water splitting would have (a) high efficiency of light absorption (low-intensity light would be sufficient to run the reaction and all photons absorbed would be used to split water), (b) efficient electron-hole separation process (if this separation process is not efficient then electron and hole recombine fast and do not liberate H_2 and O_2 respectively), and (c) high surface area of bulk material (the reaction takes place on the catalyst surface—higher surface area gives more "contact points" between the catalyst and water thus the reaction proceeds faster).

E24.31 BN > C(diamond) > AlP > InSb. See Sections 4.19, 4.20 and 24.19 for more details.

E24.32 In Na_2C_{60}, all of the tetrahedral holes are filled with sodium cations within the close-packed array of fulleride anions—in other words, it has an antifluorite structure (see Section 4.9).

In Na_3C_{60}, all of the tetrahedral holes and all of the octahedral holes are filled with sodium cations within the fcc lattice of fulleride anions.

E24.33 **a) Surface areas.** 1.256×10^3 nm^2 versus 1.256×10^7 nm^2 (a factor of 10^4).

b) Nanoparticles based on size. Recall that nanoparticles are particles or layers with at least one dimension in the 1–100 nm range. Thus, the 10 nm particle can be considered a nanoparticle, whereas the 1000 nm particle cannot.

c) Nanoparticles based on properties. A true nanoparticle should have a localized surface plasmon without characteristic momentum and of high intensity.

E24.34 **a) Top-down versus bottom-up.** The "top-down" approach requires one to "carve out" nanoscale features from a larger object. The "bottom-up" approach requires one to "build up" nanoscale features from smaller entities. Lithography is a standard approach to "top-down," and thin film deposition of quantum wells or superlattices is a standard approach to "bottom-up."

b) Advantages and disadvantages. Top-down methods allow for precise control over the spatial relationships between nanoscale entities, but they are limited to the design of rather large nanoscale items. Bottom-up methods allow for the precise spatial control over atoms and molecules relative to each other, but the long-range spatial arrangement is often difficult to realize.

E24.35 **a) Steps in solutions synthesis of nanoparticles.** All reacting species and additives must first be solvated; then stable nuclei of nanometre dimensions must be formed from solution and finally growth of particles to the final desired size will occur.

b) Why should the last two steps occur independently? The last two steps should be independent so that nucleation fixes the total number of particles and growth leads to a controlled size and a narrow size distribution.

c) Role of stabilizer molecules. Stabilizers prevent unwanted Ostwald ripening, a process in which smaller particles present in the solution dissolve and the re-solvated material from these particles precipitate again but on the surface of larger particles. The Ostwald ripening decreases the number of particles in the solution and increases size distribution and thus is an unwanted process during nanoparticle synthesis.

E24.36 Ostwald ripening is a process in which smaller particles present in the solution dissolve and this redisolved material precipitates again but on the surface of existing particles. The Ostwald ripening decreases the number of particles in the solution and increases size distribution and thus is an unwanted process during nanoparticle synthesis.

E24.37 **a) Core-shell nanoparticle diagram.** A schematic of the core and shell of the nanoparticle is shown below.

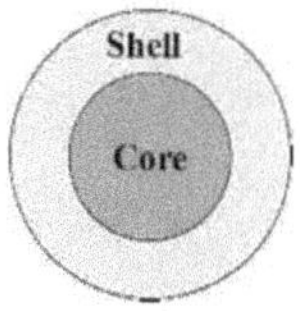

b) Production. i) *Solution based method*: first grow the core in solution and then add the additive and material for the growth of shell. ii) *Vapour-deposition method*: first deposit the material for the core and then the material for the shell.

c) Use. In biosensing, the dielectric property of the shell can control the surface plasmon of the core, while the shell can be affected by the environment. In drug delivery, the shell could react with a specific location and the core could be used as a treatment (drug).

E24.38 Examples include:

- Hexagonal boron nitride: layers contain strong B-N
- Sheet silicates (mica and clay minerals): strong Si-O bonds within sheets, weaker ionic or (less common) van der Waals interactions
- Molybdenum sulfide, MoS_2 (molybdenite): strong Mo-S bonds within layers, weak van der Waals interaction between layers.

E24.39 Special measures are required in heterogeneous catalysis to ensure that the reactants achieve contact with catalytic sites. For a given amount of catalyst, as the surface area increases, the number of catalytically active sites increases. A thin foil of platinum-rhodium will not have as much surface area as an equal amount of small particles finely dispersed on the surface of a ceramic support. The diagram shows this for a "foil" of catalyst that is 1000 times larger in two dimensions than in the third dimension (i.e., the thickness is 1; note that the diagram is not to scale). If the same a mount of catalyst is broken into cubes that have side length of 1 unit, the surface area of the catalyst is increased by nearly a factor of three.

total area = 2,004,000   total area = 6,000,000

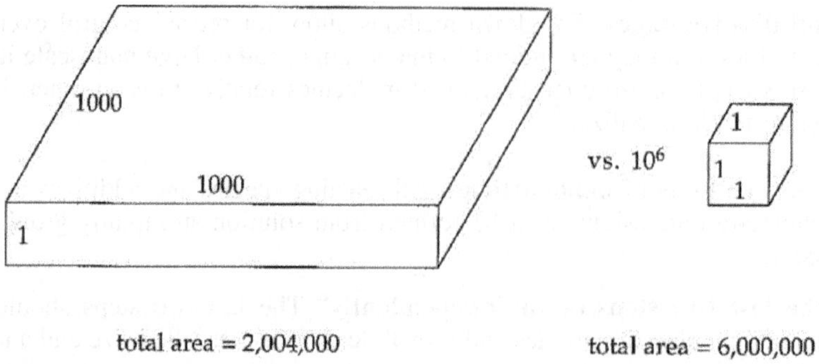

E24.40 Two observations must be explained in this catalytic deuteration of C–H bonds. The first is that ethyl groups are deuterated before methyl groups. The second observation is that a given ethyl group is completely deuterated before another one incorporates *any* deuterium. Let's consider the first observation. The mechanism of deuterium exchange is probably related to the reverse of the last two reactions in Figure 24.104, which shows a schematic mechanism for the hydrogenation of an olefin by D_2. The steps necessary for deuterium substitution into an alkane are shown below and include the dissociative chemisorption of an R–H bond, the dissociative chemisorption of D_2, and the dissociation of R–D. This can occur many times with the same alkane molecule to effect complete deuteration.

$$— Pt — Pt — Pt — Pt — \quad + \; RH \;\longrightarrow\; \overset{\overset{H}{|}\;\;\overset{R}{|}}{— Pt — Pt — Pt — Pt —}$$

$$\overset{\overset{H}{|}\;\;\overset{R}{|}}{— Pt — Pt — Pt — Pt —} \quad + \; D_2 \;\longrightarrow\; \overset{\overset{H}{|}\;\;\overset{R}{|}\;\;\overset{D}{|}\;\;\overset{D}{|}}{— Pt — Pt — Pt — Pt —}$$

$$\overset{\overset{H}{|}\;\;\overset{R}{|}\;\;\overset{D}{|}\;\;\overset{D}{|}}{— Pt — Pt — Pt — Pt —} \;\longrightarrow\; RD \;+\; \overset{\overset{H}{|}\quad\quad\quad\overset{D}{|}}{— Pt — Pt — Pt — Pt —}$$

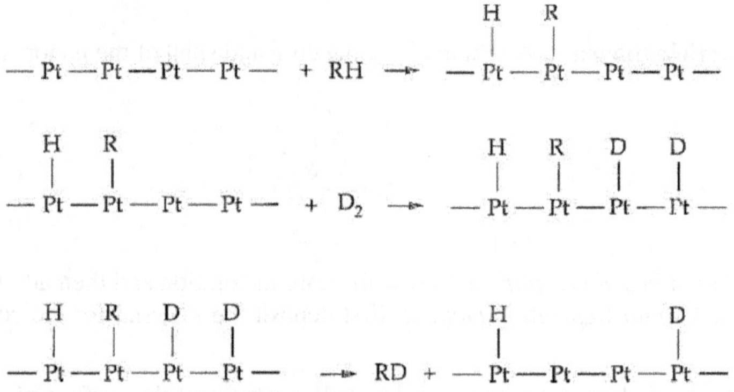

Since the $–CH_2CH_3$ groups are deuterated before the $–CH_3$ groups, one possibility is that dissociative chemisorption of a C–H bond from a $–CH_2–$ group is faster than dissociative chemisorption of a C–H bond from a $–CH_3$ group. The second observation can be explained by invoking a mechanism for rapid deuterium exchange of the methyl group in the chemisorbed $–CHR(CH_3)$ group $(R = C(CH_3)_2(C_2H_5))$. The scheme below shows such a mechanism. It involves the successive application of the equilibrium shown in Figure 24.104. If this equilibrium is maintained more rapidly than the dissociation of the alkane from the metal surface, the methyl group in question will be completely deuterated before dissociation takes place.

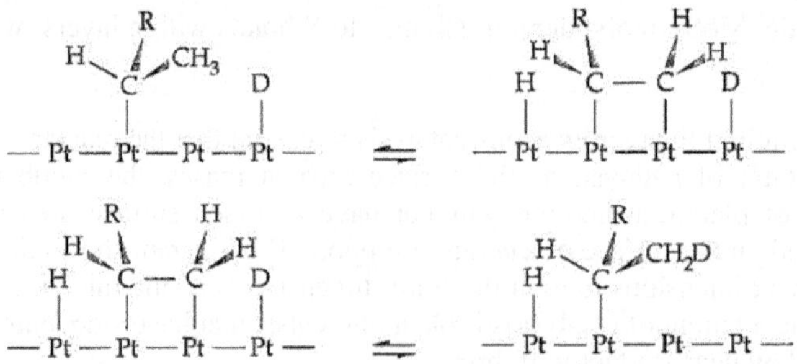

Another possibility, not shown in the scheme above, is that the terminal $–CH_3$ groups undergo more rapid dissociative chemisorption than the sterically more hindered internal $–CH_3$ groups.

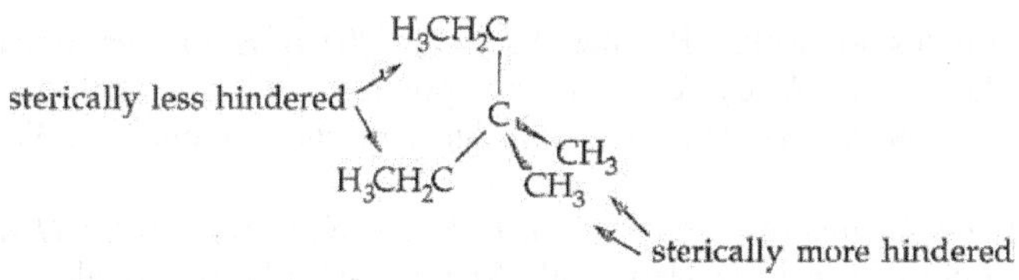

E24.41 The reduction of hydrogen ions to H_2 on Pt surface probably involves the formation of the surface hydride species. Dissociation of H_2 by reductive elimination would complete the catalytic cycle. However, platinum not only has a strong tendency to chemisorb H_2, but it also has a strong tendency to chemisorb CO. If the surface of platinum is covered with CO, the number of catalytic sites available for H^+ reduction will be greatly diminished and the rate of H_2 production will decrease.

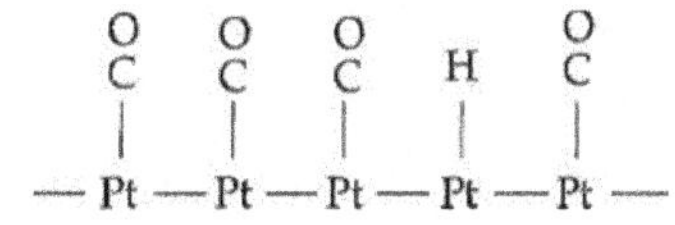

E24.42 Electrocatalysts are compounds that are capable of reducing the kinetic barrier for electrochemical reactions (barrier known as overpotential). While platinum is the most efficient electrocatalyst for accelerating oxygen reduction at the fuel cell cathode, it is expensive. Current research is focused on the efficiency of a platinum monolayer by placing it on a stable metal or alloy clusters; your book mentions the use of the alloy Pt_3N. An example would be a platinum monolayer fuel-cell anode electrocatalyst, which consists of ruthenium nanoparticles with a sub-monolayer of platinum. Other areas of research include using tethered metalloporphyrin complexes for oxygen activation and subsequent reduction.

Guidelines for Selected Tutorial Problems

T24.1 One of the most popular lambda sensors contains zirconia (ZrO_2) as a solid electrolyte. This material has a high conductivity for O^{2-} ions at elevated temperatures. The cell responds to the changes in O_2 partial pressure in exhaust gas. The reduction potential for $O_2(g) + 4e^- \rightarrow 2O^{2-}(s)$ is proportional to p(O_2, reference)/p(O_2, exhaust gas), with partial pressure of O_2 in the air as the reference. (See Section 24.4(b) *Solid anionic electrolytes* and refer to Figure 24.9.)

T24.3 Recall from Section 24.8(a) *Nitrides* that heating complex oxides under a stream of ammonia can lead to the partial reduction of metal oxide. It can also lead to the nitridation of the oxide forming oxide-nitrides and eventually, with high excesses of ammonia, nitrides. In this case, partial nitridation occurs:

$$2SrWO_4 + 2NH_3 \rightarrow Sr_2W_2O_5N_2 + 3H_2O.$$

$Sr_2W_2O_5N_2$ adopts the pyrochlore structure.

T24.5 Delafossite is a mineral with chemical composition $CuFeO_2$. Its structure has been described in Pabst, A. (1938) *American Mineralogist*, **23**, 175–176, and more recently revisited (with a look at electronic properties of this material) in Marquardt, M., Ashmore, N., & Cann, D. (2006). *Thin Solid Films*, **496**(1), 146–156. Connect this research with the fact that Fe_2O_3 can also be used as a photocatalyst, as mentioned in Section 24.18 *Photocatalysts*.

T24.7 The earliest reports of a superconducting material based on LnFeOAs structure were Kamihara, Y. et al. (2008). *J. Am. Chem. Soc.* **130**, 3296, and Takahashi, H. et al. (2008). *Nature* **453**, 376, both reporting the structure and superconducting properties of La[$O_{1-x}F_x$]FeAs. Sm, Ce, Nd, Gd, and Pr analogues were all reported in the same year as well. Some useful references are: Chen, X. H. et al. (2008). *Nature* **354**, 761 (for Sm), Chen, G. F. et al. (2008). *Phys. Rev. Lett.* **100**, 247002 (Ce analogue); Ren, Z. A. et al. (2008). *Europhysics Lett.* **82**, 57002, (Nd analogue); Ren, Z. A. et al. (2008). arXiv:0803.4283. (Pd analogue). These are historically interesting, and you should look up more current sources as well.

T24.9 Two relatively recent sources are useful: Hischer. M. (ed.). (2010). *Hydrogen Storage: New Materials for Future Energy Storage*. Hoboken: John Wiley & Sons, Incorporated; and Godula-Jopek, A. et al. (2012). *Hydrogen Storage Technologies: New Materials, Transport, and Infrastructure*. Weinheim: Wiley-VCH.

T24.11 A very good starting point for this research is Guo, J., Chen, X. (2012). *Solar Hydrogen Generation: Transition Metal Oxides in Water Photoelectrolysis*. New York: McGraw-Hill Professional.

Chapter 25 Green Chemistry

Self-Tests

S25.1 To decide which type of industry better tolerates the reaction waste, we must first calculate the E-factor for the reaction and then compare it with the E-factors listed in Table 25.1.

The E-factor for the reaction is:

E-factor = (mass waste)/(mass product) = 40 kg/1 kg = 40.

This E-factor would be tolerated the best by the pharmaceutical industry.

S25.2 The atom economy of a process (or reaction) can be calculated using Equation 25.2 in your textbook:

$$\text{Atom economy (\%)} = \frac{\text{Molecular mass of desired product}}{\text{Molecular mass of all reactants}} \times 100\%.$$

Applied on the reaction $SiO_2(s) + 2C(s) \rightarrow Si(s) + 2CO(g)$:

$$\text{Atom economy (\%)} = \frac{Mr(\text{Si})}{Mr(\text{SiO}_2) + 2Mr(C)} \times 100\% = \frac{28.09\,\text{g mol}^{-1}}{60.09\,\text{g mol}^{-1} + 2 \times 12\,\text{g mol}^{-1}} \times 100\% = 33.4\%$$

S25.3 The copolymerization follows the reaction:

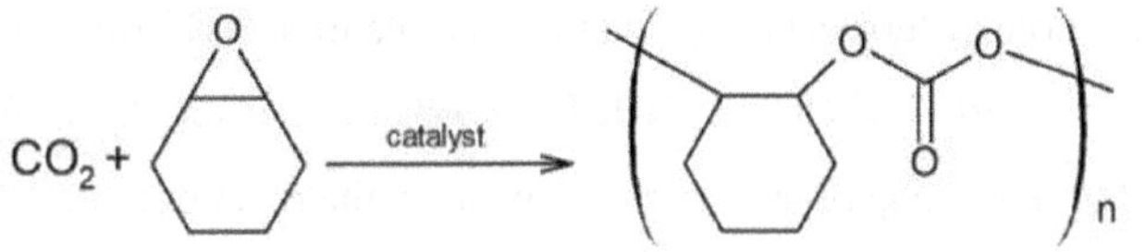

The molecular mass of one repeat unit is 142 g mol^{-1} of which 44 g mol^{-1} is due to CO_2. Thus:

(44 g mol^{-1})/(142 g mol^{-1}) = 0.31 or 31%.

S25.4 If you look at the reactions provided in Example 25.4, you will see that Rh catalyst is involved in the water gas shift reaction. This side reaction "ties up" some of the catalyst. If Cativa's process does not suffer from this side reaction, that means all Ir catalyst is involved in the desired catalytic cycle. Therefore, the relative activity of the catalyst is increased.

Exercises

E25.1 Using Equation 25.1 and following the procedure outlined in Example 25.1 and Self-Test 25.1 we get an E-factor of 3.5.

E25.2 Using Equation 25.1 and following the procedure outlined in Example 25.1 and Self-Test 25.1 we get an E-factor of 7.

E25.3 The reaction

$$2PbCl_2 + 4CH_3MgBr \rightarrow (CH_3)_4Pb + Pb + 4MgBrCl$$

shows that two moles of $PbCl_2$ are required for one mole of product. First, we have to calculate the number of moles of $PbCl_2$ in 4.0 g:

$$n(\text{PbCl}_2) = \frac{m(\text{PbCl}_2)}{Mr(\text{PbCl}_2)} = \frac{4.0\,\text{g}}{278.11\,\text{g mol}^{-1}} = 1.4 \times 10^{-2}\,\text{mol}.$$

Theoretically, only half of this is converted to $(CH_3)_4Pb$:

$$n((CH_3)_4Pb) = ½ \times 1.4 \times 10^{-2} \text{ mol} = 7 \times 10^{-3} \text{ mol}.$$

This can be converted to mass to give us theoretical yield of the reaction:

$$m((CH_3)_4Pb) = 7 \times 10^{-3} \text{ mol} \times 267.34 \text{ g mol}^{-1} = 1.92 \text{ g}.$$

Now we can calculate actual yield:

$$\text{yield}(\%) = \frac{1.54\,\text{g}}{1.92\,\text{g}} \times 100\% = 80\%$$

The atom economy of the reaction can be calculated using Equation 25.2 and following the procedure outlined in Example 25.2 and Self-Test 25.2:

$$\text{Atom economy}(\%) = \frac{\text{Molecular mass of desired product}}{\text{Molecular mass of all reactants}} \times 100\%$$

$$= \frac{267.34\,\text{g mol}^{-1}}{2 \times 278.11\,\text{g mol}^{-1} + 4 \times 119.24\,\text{g mol}^{-1}} \times 100\% = 25.9\% \cdot$$

The reaction yield is quite acceptable 80%. However, this reaction has a very poor atom economy.

E25.4 First we have to determine the number of moles in 1.5 g of ferrocene:

$$n(\text{Cp}_2\text{Fe}) = \frac{m(\text{Cp}_2\text{Fe})}{Mr(\text{Cp}_2\text{Fe})} = \frac{1.5\,\text{g}}{186.3\,\text{g mol}^{-1}} = 8 \times 10^{-3}\,\text{mol}.$$

Theoretically, for each mole of ferrocene we obtain one mole of acetylferrocene, thus:

$$n((C_5H_4COCH_3)(C_5H_5)Fe) = n(\text{Cp}_2\text{Fe}) = 8 \times 10^{-3} \text{ mol}.$$

This can be converted to mass to give us theoretical yield of the reaction:

$$m((C_5H_4COCH_3)(C_5H_5)Fe) = n((C_5H_4COCH_3)(C_5H_5)Fe) \times Mr((C_5H_4COCH_3)(C_5H_5)Fe)$$
$$= 8 \times 10^{-3} \text{ mol} \times 228.07 \text{ g mol}^{-1} = 1.82 \text{ g}.$$

Now we can calculate actual yield:

$$\text{yield}(\%) = \frac{1.15\,\text{g}}{1.82\,\text{g}} \times 100\% = 63\%.$$

The atom economy of the reaction can be calculated using Equation 25.2 and following the procedure outlined in Example 25.2 and Self-Test 25.2:

$$\text{Atom economy}(\%) = \frac{\text{Molecular mass of desired product}}{\text{Molecular mass of all reactants}} \times 100\%$$

$$= \frac{228.07\,\text{g mol}^{-1}}{186.3\,\text{g mol}^{-1} + 102.09\,\text{g mol}^{-1}} \times 100\% = 79\% \cdot$$

This reaction has quite acceptable yield and atom economies, although both could be higher (particularly the yield).

E25.5 a) **Turnover frequency** is the *amount* of product formed per unit time per unit *amount* of catalyst. In homogeneous catalysis, the turnover frequency is the rate of formation of product, given in mol L^{-1} s^{-1}, divided by the concentration of catalyst used, in mol L^{-1}. This gives the turnover frequency in units of s^{-1}. In heterogeneous catalysis, the turnover frequency is typically the amount of product formed per unit time, given in mol s^{-1}, divided by the number of moles of catalyst present. In this case, the turnover frequency also has units of s^{-1}. Since one mole of a finely divided heterogeneous catalyst is more active than one mole of the same catalyst with a small surface area, the turnover frequency is sometimes expressed as the amount of product formed per unit time divided by the surface area of the catalyst. This gives the turnover frequency in units of mol s^{-1} cm^{-2}. Often one finds the turnover frequency for commercial heterogeneous catalysts expressed in rate per gram of catalyst.

b) Selectivity is a measure of how much of the desired product is formed relative to undesired by-products. Unlike enzymes (see Chapter 26), man-made catalysts are rarely 100% selective. The catalytic chemist usually has to deal with the often difficult problem of separating the various products. The separations, by distillation, fractional crystallization, or chromatography, are generally expensive and are always time-consuming. Furthermore, the by-products represent a waste of raw materials. Recall that an expensive rhodium hydroformylation catalyst is sometimes used industrially instead of a relatively inexpensive cobalt catalyst because the rhodium catalyst is more selective.

c) A catalyst is a substance that increases the rate of a reaction but is not itself consumed. This does not imply that the added catalyst does not change during the course of the reaction. Frequently, the substance that is added is a *catalyst precursor* that is transformed under the reaction conditions into the active catalytic species.

d) The catalytic cycle is a sequence of chemical reactions, each involving the catalyst, that transform the reactants into products. It is called a cycle because the actual catalytic species involved in the first step is regenerated during the last step. Note that the concept of a "first" and "last" step may lose its meaning once the cycle is started.

e) Catalyst support. In cases where a heterogeneous catalyst does not remain a finely divided pure substance with a large surface area under the reaction conditions, it must be dispersed on a support material, which is generally a ceramic-like γ-alumina or silica gel. In some cases, the support is relatively inert and only serves to maintain the integrity of the small catalyst particles. In other cases, the support interacts strongly with the catalyst and may affect the rate and selectivity of the reaction.

E25.6 The key difference between two reactions is that HI would hydrolyse MeCOOMe and produce MeI and MeCOOH (instead of water). The reaction of the ethanolic acid with the acetyl iodide (produced at the end of the catalytic cycle) leads to ethanoic anhydride. See the catalytic cycle in Figure 25.3.

E25.7 The two terms are related. **Hazard** refers to anything that could pose a harm to people, property or the environment. **Risk** is a measure of probability the hazard will do harm.

E25.8 The safety principle is consideration of safety of a reaction, process or design by bearing in mind known hazards and risks associated with them.

E25.9 Real-time, in-process monitoring provides timely information on a reaction's progress, state of a catalyst, formation of any by-products, etc. This information can alert to any problems that arise during the process (for example, uncontrolled increase in the temperature of the reactor, unusually high rate of by-product formation, etc.) and thus help in prevention of serious accidents to people, property or the environment.

Guidelines for Selected Tutorial Problems

T25.1 It is interesting that Paul Hermann Müller won the 1948 Nobel Prize in Physiology or Medicine for his discovery of DDT's pesticide properties. To get a good idea on DDT's regulatory history up to 1975 you can check the following website http://www.rst2.edu/ties/ddts/university/docs/BackgrSurvey.pdf that contains an EPA report on this topic. A good source is the Stockholm Convention website http://chm.pops.int/ that covers persistent organic pollutants. The site can be searched for DDT, and current status of this compound retrieved. Of somewhat broader interest is a recent book "Banned: A History of Pesticides and the Science of Toxicology" by Frederick Rowe Davis that covers DDT history as well (published by Yale University Press in 2014).

T25.3 A good starting point is the review article M.E. Franks et al. "Thalidomide" *The Lancet* **2004** 363, p. 1802–1811. This article focuses on historical and current uses of this compound. The references cited in the article as well as article citing this review are also useful.

T25.4 Here is an example as a starting point for discussion. Collecting discarded, empty aluminium pop cans, melting them and using this as a raw material to make, say, parts for automobile industry is an example of recycling. Green chemistry would analyse the steps involved in this recycling process and look for the answers to questions such as:

- Have all hazards in the process been recognized?
- Have risks been properly assessed?
- Can we find a new process that would reduce the waste, reduce greenhouse gas emissions, use less energy, etc?
- If there are potentially harmful chemicals used, can we find less hazardous alternatives?

T25.5 Early refrigerants were NH_3, CH_3Cl and SO_2. Based on the previous material in the textbook you can discuss the hazards associated with these chemicals. CFCs were designed to replace these three substances. CFCs have been implicated in the destruction of the ozone layer. Considering the importance of this environmental topic, many sources are easily accessible.

T25.7 There are three major industrial processes for the production of acetic anhydride, described very briefly as a starting point:

1) The *ketene process* involves thermal cleavage of acetic acid to form ketene and water:

$$CH_3COOH \xrightarrow{700-750°C} H_2C{=}C{=}O + H_2O.$$

This step requires phosphoric acid as a catalyst. Formed ketene must be rapidly removed from water (to avoid backward reaction) and reacted with glacial acetic acid to produce the anhydride:

$$H_2C{=}C{=}O + CH_3COOH \rightarrow (CH_3CO)_2O.$$

The produced anhydride is distilled at the end.

2) The second process starts with oxidation of acetaldehyde with oxygen to produce peracetic acid:

$$CH_3CHO + O_2 \rightarrow CH_3COOOH.$$

The peracetic acid reacts with another molecule of acetaldehyde to produce acetic anhydride:

$$CH_3COOOH + CH_3CHO \rightarrow (CH_3CO)_2O.$$

The reactions take place in the solution without isolation of peracetic acid (a one-pot reaction). The best solvent proved to be ethyl acetate. The oxidation step requires a catalyst which is, most commonly, a mixture of cobalt acetate and copper acetate.

3) Carbonylation of methyl acetate is a process similar to the Monsanto acetic acid synthesis. It uses methyl acetate and CO as starting materials, that, in the presence of a catalyst produce acetic anhydride:

$$CH_3COOCH_3 + CO \xrightarrow{\text{catalyst}} (CH_3CO)_2O.$$

The most efficient catalysts are Ir complexes but some Ni complexes have been in use as well. Both require addition of an iodide (commonly CH_3I, LiI or even I_2) for catalytic activity.

Chapter 26 Biological Inorganic Chemistry

Self-Tests

S26.1 Uncomplexed Fe^{2+} is present at very low concentrations (approximately 10^{-7} M). By contrast, Fe^{3+} is strongly complexed (relative to Fe^{2+}) by highly specific ligands such as ferritin and by smaller polyanionic ligands, particularly citrate. Free Fe^{2+} is therefore easily oxidized.

S26.2 The protein's tertiary structure can place any atom or group in a suitable position for axial coordination. Thus, the protein folding is responsible for bringing the unusual methionine sulfur atoms (recall that sulfur is a soft donor) in axial sites. This prevents water molecules (which would be the natural choice for Mg^{2+}, a hard cation) from occupying the axial sites.

S26.3 Blood plasma contains 0.1 M Na^+ but very little K^+; the opposite is true inside cells (see Table 26.1). This differential results in an electrical potential that is used to drive reactions. If a patient was given an intravenous fluid containing KCl instead of NaCl, the potential across the cell membrane would collapse, with severe consequences.

S26.4 Calmodulin does not bind to the pump unless Ca^{2+} is coordinated to it. As the Ca^{2+} concentration increases in the cell, the Ca-calmodulin complex is formed. The binding of this Ca-calmodulin complex to the pump is thus a signal informing the pump that the cytoplasmic Ca^{2+} level has risen above a certain level and to start pumping Ca^{2+} out.

S26.5 A possible reaction sequence is shown below. Starting from an Fe(II) porphyrin complex (L is a neutral axial ligand throughout the sequence) we first obtain a peroxo bridged Fe(III) porphyrin dimer. Peroxo bridge can oxidatively cleave producing two equivalents of oxido Fe(IV) porphyrin monomers. This rather reactive species (recall that +4 is not a stable oxidation state for Fe and that Fe compounds with oxidation states above +3 are strong oxidizing agents) can react with another equivalent of the starting Fe(II) porphyrin complex producing an oxido-bridged Fe(III) porphyrin dimer. Note that all complexes are neutral species.

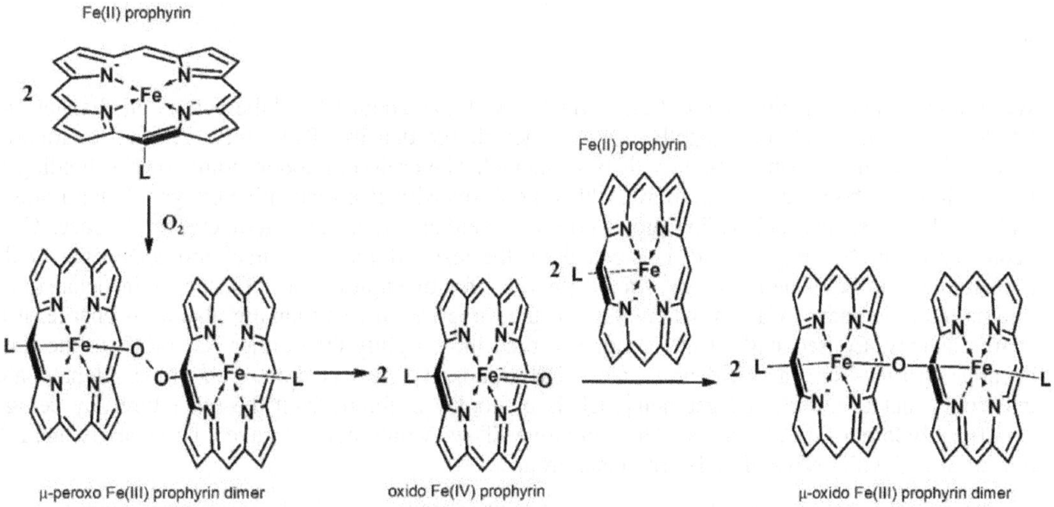

S26.6 Hydrogen bonding and replacement of one cysteine with a histidine should lead to an increase in the reduction potential of an Fe-S cluster. Hydrogen bonding should stabilize the more reduced, electron-rich oxidation level of an Fe-S cluster. Two His-for-Cys substitutions raise the reduction potential in the Rieske centres because imidazole is a better ligand to Fe(II) than Fe(III) sites. Surrounding a cluster with negatively charged protein residues would likely decrease the cluster's reduction potential because it will be more difficult to add an electron to a site that is already negatively charged in a low dielectric medium.

S26.7 There is greater covalence in blue Cu centres than in simple Cu(II) compounds. The unpaired electron is delocalized onto the cysteine sulfur and spends more time away from the Cu nucleus, decreasing the coupling. In most of the Cu blue centres, the reduced form is Cu(I), which is a soft acid, so it makes sense that it bonds well with sulfur-containing ligands such as a cysteine residue.

S26.8 Cu(III) is a d^8 metal centre. It is likely to be highly oxidizing, and probably diamagnetic with a preference for square planar geometry, a favoured geometry for the d^8 electronic configuration.

S26.9 The way to obtain a better model for Fe(IV) oxo centres is to force the octahedral geometry at Fe(IV). This would produce a complex molecular orbital diagram as shown in the Example 26.9. For non-haem Fe proteins the ligands should contain N and O donor atoms (hard donors).

S26.10 Species such as CH_3Hg^+ and $(CH_3)_2Hg$ are hydrophobic and can penetrate cell membranes. Cobalamins are very active methyl-transfer reagents that can methylate anything in the cell, which is bad news for the human body. Unlike many other mercury compounds, methylmercury is soluble in water. Methylation of Hg^{2+} thus increases concentration of this cation in body fluids.

S26.11 Spectroscopic measurements that are metal specific, such as EPR, which detects the number of unpaired electrons, could be used on both enzyme and isolated cofactor. Attempts could be made to grow single crystals from the solution and perform single-crystal X-ray diffraction and EXAFS to reveal molecular geometry, bond distances, and angles between Fe or Mo and the sulfur ligands. Both these techniques can be carried out on the enzyme and cofactor dissolved in DMF. Of these, single-crystal X-ray diffraction is our most powerful technique for determining structure. Mössbauer spectroscopy could also be useful in determining (average) oxidation states and geometries at Fe centres.

S26.12 Cu(I) has an ability to undergo linear coordination by sulfur-containing ligands. The only other metals with this property are Ag(I), Au(I), and Hg(II), but these are not common in biology. Cu(I) can also be found in trigonal planar and tetrahedral coordination environments. Binding as Cu(II) would be less specific because it shows a strong preference for square planar or tetragonal geometries. Also, importantly, Cu(II) is able to oxidize thiolates ($R–S^-$) to RS–SR.

Exercises

E26.1 We should compare the ions in question with what we know about K^+ and the potassium channel. In comparison to K^+, Na^+ is smaller and consequently has a higher charge density. Thus, for a channel or ionophore to be Na^+ specific it must have a smaller cavity than K-channel. The preferred donor atoms for Na^+ binding are the same as for K^+ binding—both atoms form more stable complexes with electronegative oxygen donor atoms. Since both are s-block elements, and lack LFSE (no d orbitals), neither has a clear geometry preference. Consequently, the geometry of an Na^+ binding site is dictated by the size—if the K-channel provides eight O donors in cubic arrangement, an equivalent for Na^+ should provide (for example) 6 or 7 O donors in octahedral or pentagonal bipyriamidal geometry. Ca^{2+} is the smallest of the three cations, and having also a 2+ charge, it has the highest charge density. Consequently, Ca^{2+} is going to bind more tightly than either Na^+ or K^+ to the residues that have negative charge—for example deprotonated –OH groups. It is also a cation of an s-block element and has no LFSE that could dictate a preferred geometry. Cl^- is obviously different from the other three by being an anion (and significantly larger than all others). Thus, to move Cl^- we would need to have a large cavity and rely on hydrogen bonds formed with –NH and –OH functional groups.

E26.2 Calcium-binding proteins can be studied using lanthanide ions (Ln^{3+}) because, like calcium ions, they are hard Lewis acids and prefer coordination by hard bases such as anionic oxygen-containing ligands like carboxylates. The larger size of lanthanides is compensated for by their higher charge, although they are likely to have higher coordination numbers. Many lanthanide ions have useful electronic and magnetic properties. For example, Gd^{3+} has excellent fluorescence, which is quenched when it binds close to certain amino acids such as tryptophan.

E26.3 Co(II) commonly adopts distorted tetrahedral and five-coordinate geometries typical of Zn(II) in enzymes. When substituted for Zn(II) in the parent enzyme, the enzyme generally retains catalytic activity. Zn(II) is d^{10} and therefore colourless; however, Co(II) is d^7, and its peaks in the UV–Vis spectrum are quite intense (see Chapter 20) and report on the structure and ligand-binding properties of the native Zn sites. Furthermore, it is almost impossible to use ^{1}H NMR to study the active site of zinc enzymes because of all the interferences and overlaps one gets with the amino acids themselves. Fortunately, Co(II) is paramagnetic, enabling Zn enzymes to be studied by EPR after substitution.

E26.4 Acidity of coordinated water molecules is in the order Fe(III) > Zn(II) > Mg(II). Ligand-binding rates are Mg(II) > Zn(II) > Fe(III). Mg(II) is usually six-coordinate, so it can accommodate more complex reactions (such as rubisco). It is also the weakest Lewis acid of the three, has the highest mobility, and has a clear preference for hard donors. Zn(II) has low mobility because it binds strongly to amino acid residues in proteins but has excellent polarizing power, fast ligand exchange rate, and no preference for any coordination geometry. These features, combined with no redox activity, make Zn(II) a perfect choice for substrates that are not readily activated but should not be reduced or oxidized. The redox activity of Fe(III) makes its use in metalloenzymes relatively limited (in comparison to Zn(II)).

E26.5 Since iron(V) would have the electron configuration of $[Ar]3d^3$, the ion would have three unpaired electrons. The logical choice would be either EPR (electron paramagnetic resonance) or Mössbauer spectroscopy. See Chapter 8, *Physical Techniques in Inorganic Chemistry*, for more discussion on both instruments.

E26.6 Ferredoxins contain Fe-S clusters, see structure **25** in the textbook. The oxidized spectrum is consistent with all iron atoms in the cluster being Fe(III)—there are two peaks (because of quadrupole coupling) centred at about 0.3 mm/s with small coupling. The reduced spectrum has four lines, two of which are almost identical to the lines observed for the oxidized form and two new ones on each side. The two central lines still correspond to Fe^{3+}. The two new ones show a larger quadrupole coupling and a higher isomer shift, characteristics consistent with high-spin Fe^{2+}. Since the signals from Fe^{2+} and Fe^{3+} in the reduced form are well separated, the electron received during the reduction is localized on one iron atom only; that is, there is no delocalization.

E26.7 The structural changes accompanying oxidation of the P-cluster raise the possibility that the P-cluster may be involved in coupling of electron and proton transfer in nitrogenase. Among the important changes in the structures is the change in coordination of the cluster. The oxidized form of the P-cluster has a serine and the amide nitrogen of a cystine residue coordinated to Fe atoms which dissociate in the reduced form. Since both ligands might be protonated in their free states and may be deprotonated in their bond states, this raises the possibility that two-electron oxidation of the P-cluster simultaneously releases two protons. Transfer of electrons and protons to the FeMo-cofactor active site of nitrogenase needs to be synchronized; the change in structure suggests that the coupling of proton and electron transfer can also occur at the P-cluster by controlling protonation of the exchangeable ligands.

E26.8 Transfer of a $-CH_3$ group (transfer as CH_3^+, binding as CH_3^-) is expected to involve Co(I) and Co(III), which is found in the active site of cobalamin. Transfer and activation of CO is expected to involve an electron-rich metal such as the bioactive metals Fe(II) or Zn(II). Most likely the mechanism involves an insertion of a coordinated carbonyl into a metal–alkyl bond.

E26.9 The discovery of substantial amounts of O_2 would indicate the presence of photosynthesis and consequently some life-form that is capable of photosynthesis. On Earth, the plants are organisms capable of photosynthesis, but this does not preclude a possibility of some other photosynthetic life-form.

E26.10 There are several possibilities. For example, EPR spectroscopy is a very good method to follow the changes in oxidation states—it can be used to distinguish between different electronic states of the Mo centre. Use of isotopically labelled water molecules, for example $H_2{}^{17}O$, can be combined with Raman and IR spectroscopies to monitor the movement (transfer) of oxygen atoms.

E26.11 One possible mechanism is shown on the following figure. The first step is formation of halogen bond between electron rich Co(I) (a Lewis base) and X-R (a σ^* orbital of RX acts as a Lewis acid). This Co(I)–X–R fragment is linear (see Section 5.7f *Group 17 Lewis Acids*). Next, Co(I) is oxidized to Co(III) and Co–X bond is formed. Note that this step is not an oxidative addition, although it might appear as such, because the coordination number at Co is increased only by one, not two. The conserved tyrosine residue protonates R^- and forms R–H product. This removes a very nucleophilic reagent that is also capable of bonding to Co. The Co(III) is next reduced twice with electrons supplied through two 4Fe–4S clusters present in the structure of this protein. This regenerates Co(I) active site and releases chloride as the second product of the reaction.

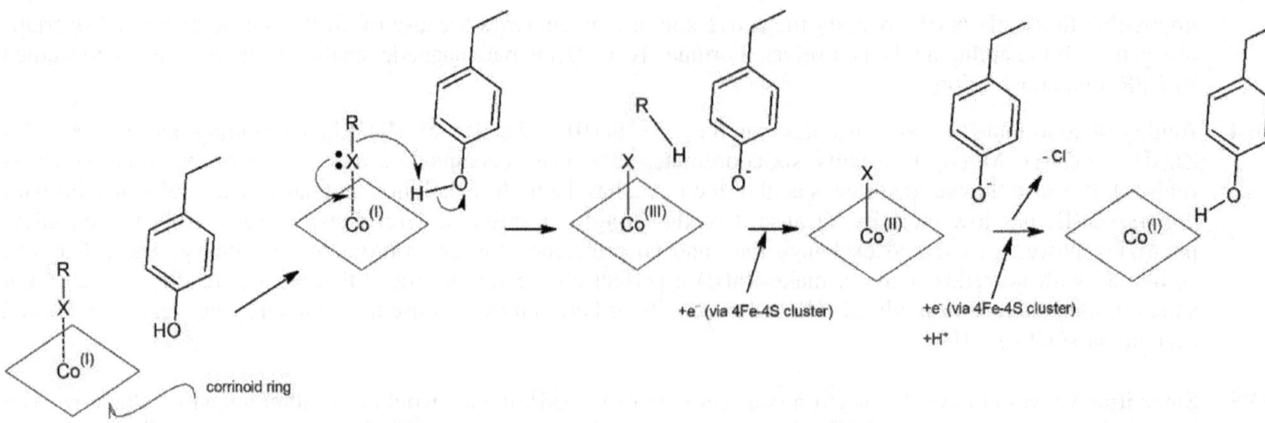

Guidelines for Selected Tutorial Problems

T26.2 Consider what is meant by the term "efficiency" in catalysis (you might want to consult Chapter 25 as well). In terms of mechanistic questions, consider the complexity of the process for production of NH_3 from N_2 as well as the complexity of the nitrogenase cofactor. For example, can the 3D structure help to decide on the binding site of N_2? Does it reveal much about the electron flow within the cofactor? What happens to a produced NH_3 molecule (how is it removed from the active site)? Consider also the details about the active site structure we are missing.

Chapter 27 Inorganic Chemistry in Medicine

Self-Tests

S27.1 A general structure of semithiocarbazide ligand is shown below. The easiest and most straightforward way to fine-tune the reduction potentials of Cu(II) complexes is by modifying R substituents (R^1–R^5) on the ligand backbone. The Cu semithiocarbazide complexes used in medicine usually contain tetradentate versions of the general structure shown.

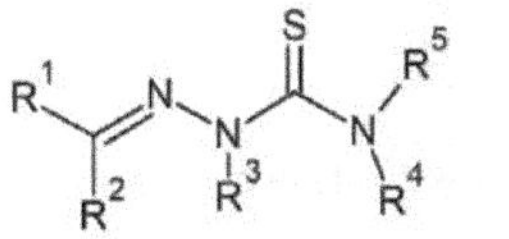

Exercises

E27.1 Au(III) is in many ways similar to Pt(II), with similarities mostly arising because the two are isoelectronic species with d^8 electronic configuration. Thus, both prefer square planar geometry that seems to be necessary for the activity of Pt(II) anticancer drugs. Au(III) is also a soft Lewis acid, meaning that it has low preference for Cl^- and O-donor ligands. The major difference between Pt(II) and Au(III) is that Au(III) is easily reduced under hypoxic conditions to Au(I), which prefers linear geometry as in $[AuCl_2]^-$. Au(I) can also disproportionate to Au(0) and Au(III).

E27.2 Although copper is a relatively inert element, it does slowly oxidize in air, producing copper oxides (and upon longer exposure to the atmosphere, basic copper carbonates malachite and azurite). Both Cu_2O and CuO are basic oxides. Epidermis, the outermost layer of skin, is slightly acidic (pH about 5). When oxidized parts of the bracelet get in touch with acidic epidermis, a small amount of oxides dissolve and produce Cu^{2+}, which is water soluble and can pass through the skin.

E27.3 Likely products of the $[H_3BCO_2]^{2-}$ decomposition are $B(OH)_3$, H_2, CO, and OH^-.

By analogy to carbonate, you can expect boranocarbonate to be a rather good base. Consequently it can be stable in a basic medium (just like carbonate), but in neutral or mildly acidic medium it would be protonated to produce hydrogenboranocarbonate:

$$[H_3BCO_2]^{2-} + H_2O \rightarrow [H_3BCOO_2H]^- + OH^-$$

Hydrogenboranocarbonate can eliminate OH^- to produce $[H_3BCO]$ adduct (boranocarbonyl):

$$[H_3BCO_2H]^- \rightarrow [H_3BCO] + OH^-.$$

$[H_3BCO]$ is unstable when not under CO atmosphere and CO can be easily displaced by water molecules. The intermediate aquo adduct $[H_3B(OH_2)]$ rapidly decomposes to give H_2BOH and H_2. H_2BOH reacts rapidly with another two water molecules to produce the final products $B(OH)_3$ and two more equivalents of H_2. The overall reaction is:

$$[H_3BCO] + 3H_2O \rightarrow B(OH)_3 + CO + 3H_2$$

E27.4 Choose a trace element (not CHNOPS); list what its important chemical properties are, including aqueous chemistry, redox chemistry; identify proteins in which such an element appears; discuss how the element's chemical properties suit it for its task.

E27.5 Gallium(III) can bind to transferrin and lactoferrin, and can even be incorporated into ferritin (a biological "iron storage"). These enzymes use nitrogen and oxygen donors to bind iron. Since Ga(III) also has a good affinity for the same donors, it can compete with Fe(III) for the binding sites. One major difference between Fe(III) and Ga(III) is their redox chemistry. While Fe(III) can be reduced to Fe(II), Ga(III) cannot. This is an important difference because the reduction is used to liberate Fe from ferritin—Fe(II) is more labile and hence can easily be released from the complex. Ga(III) cannot be reduced and would remain bound in ferritin.

E27.6 Bismuth(III) compounds frequently used in treatment of gastric ailments are bismuth subsalicilate and bismuth subcitrate. In the strongly acidic environment (pH ~ 3) that abounds in organic acids, these Bi(III) compounds form polymeric species and clusters that can form a protective layer (or coating) on the stomach. The Bi(III) polymers and clusters slowly release Bi^{3+}, which pathogens can uptake. Once in the system, Bi(III) interferes with many bacterial proteins (e.g., transferrin and urease) and inhibits many important cellular functions resulting in the pathogen's death.

Guidelines for Selected Tutorial Problems

T27.2 Some useful sources (and references cited within) are:

Bratsos, I., Gianferrara, T., Alessio, E., Hartinger, C. G., Jakupec, M. A. and Keppler, B. K.. (2011). Chapter 5: Ruthenium and other non-platinum anticancer compounds. In Alessio, E. (ed.), *Bioinorganic Medicinal Chemistry*. Weinheim: Wiley-VCH Verlag & Co.

Gasser, G., Ott, I., Metzler-Nolte, N. (2011). Organometallic anticancer compounds. *Journal of Medicinal Chemistry 54*(1), 3–25.

Hartinger, C. G., Dyson, P. J. (2009). Bioorganometallic chemistry—from teaching paradigms to medicinal applications. *Chemical Society Reviews 38*(2), 391–401.